Thematic mapping

Other books by Kenneth Field

Landmarks in Mapping, ed.
Cartography.

Thematic mapping

101 inspiring ways to visualise empirical data

Kenneth Field

This is my truth, tell me yours.

Esri Press
REDLANDS | CALIFORNIA

Esri Press, 380 New York Street, Redlands, California 92373-8100
Copyright © 2021 Esri
All rights reserved.
Second printing 2022

ISBN: 9781589485570
Library of Congress Control Number: 2020950027

For Linda, and Wisley.

Contents

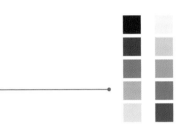

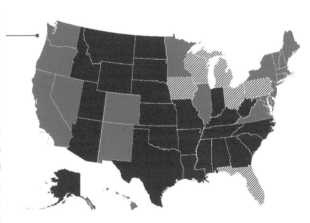

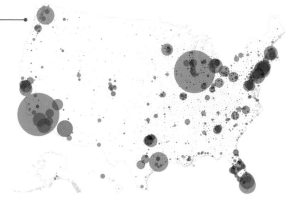

Contents

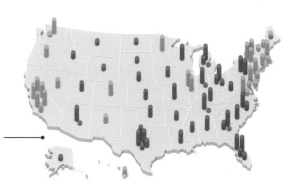

Foreword: Wandering multiplicity

Alberto Cairo

I could begin a foreword to Ken's splendid *Thematic Mapping: 101 Inspiring Ways to Visualise Empirical Data* by referring to master mapmaker and educator Mark Monmonier's admonitions about all maps being white lies or Alfred Korzybski's warning that maps aren't territories; as for a representation to be identical to the reality it stands for, it'd need to be that reality itself. Both sayings are true but have become tired clichés through misuse, overuse, and obliviousness to what their authors wrote before and after them. We live in a time when brief slogans and witticisms often are considered deep wisdom. Devoid of their context, even the words of great thinkers become hollow and commoditized in that manner.

'This is my truth, tell me yours' is no hollow slogan. On the contrary, it's one of the most extensive, systematic, and thorough demonstrations I've seen of what cartographers and information designers have said for decades: there's no perfect graphic, and not only because a map can't be a territory because of the territory's endless complexity or because of the obvious limitations of our helplessly imperfect maps. It's also because the effectiveness of any visual display of information doesn't depend just on the quality of the data it contains or on the honesty of the adeptness of its creator but also on who views it. A picture isn't worth a thousand words to audiences who don't know how to read it.

Moreover, there is often a multiplicity of possible interpretations of any single phenomenon. Shouldn't then the graphical representations of that phenomenon also be multiple, as each graphic might cast light on the phenomenon in a unique manner? Ken takes all these ideas seriously, so *Thematic Mapping* offers a multifaceted grand tour of alternative maps of the same topic.

This book is also a lesson on cartographic reasoning and decision-making. All its examples illuminate and exemplify varied aspects of the craft. Ken's cheerful—and occasionally ironic—notes paired with each map are of great help. I easily can see this book adopted as an introduction to mapmaking courses. It'll help students understand the relative strengths and weaknesses of various types of thematic maps—which are always dependent on the nature of the information displayed and their goals—and the many trade-offs designers face in their daily practice.

This foreword would be a disservice to you if I didn't emphasise how beautiful the maps are that this book contains. It seems that people who visualise or write treatises about it tend to downplay the joys that great graphics provoke. I write 'beautiful', and not uppercase 'Beautiful', because I deeply distrust idealism in philosophy, the ideology that makes 'Beauty' a vaporous entity or quality independent from its observers.

I have no problem with the circular argument that 'beautiful' is simply whatever a human being—or several human beings—perceives, feels, and describes as beautiful. Sensual pleasure often is looked down upon by stern prophets of visualisation orthodoxy who, lost in their platonic caverns, seem to believe that sheer enjoyment has no value on its own and, therefore, ought to be either ignored or always secondary to bare clarity.

Most of the maps in *Thematic Mapping* are indeed clear, enabling insights and permitting exploration of data, but others are playful or quirky experiments, and they aren't lesser works because of that; they are experimental wanderings and, as J. R. R. Tolkien wrote, not all those who wander are lost. *Thematic Mapping* teaches us to appreciate and embrace multiplicity and wandering, and it's a wondrous book because of it.

Professor Alberto Cairo is the Knight Chair in Visual Journalism at the School of Communication of the University of Miami, where he teaches courses on infographics and data visualisation. He has authored several best-selling books, including, most recently, *How Charts Lie* (2019). He has held senior positions at various high-profile news media outlets and won numerous awards during his time leading the Interactive Infographics Department at *El Mundo*. He can be found on Twitter at @albertocairo.

Preface

I'm not writing another book; I'm done

Sometime in the early part of 2018, Professor Alberto Cairo visited Redlands, California, and I was keen to show him the final draft of my first book, *Cartography*. It had taken me several years to write, and I poured everything I had into it. It was designed to contain all I knew and had learnt about cartography, and, frankly, I was exhausted. Yet Alberto stumped me with a question I'd not even considered: 'What are you going to do for your second book?' My response was brief: 'I'm not writing another book. I'm done.' He said, 'You will.' What you hold in your hand is my second book on cartography. More recently, Alberto has said to me, 'I told you.'

The idea for this book emerged out of part of my work in exploring ways in which you can make different maps out of the same data. I've always had an interest in thematic maps, and when I moved from the United Kingdom to California in 2011, I inevitably started making many more maps of United States data than maps of United Kingdom data. Elections always make for fascinating events, and with their frequency, there's always scope to make new maps. And for someone interested in the political machinations of where I live, making maps of political data helps me scratch two itches at the same time. In the United States, election data is broadly open and available (though not always in a nice, clean, orderly format), and it provided me with a way to better understand the political geography of the country that was now my native coordinate.

A few maps turned into a few more. Barack Obama was coming to the end of his first term as president, and the 2012 election was on my radar. I made many maps of the 2012 election, in which Obama beat Mitt Romney to gain a second term. I started teaching workshops with these maps, and since 2013 at the annual Esri User Conference, I've taught technical workshops on thematic mapping using the 2012 election data as a case study. But no map is ever done, and with Donald J. Trump's victory in the 2016 presidential election, the maps were updated, I made a few more, and the workshops continued to grow. Without even knowing it, I had the basis of a second book, focused on thematic mapping.

Thematic maps are such a rich source of interest, and in the cartographic community, there have been plenty of fine academic texts over the years that explore the detail of their construction. As a student, one of my favourite books was *Thematic Maps*, by David Cuff and Mark Mattson, published in 1982 by Methuen in the United Kingdom. It became one of the core texts of my undergraduate cartography degree at Oxford Polytechnic, and I continued to use it once I began my own career as an academic. Cuff and Mattson had based their text on the work of a wide range of United Kingdom, United States, and Canadian academic cartographers, and crucially, also that of practising professional cartographers working in the journalistic realm. This recognition that cartography was not, and never has been, simply an academic pursuit is as important today as it ever has been.

Many more maps are made by people who have little cartographic training than by those who have a background in cartography. Visual journalism has blossomed over the last decade. The technology we have at our disposal means the entry point to making a map of an interesting dataset keeps getting lowered. This is all great news for thematic maps, and we see them everywhere. But having a source text that organises traditional concepts in a fresh way also has never been more necessary. That is how I see the contribution of this book. It's not a lengthy treatment of the nuts and bolts of cartography. For that, you might want to look at my first book. But it's a complementary and updated book to Cuff and Mattson, and to many others such as Francis J. Monkhouse and Henry R. Wilkinson's wonderful *Maps and Diagrams* (1973); Borden D. Dent's *Cartography: Thematic Map Design*, 6th ed. (2008); and Jacques Bertin's deep-dive text, *Semiology of Graphics: Diagrams, Networks, Graphs* (reprinted by Esri Press in 2010).

Away from the purely cartographic realm, there are plenty of excellent books that explore the role and design of information graphics more widely, such as Cairo's own books. His latest, *How Charts Lie*, encourages people to become smarter about visual information. Most similarly themed books attempt to do the same, such as Andy Kirk's *Data Visualisation* (2nd ed., 2019) and Edward Tufte's range of books on displaying quantitative information. I also have this aim in mind, for it isn't just about sharing knowledge and advice on how to make a better map but also better enabling people to read maps.

This book focuses more on the portrayals that apply to data of different types, such as quantitative data for areas, points, and lines. It explores the display of empirical data through illustrations and attempts to do so comprehensively, with as few words as possible. The intent is not to set out detailed layout, design, and production content because most will be working in an environment of specifications and software that largely is dictated. And tastes differ.

Coming up with a list of 101 maps, graphs, and charts proved to be straightforward for the first 70 to 80. Many common techniques had to be included, as well as a wide array of techniques that are seen less commonly, largely because of difficulties either in construction or interpretation. Many of those in the latter category are excellent alternatives, and hopefully this book will help lead to seeing more examples in the media. I also have included some obscure mapping techniques. Not because they are less relevant, but they perhaps have fallen out of favour as software tends to lead people down a path of least resistance using common defaults. But again, you might feel it worth the extra effort to try something a little different. I also have included remakes of some of the most impressive work seen in the media over the last couple of decades. Visual journalism often innovates more than academia, perhaps to try to appear different and engaging. Finally, there are a few curiosities—maps that usually are made as one-offs for fun or to be provocative. But it's worth remembering that there are only 101 examples in this book. The maps included are only a subset of a longer list of potential maps, charts, and graphs that were considered. There likely are as many more that might be made, though they also likely will be removed even further from conventional practice, and the book certainly wasn't simply about showcasing the novel or the bizarre. But maybe you can take inspiration and come up with something altogether new.

My hope is that this book will support not only everyday mapmakers by giving them a visual glossary of the different ways they might map statistical data, but it also might be useful to those specifically making political maps. And it shouldn't matter which country's political system you're mapping or what political persuasion you bring to the mapmaking process. It should be suitable for any introductory course on thematic mapping or as a text that supports that content in a broader course on cartography, geography, or GIS. It also provides a way for political scientists to see election data through different lenses as a way to encourage them to use maps to delve into the detail of election data and to illustrate how maps mediate the message. Plenty of more detailed cartographic theory and concept deliberately are missing from this book, but I'd direct students and interested parties to many other excellent references in the field. I also have included an appendix to the book that details prior art in the maps and other visuals described in the book. I do this for two reasons. Firstly, as a reference so you can explore early examples of the map types featured. Old is always new again, and we can learn as much from the maps that have come before as from our modern efforts. Secondly, as a way of recognising inspiration behind the work I've included because we all follow in the footsteps of previous mapmakers.

Finally, as I have come to learn, writing a book is not an easy task, and it relies on so many friends, family, and colleagues to get the job done. I am indebted to my close colleagues at Esri for giving me the time, space, and support to see this project through. Thanks especially to Clint Brown for believing in this project, and to Esri Press for steering it to publication. I work with some truly talented people who are never short of cartographic advice and critique. Chief among them are John Nelson, Wes Jones, Edie Punt, Nathan Shephard, David Watkins, and Craig Williams, who all have contributed in many ways to the content in the book, as well as general advice to make my work better.

Early manuscripts of the book were reviewed by many colleagues at Esri, as well as external experts Alberto Cairo, Lauren Tierney, Rosemary Wardley, Anthony Robinson, Amy Rock, RJ Andrews, and Amy Griffin. All offered suggestions and comments that have led to substantial improvements. It's important to acknowledge the wonderful community of cartographers who help shape work like this, in which collaboration leads to much stronger work. That said, as a Brit trying to navigate the vagaries of the American political system as well as map it, all errors are solely down to me.

Finally, my close friends and family all get put through a torrid time when I'm in the zone. I always say that maps are both my passion and my profession, and sometimes the life-work balance is blurred and lopsided. Without the love and support of my wonderful partner, Linda Beale, my family back in the UK, my good friends and colleagues around the world, and our faithful dog, Wisley, writing a book would not be possible. In fact, doing any work of substance, based on passion, wouldn't happen. I am deeply grateful for all the unseen support I receive, and I appreciate it from everyone.

Just don't ask me what the third book will be about.

Prologue

Speaking to different truths

Maps are ubiquitous, yet maps are not made equally. They are not read equally either. They are constructs of many decisions, opportunities, constraints, the people who make them, and the people who read them. Yet maps are one of the most often used and trusted mechanisms for displaying data that has a people component, and by that, I mean data in which people are on the map. This might be population density, levels of poverty or income, or any pattern of human activity. Data collects what we do, how we do it, and, crucially, where we do it. This might be through routine census surveys, market research, or to show the outcome of how we vote in elections. Data about us is plotted every day in official publications, atlases, and news media. In fact, we forever are looking at maps in which we are part of the map.

In April 2017, as President Donald J. Trump celebrated his first 100 days in office after his election victory in November 2016, he showed one such map, shown at right. In one of his first press conferences in the Oval Office, he shared a small printed map with the assembled media. *Reuters* quotes Trump as saying, 'Here, you can take that. That's the final map of the numbers. It's pretty good, right? The red is obviously us.' Trump was using a map to show clearly how he'd turned the United States red, but it sparked a debate about the veracity of the map and the extent to which it might be seen to obfuscate the election results through its design.

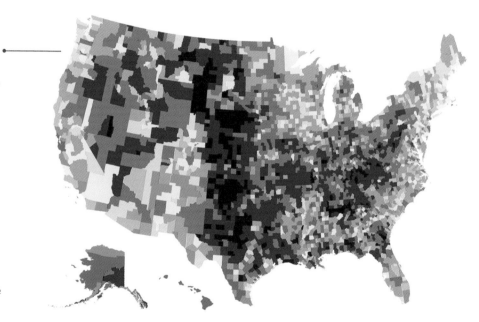

Despite many having criticised Trump's map for being wrong or misleading, it wasn't. It was just as accurate and precise as many other maps that were made to show the election results across the media. It was also the map of the results used by *Fox News*, *Breitbart*, and *InfoWars*. But it was a map that painted a specific picture of the results through the choices the mapmaker had made. It was a map that had a lot of Republican red on it and which perfectly illustrated the apparent strength and extent of his win using a certain thematic mapping technique (a choropleth map) coupled with a version of the data that emphasised his support (percentage share of the vote by county). Any Republican victor would show the same type of map with the same design choices.

Had Hillary Clinton won, or had the voting system in the US been based on the popular vote, which would have secured her victory, she would have shown broadly the same election data, but the map she would have shown would have been much bluer. It likely would have used a different thematic mapping type and manipulated the data to speak to a Democratic Party win. People would have had the same conversation about a misleading map but from a different political perspective.

How can this be the case? Surely, a map is a map, particularly when the data it is based on is the same, and the opportunity to manipulate it to show different versions of the results can't be possible. It doesn't matter whether we're talking about election maps or some other map of population-based data, the same opportunity for the map to put on many masks is possible and likely to occur to a much greater extent than a general audience might know or imagine. This book explores and explains this underappreciated fact of thematic mapping and invites mapmakers to focus upon the decisions they make in designing a map, as well as encouraging map readers to look with a critical and unassuming eye.

This entire book is based on the fundamental idea that there are few maps that can be thought of as right or wrong. They all tell different truths and different shades of the truth, and what represents one person's truth well may not chime with another person's truth.

The world now recognises the term *post truth* as the circumstances in which objective facts are less influential in shaping public opinion than appeals to emotion and personal belief. Maps always have been purveyors of post truth. They are designed to deliver messages which are often shaped to tell a story. This can be summed up by looking to past writers as much as it can to President Trump and his election map. In 1946, George Orwell, in his novel *1984*, stated, 'Political language is designed to make lies sound truthful and murder respectable, and to give an appearance of solidity to pure wind.' We might update this to 'Maps are designed to make lies appear truthful, misinformation respectable, and to give an appearance of fact to pure illusion.'

Aneurin Bevan, a Welsh socialist, stated in 1948 that the idea of alternative truths could be summed up with the phrase, 'This is my truth, tell me yours.' This concept introduces another actor, the consumer, to the production of information. For a map, this is the reader, who brings their own knowledge, understanding, and biases to how they see and interpret a map. For many, Trump's map speaks the truth. For others, it was, at best, a poor way to represent the results, and at worst,

'Maps are designed to make lies appear truthful, misinformation respectable, and to give an appearance of fact to pure illusion'

a complete lie. Facts are irrefutable, yet the way we represent them through maps allows us to paint different pictures, to make them more subjective, and to offer opinions through the way we manipulate them. Your view of a map is going to be predicated by what you already believe in and who, and what, you trust. For a political map, you well may trust a map shown on a right-leaning media outlet if you're a Republican Party voter. And yet there probably will be just as many Democratic Party voters who trust the version shown on a left-leaning media outlet. This sort of false belief in what we see is simply confirmation bias.

Statistician and artist Professor Edward Tufte paraphrases a quote by the late US senator Daniel Patrick Moynihan as 'Everyone is entitled to their own opinions, but they are not entitled to their own facts.' Yet when you see a map that backs up your preconceptions, you likely process that as fact, to the exclusion of other versions of the same facts.

Maps are good at showing facts in different and compelling ways, but as Darrell Huff wrote in *How to Lie with Statistics* (1954), 'If you torture the data long enough, it will confess to anything.' So to believe in truth, you first must be willing to be proven wrong. I hope that by challenging you to see the many varied faces that the maps in this book take on, you will begin to appreciate that the version you are presented by any one person or outlet is simply one version of the truth, however well constructed, beautifully designed, and believable it appears. When you make your own maps, it should act as a guide to the rich and varied ways in which you can shape your own truths.

Each map in this book looks different from the next because of the way the map type leads to an image on the page and, in your mind, the way data has been manipulated, and the choices of colours, symbols, and other cartographic detail. This is both the beauty and curse of cartography because there simply isn't one objective and immutable way to make the map. Even though the result of an election is an incontrovertible fact, the many ways you can make a map shows that you still can manage to bend the truth and sway public opinion in multiple ways. The maps show you the benefits and drawbacks of different techniques. Each map offers advice on how to optimise it and how people may read it. You more than likely will see maps that appeal to you and some that you have a viscerally negative reaction toward. This

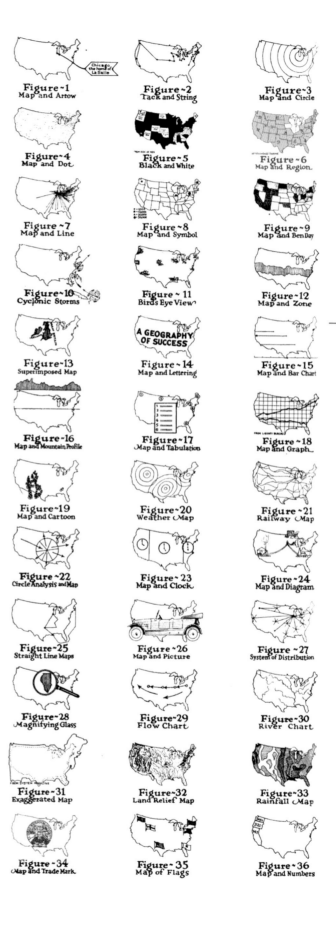

is a natural experience, and along the way, it should become clearer which maps do which jobs, when to make different choices, when to go beyond the defaults, and also when you might want to extend the bounds of your mapping to persuade or provoke. The book does this by presenting 101 maps of the same data, the 2016 presidential election data, in which Trump's map is one version that we'll deconstruct.

The idea to showcase many maps of the same data isn't new. In 1922, E. P. Hermann's *Maps and Sales Visualization* showed 36 ways to make maps of data, shown in summary to the left. What I hope to achieve by taking a similar approach is to make you think about the rich palette of maps you can develop of population-based data, whether of an election or otherwise. You might use the examples herein as a pick 'n' mix, or even as a starting point for alternative ways to make a map. But I also hope you'll use it to better understand how some of the more traditional, perhaps conventional, and widely used techniques can be made better.

In 1983, French cartographer Jacques Bertin wrote in his classic *Semiology of Graphics* that 'for a graphic to be "useful", it must be "efficient". The rules governing graphic efficiency stem from the properties of visual perception'. This concept embodies a key approach in this book, that to understand how a map works, you first must understand how it is seen and understood by a reader. Bertin demonstrated the basic graphical problem of deciding the appropriateness of using a graphic to represent a phenomenon using a dataset of the work force in the primary, secondary, and tertiary sectors for 90 French departments. That's 270 pieces of data. He illustrated 100 different graphics for the same information, a small selection of which are shown to the right.

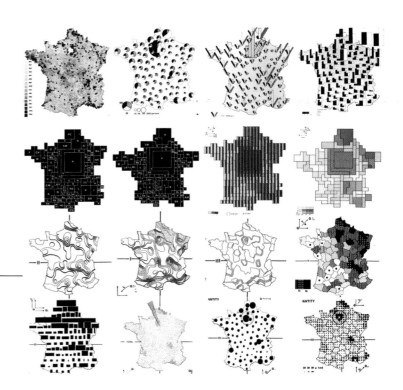

The principal purpose was to show how to transcribe the data graphically and to evaluate the properties of each approach, its efficiency, and the decisions over your chosen approach as a function of habits, personal aptitudes, and even fashion. He proposed that negative decisions often were made because of time constraints, what we might refer to as quick and dirty maps to paint a picture rapidly but which end up perhaps painting an erroneous picture. The decision on which map to make, considering an almost unlimited combination of graphical sign-systems, is to determine what suits the purpose and the audience. But it is also about appreciating how Bertin's retinal variables, otherwise known as *visual variables*, operate perceptually and cognitively. Or understanding how position, size, shape, value (lightness), hue, orientation, and texture work. These characteristics, and more, are all ways you can vary graphical marks on a page or screen, and they all encode meaning. Understanding how to vary meaning by varying the graphical marks helps identify what map or maps might be more effective than others in delivering a message.

The 101 maps herein show how an understanding of their construction supports their utility, and by appreciating how they are read, it helps you make a more informed choice about what map to make yourself. Each map creates a unique visual lens into the data. It brings elements into focus and makes others recede. Each map is as accurate as the next, but they all manipulate the data and the graphical sign language in multifarious ways. Some maps are a little more deceitful than others, it's true, and some are ambiguous, but none are wrong. They all have the appearance of being objective and precise. They all are designed to look attractive and inviting, to pull you in to encourage you to ask questions of the data and seek answers. Whether you get what you want from the map will vary considerably from one map to another. Even beautifully designed maps, made as objectively as possible, can be misleading.

The maps represent a wide variety of cartography, from classic and well-worn approaches to novel techniques seen less often or which are genuinely new and innovative. But as with many maps, old is often new again, and we can trace mapping techniques to many others who have gone before. Chief among the conventional techniques are choropleth maps, which can be traced back to the work of Pierre Charles Dupin, who published what is regarded as the first choropleth map, in 1826. It illustrated levels of illiteracy in France using shading from black (least literate) to white (most literate). The first example of the use of a choropleth map for an election in the United States is the example to the right from a US Census Bureau atlas published in 1883. This map of the 1880 presidential election is effectively a choropleth of the popular vote.

It's a beautiful map in many ways. It uses an appropriate projection, sensibly shaded colours that relate to the two parties (albeit, the inverse of red = Republican Party and blue = Democratic Party that we're used to today considering the switch in the 1980s). It has an aggregated inset map to

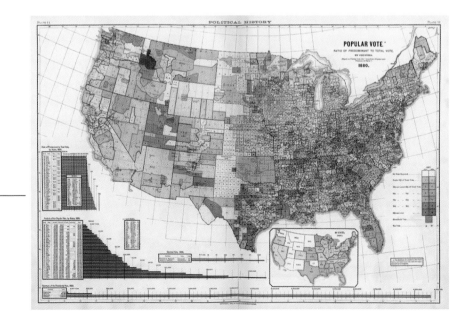

show the outcome per state, and it incorporates a range of additional graphs and charts to explain the results. This is key to many good maps. In addition to presenting only a map, providing context can help the map reader better navigate the map and enables them to interpret it better. Crucially, a good map tunes to the signal, omits the noise, and allows people to ask good questions and recover meaningful answers.

Even maps that you might think of as more modern often have a longer history. The first, and possibly the first attempt to make a map of election results that disregarded conventional physical geography, was a cartogram published in *The Times* (London) in 1895. The development of the cartogram as a technique for thematic mapping can be traced back to early 19th-century atlases in the United States but cartograms were not described as such until 1870. *The Times'* cartogram was referred to as the copyright 'simplex chart', which showed the various parliamentary geographies represented as tessellating squares of equal size. Additional internal symbology and a black/white distinction gave rise to the use of the multivariate symbol to show whether a seat was an urban borough, a metropolitan borough, or a university seat (members of certain universities got to vote twice in 1895 all the way to 1945, after which university seats were abolished).

Cartograms have become a popular mechanism for portraying elections from 1895 onward, not just in the UK but in many other countries. Cartograms routinely have been used in German election mapping since at least 1903 and in US election reporting by the *New York Times* for the last decade and almost certainly earlier. In the UK, cartograms traditionally have been most popular when the Conservative and Unionist parties have been least popular. In both 1964 and 1966, a topologically correct cartogram of the UK (now without most of Ireland) appeared. Both were created by Thomas Henry Hollingsworth and would have taken many days, if not weeks, to produce by hand. These are all early evidence of the desire to equalise our geographical view of the world to present the relative importance of political units. The exclamation in the top right that this is 'a new kind of election map' well may be correct. I have no alternative facts to challenge the assertion, but it is also a well-worn phrase we often hear today, even when there is irrefutable evidence that it might not be the case.

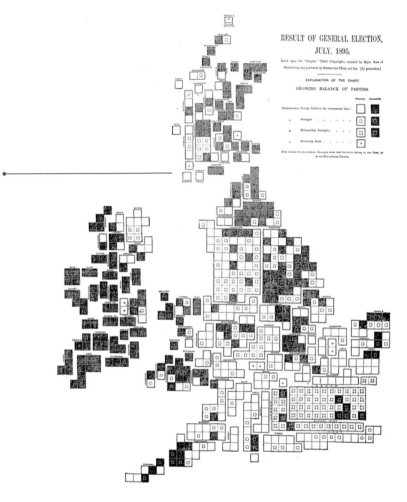

This book is organised into nine chapters, each encompassing a major category of visualisation. It begins with the most common, maps that use areas to report data, and then moves through maps that use point symbols, line symbols, and so on. This structure is convenient, but, of course, some maps straddle broad categories so it's also a fairly loose structure, and you'll discover maps that easily might have been placed in a different chapter. Within each chapter, a map gets its own spread with an accompanying discussion of the map type, the main considerations for designing it effectively, and notes on how it is read by the audience. Along the way, it signposts the things you should be aware of in making the map and how it will be interpreted.

You'll also see handwritten annotations and callouts scribbled around the maps. These writings comprise small comments that pick out aspects of the map's design and how it impacts a specific component of the map. They are the sort of comments you might see on a marked-up map, in which a cartographic editor has scribbled comments in the margins. It gives you an insight into how people with an eye for cartographic design often make small decisions in the art and science of the display that have a large consequence on how the finished map looks and what it achieves in terms of communicating the data.

Although many of the maps in the book take a similar style, to support the idea that you have an atlas of maps in your hand, each map demands a unique treatment. Many small differences exist between different map types and the way data is arranged, which means you simply cannot design by template. For instance, where you position the legend and how labels are designed, included, abbreviated, or excluded are all governed by the map you're making. The callouts express where such changes have been made and the reasons behind them. You may not agree with all these decisions because cartography is partly an artistic expression, but by seeing the thought processes for the decisions taken, you at least will understand why things are as they are.

You also hopefully will appreciate that, in book form, the maps you see are inevitably static. Cartography is an art, science, and technology. This book approaches thematic mapping by showing and discussing the art and science. The technology is analogue. It's primarily a print publication, and the medium we're working in is paper (even in e-book format). It's a high-resolution medium, and the maps are designed to be viewed in static mode.

Many people annotate books, underlining key points or filling passages of text with a multicoloured marker pen. The idea is to pick out salient points that resonate, or facts that can easily be identified later when flicking through.

The annotated scribbles you see on the maps are comments I've made, in my hand, that provide additional context for some of the smaller decisions made in designing the map, what it achieves, and how it leads to a certain message in the map.

It's written in my handwriting. Yes, this is a digital version of my handwriting so I hope you can read it and find the comments useful, and unobtrusive!

But the world is not static, and many maps you see these days are delivered using different technology through a digital medium such as a web map or a mobile app. In many ways, the ideas you see in this book are equally applicable if you made the map for a digital medium. Basic cartographic ideas work across many technologies. Where a digital medium differs is in the additional ways in which you can interact with the map. A digital map might be static but slippy, allowing you to pan and zoom. Such a map loses its edges as you use a screen as a window into the map. Additionally, interaction can be optimised so you can retrieve information through rollovers or clicks or turn data layers on or off. Even contextual layers, such as labels, might be off by default but available through an optional setting.

The ways in which you might modify the map's design for digital delivery are discussed throughout, especially in sections in which digital delivery would be the preferred medium, such as for the 3D maps. Paper is an insufficient medium to support some of the maps you'll see because interactivity is an important component for optimum viewing. This book is supported by digital versions of many of the maps you see. They are designed and optimised for web delivery and make use of additional capabilities, such as the multi-scale environment you get in a web map. The gallery of maps from this book, in their web map form, can be seen at esriurl.com/election2016.

This book is apolitical. Yes, you'll see maps that tend toward red, and you'll see maps that tend toward blue. And some will be purple. Election maps are bound to pique interest precisely because people are invested in the outcome of an election. They care about the consequences as individuals and for how the society and economy they live in is administered. The choice of this dataset for this book is simply to engage. The ideas and examples are applicable beyond simply mapping the results of a US election—or any election, for that matter.

Apply the ideas to your own election datasets. More than that, apply them to any population-based data, any empirical data, and any data about any phenomenon. This book is about thematic mapping. It shows you 101 versions of the truth of the 2016 presidential election. Whatever my truth is, you'll find maps in here that speak yours.

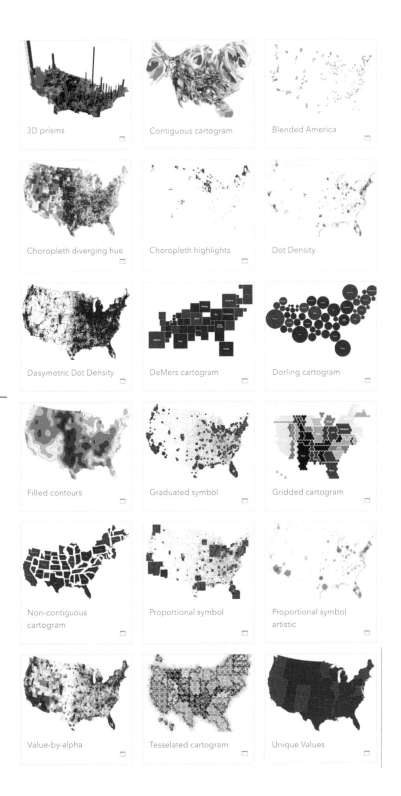

3D prisms

Contiguous cartogram

Blended America

Choropleth diverging hue

Choropleth highlights

Dot Density

Dasymetric Dot Density

DeMers cartogram

Dorling cartogram

Filled contours

Graduated symbol

Gridded cartogram

Non-contiguous cartogram

Proportional symbol

Proportional symbol artistic

Value-by-alpha

Tesselated cartogram

Unique Values

1. Preparation

As every politician knows, it's all about playing to the base, and the same is true in cartography.

Before pouring data into a computer and allowing it to help make a map, it's worth spending a while thinking about the base upon which the map will sit. This includes a whole range of contextual data, map reference material, and overarching stylistic decisions. Every map requires preparation, thinking about what the map is going to show and how it will achieve it through design.

The map's message will be mediated by decisions made on the projection used, the design choices employed, and the colours chosen. Working out at the outset what's going to work will provide a blueprint for success, saving time and making a better map for the reader.

This chapter focuses purely on the geography of the area being mapped, including some of the things that can be changed to make a better map and some of the constraints which simply must be accepted but still might be mitigated against.

From geography to map

Selecting a projection that works

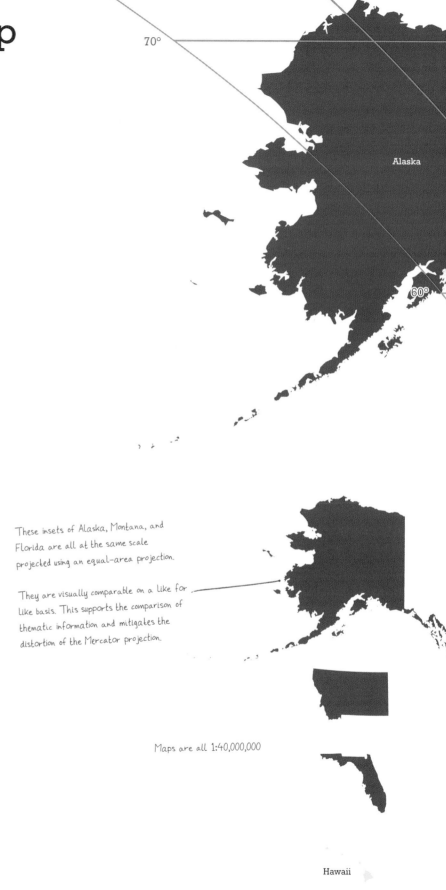

Whatever map you make, you must account for the geography of the place itself because it impacts your design choices. The geography of the maps used in this book is the United States. Its extent and spatial configuration mean you must deal with its character to make meaningful maps.

For election mapping, the US consists of 48 contiguous states, plus the remote islands of the state of Hawaii, the disconnected state of Alaska, and the small but significant District of Columbia. That's 51 areas of differing size and with a spatial configuration that causes difficulties if you want them all in their correct position on the same map. Firstly, you'll almost never show Alaska and Hawaii in their correct positions. You'll use insets instead. Secondly, let's look at the impact of the choice of projection on the way geography is represented.

By default, many maps use a Mercator projection, whether it's the one that is most familiar to you or the default for your mapping software. In a Mercator projection, scale varies across the map and is consistent only with other places for the same line of latitude. This has the effect of massively distorting the size of mapped objects the farther north or south you are from the equator. Look at the difference in size of the states of Alaska, Montana, and Florida on the main map. Their latitudes fundamentally define their size relative to one another, making Alaska disproportionately larger by comparison. Regardless of the real differences in the size of areas, this distortion propagates into how the map is seen, perceived, and cognitively processed because larger areas tend to be seen more prominently.

For thematic mapping, you want to make the best use of space and use a consistent framework from which people can visually compare one place to another. So the first major task in representing the geography of the US is selecting a projection that does the job. To maximise the use of the page or screen, you also likely will want to adjust the position of the 'problematic' states of Alaska and Hawaii by using insets. This will allow you to make the best use of the real estate on your map. The choice of projection is governed by what's sensible for the mapping task at hand—and you always will want to preserve area for thematic mapping. Moving Alaska and Hawaii is a compromise for this geography but one which brings other benefits so, on balance, it is a pragmatic decision.

Thus, this first page is designed deliberately to show you what not to do. **There's no small- or medium-scale thematic map for which Mercator (or Web Mercator) is a good choice.** We'll make some corrections on the next spread.

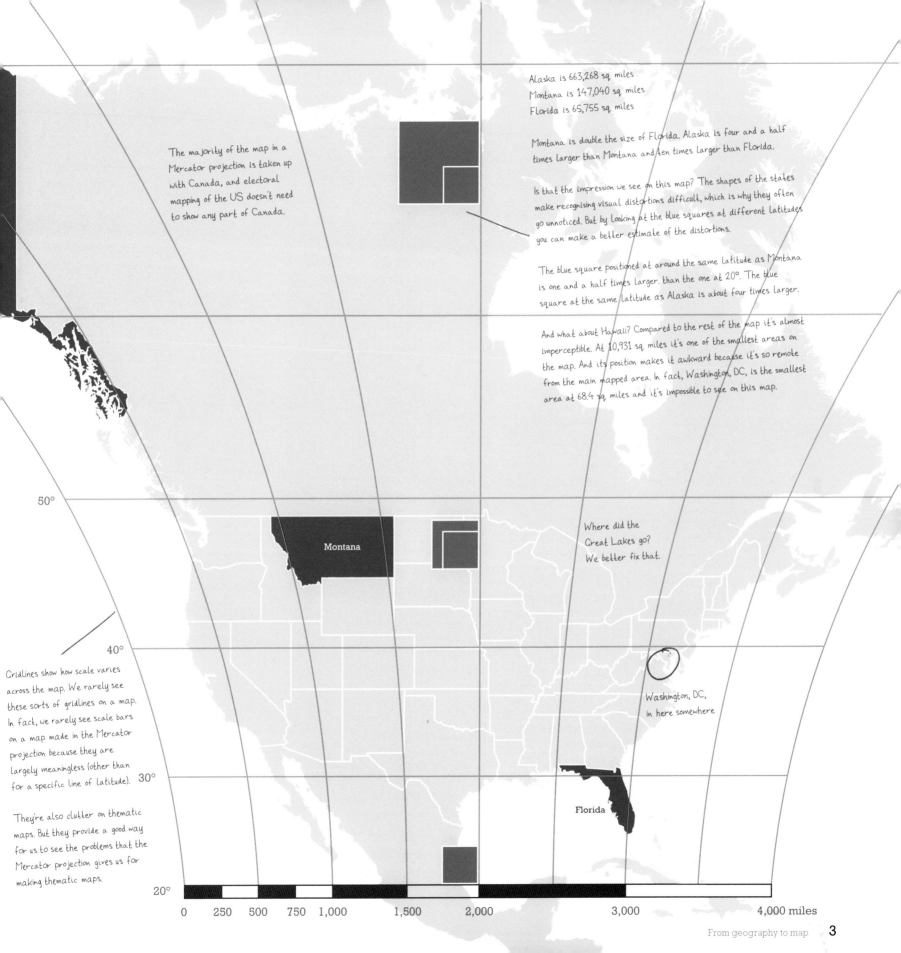

Alaska is 663,268 sq miles
Montana is 147,040 sq miles
Florida is 65,755 sq miles

Montana is double the size of Florida. Alaska is four and a half times larger than Montana and ten times larger than Florida.

Is that the impression we see on this map? The shapes of the states make recognising visual distortions difficult, which is why they often go unnoticed. But by looking at the blue squares at different latitudes you can make a better estimate of the distortions.

The blue square positioned at around the same latitude as Montana is one and a half times larger than the one at 20°. The blue square at the same latitude as Alaska is about four times larger.

And what about Hawaii? Compared to the rest of the map it's almost imperceptible. At 10,931 sq miles it's one of the smallest areas on the map. And its position makes it awkward because it's so remote from the main mapped area. In fact, Washington, DC, is the smallest area at 68.4 sq miles and it's impossible to see on this map.

The majority of the map in a Mercator projection is taken up with Canada, and electoral mapping of the US doesn't need to show any part of Canada.

Where did the Great Lakes go? We better fix that.

Montana

Washington, DC, in here somewhere

Gridlines show how scale varies across the map. We rarely see these sorts of gridlines on a map. In fact, we rarely see scale bars on a map made in the Mercator projection because they are largely meaningless (other than for a specific line of latitude).

They're also clutter on thematic maps. But they provide a good way for us to see the problems that the Mercator projection gives us for making thematic maps.

Florida

50°

40°

30°

20°

0   250   500   750   1,000   1,500   2,000   3,000   4,000 miles

Building your map

Laying strong foundations

A good map begins with a strong foundation. It's like any other visual form of communication: build the base first, organise all your key geographical layers, and the thematic information you want to focus on will sit effortlessly on top.

There are several key decisions to make to ensure you set up your map appropriately for the job at hand. For thematic maps, one of the most fundamental considerations is the projection you choose. The main perceptual and cognitive task you're asking of your reader is to compare different parts of the map to each other. Yet geography gets in the way because, in almost every circumstance, you'll be dealing with a map in which areas vary considerably in size. This variation already causes cognitive issues, which can be exaggerated using the wrong projection. **You don't want to imbue bias in your map through selecting an inappropriate map projection.**

Different map projections preserve different properties, such as direction, size, shape, area, and distance. **The property you must preserve for thematic mapping is area.** You should choose an equal-area projection in which comparison of one area to another is the key to understanding the map. This type of projection provides a consistent framework for people to make visual comparisons and limits the potential for making poor inferences.

Datasets often contain far more detail than required so processing polygon boundaries and linework to simplify and smooth them will pay dividends when the map is small scale. The main map of the US in this book is 1:13,000,000, and it does not need all the sinuosity of detailed linework you see at larger scales. Generalising geographical data creates a much cleaner looking map. If you're making a web map, you may decide to have generalised data at small scales, and then switch to more detailed data at larger scales in which the scale supports a finer level of geographical detail.

Make decisions on colour schemes, typefaces, and fonts early on, and you can reuse them for many of the maps you'll end up making in your series. Subtlety is often a good approach. Soft colours and clean typefaces. **Don't mix and match too much, and the map becomes pleasing to look at.**

To the right is the basic map designed to support the creation of most of the maps in this book. It's based on county equivalents. Most states have county-level reporting. Some, such as Alaska (boroughs) and Louisiana (parishes), don't have counties, but for the purposes of mapping, they are treated as equivalent. This map also can act as a reference map in an atlas, particularly when certain map types are unable to support detailed labelling.

The four-colour theorem has been used to delineate administrative areas. The use of colour is subtle for the fill and also used for emphasising state boundaries.

A gradient vignette is applied to the areas beyond the main map to signify adjacent geographies, and avoid the map simply floating in space. Blue suggests water and beige suggests land.

Insets for Hawaii and Alaska is a well-established method for dealing with dispersed geographies.

1:14 million

All labels are set in Gill Sans for consistency. Hierarchy is created by varying elements of the typeface, rather than mixing different typefaces.

State and county boundaries have different treatments to create a sense of hierarchy.

Geography naturally leads to congestion of map features. Here, labels have all been reduced in size to ensure they do not overlap and remain visible.

Subtle halos can be used between labels and the background to aid legibility of type across different colours and variations in the underlying map detail.

The map is projected using the Albers Equal Area Conic projection. It works well for the US and preserves area.

Small-scale thematic maps rarely need north arrows because they are normally of familiar areas. Regardless, the projection means north actually varies across the map anyway.

Main map
1:13 million

Scale is referred to using a representative fraction (RF). Here, one map unit = 13 million real-world units (e.g. cm). Thematic maps rarely have detailed information for scale because it's superfluous to the main theme of the map.

1:10 million

The lines are drawn

Boundaries always shape the map

Some decisions taken outside your mapmaking process will impact the way the map looks and the stories it tells. These are conditions you have little or no control over.

Maps of population data report the results of a collection and counting process that occurred within some form of area. These areas were drawn up at some point in history on the basis of criteria, such as they should contain a roughly equal number of people, be contiguous, reflect socio-economic equity, or preserve existing political communities. In election mapping, political boundaries become contentious because they can be changed, and often are changed, and any decisions about where to draw new lines have ramifications for the map, as well as how the votes stack up. The process of redrawing boundaries for political gain is referred to as *gerrymandering* or *redistricting*.

Before you make design decisions such as classifying and symbolising your data, the map you make of any election data is fundamentally, and profoundly, a function of the decisions made in drawing the boundary lines. This results in several issues, which broadly are referred to as the modifiable areal unit problem (MAUP), arising from the imposition of artificial units of spatial reporting on continuous geographical phenomena resulting in the generation of artificial spatial patterns. **MAUP always imbues your map with inherent bias because the spatial patterns you map are controlled by how the boundaries were drawn.** All other things being equal, a different set of boundaries would lead to a different set of areas and a different map.

The common first-past-the-post political system (to which the US adheres) means individual votes are aggregated from smaller areas to larger ones for the purposes of counting and determining which party and person wins. An entirely different set of areas can be created depending on the spatial resolution of the aggregated units. Additionally, the shapes can vary depending on which of your basic unit areas you combine into an aggregated area. The results on the map mean that you inevitably report not only the results but how those results have been constituted as a function of the decisions about boundaries, scale, and aggregation.

The illustrations to the right show how a hypothetical set of 100 voters can be aggregated at different scales and into different boundary configurations, which, ultimately, lead not only to a different political outcome but a different map. It's a simplification but mirrors the decisions that redistricting or redrawing boundaries have on results.

Shape shifting

The 4th Congressional District of Illinois is one of the oddest in US political geography. It has changed shape numerous times, to cover very different areas, with the latest incarnation designed to delineate the mainly Hispanic districts in Chicago. It has previously been two unconnected areas and even today is only connected to the West by a narrow band, Interstate 294.

There have been accusations that redistricting to group Latino populations is racially biased. The impact can be seen to pool votes either to ensure support, or to concede victory in favour of making gains elsewhere. Whatever the debates about the motivation, it impacts the aggregation of votes, and the map.

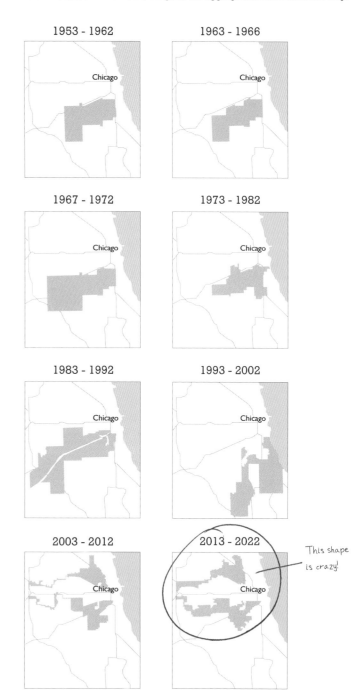

*road network as of 2018 for context

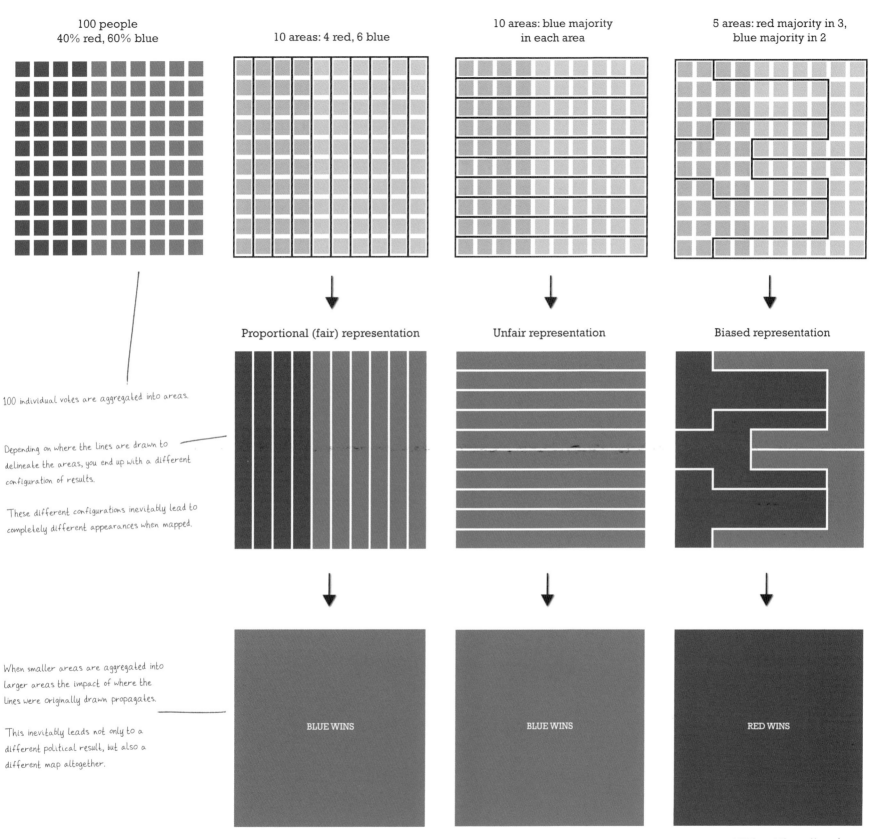

100 people
40% red, 60% blue

10 areas: 4 red, 6 blue

10 areas: blue majority
in each area

5 areas: red majority in 3,
blue majority in 2

100 individual votes are aggregated into areas.

Depending on where the lines are drawn to delineate the areas, you end up with a different configuration of results.

These different configurations inevitably lead to completely different appearances when mapped.

Proportional (fair) representation

Unfair representation

Biased representation

When smaller areas are aggregated into larger areas the impact of where the lines were originally drawn propagates.

This inevitably leads not only to a different political result, but also a different map altogether.

BLUE WINS

BLUE WINS

RED WINS

40% red and 60% blue but the result can be red simply due to where the lines are drawn!

Colouring the map

Designing colour schemes with purpose

Picking colours for a map might seem a trivial aspect of the design, but think about any picture, artwork, or patterned shirt to which you recall having a visceral reaction. Colour is emotive. Bright colours can be brash and shouty. Dull colours might be depressing. Colour also has connotations such as blue for rivers or green for forests. **People often have favourite colours and, conversely, colours they do not like. It's not a simple choice.**

In designing a colour palette for thematic maps, you should consider the topic and ask yourself if there's a colour scheme, or a key anchor colour that has some connection to what you're mapping. Often, there isn't, and so your palette can be drawn from a wider choice of hues, and you might go for your likes rather than dislikes. **When you're mapping a topic such as election data, you're obliged to consider how colour is used more generally in maps, logos, and other material that people already are familiar with.** In the US, since 2000, the two main parties are generally referenced by their colours, red for the Republican Party and blue for the Democratic Party. So these colours at least should be a good starting point for your own scheme. Perhaps check to see if there are colour specifications in cyan-yellow-magenta-black (CYMK) or red-green-blue (RGB) that reference actual colours. If so, these specifications might be good to use.

For the maps in this atlas, red and blue anchor colours were chosen that are a little softer than the reds and blues you see on placards and logos. Those are too bright for print and cause visual noise and disturbance (often as what appears to be flickering in the image at the boundaries between adjacent colours). They have a similar value (lightness) so one isn't more dominant than the other to avoid political bias. A unified palette was then developed of several lighter and darker versions of the same general colour, again keeping lightness and saturation constant at each step. This provides a good set of colours to use, particularly when it comes to mapping detail and need variations on a theme using different shades of colour.

It's also important to consider the impact of your colour choice on those with colour deficiencies, a condition which affects about 10 percent of the population. The map to the right illustrates how the chosen palette works for the most common colour deficiency, deuteranopia (no green vision). There's also a range of palettes for other deficiencies. The crucial check is that your chosen palette can be distinguished by people with colour deficiencies. The reds and blues in the colour scheme can be discerned by those with colour deficiencies. The colours also were checked by people who have colour deficiencies, and they confirmed the colours still worked for the purpose at hand.

Normal colour vision

Darker reds mean a higher Republican share of the vote and darker blues mean a higher Democratic share of the vote.

It's the relative value (or lightness) of the colours that we perceive and interpret as less to more so it's this characteristic that's important to preserve when looking at your palette through the eyes of others.

Republican   Democratic

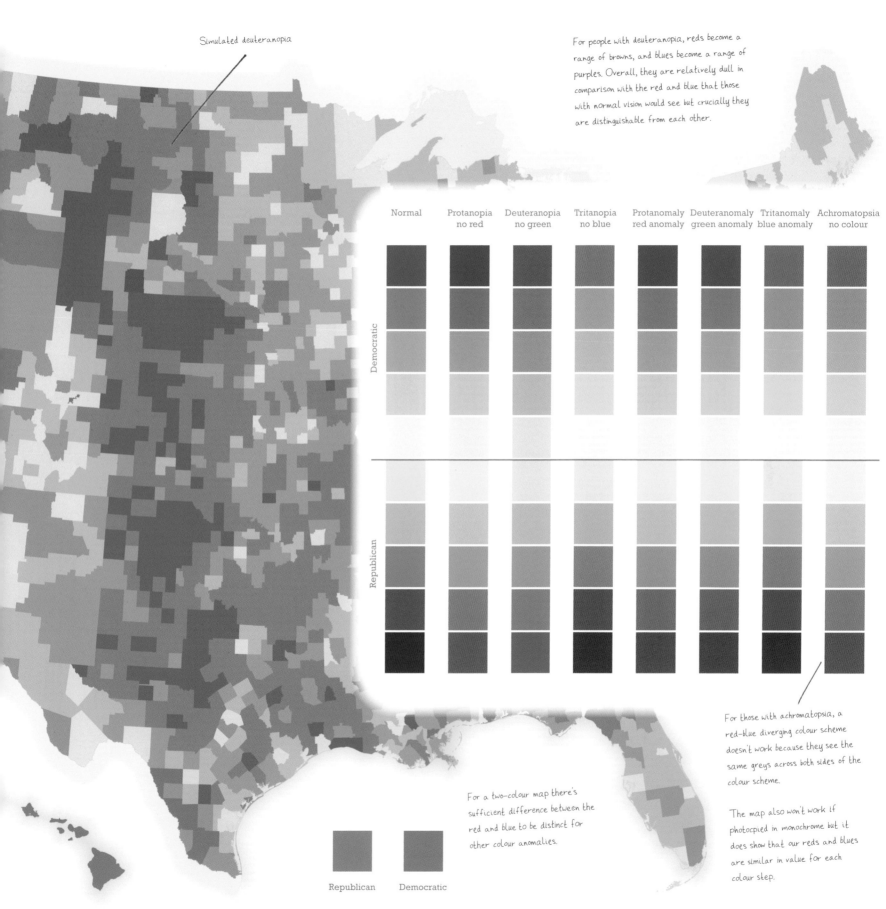

Simulated deuteranopia

For people with deuteranopia, reds become a range of browns, and blues become a range of purples. Overall, they are relatively dull in comparison with the red and blue that those with normal vision would see but crucially they are distinguishable from each other.

	Normal	Protanopia no red	Deuteranopia no green	Tritanopia no blue	Protanomaly red anomaly	Deuteranomaly green anomaly	Tritanomaly blue anomaly	Achromatopsia no colour
Democratic								
Republican								

For a two-colour map there's sufficient difference between the red and blue to be distinct for other colour anomalies.

For those with achromatopsia, a red-blue diverging colour scheme doesn't work because they see the same greys across both sides of the colour scheme.

The map also won't work if photocpied in monochrome but it does show that our reds and blues are similar in value for each colour step.

Republican   Democratic

Reference information

Adding detail that helps people interpret the map

Maps almost always require a wide range of ancillary information that helps the map reader interpret what they are reading. This is certainly true for topographic maps, which are loaded heavily with symbols and labels so map readers can see what is where and be able to interpret it in a meaningful way in terms of language. This isn't always the case with thematic maps, which traditionally have little contextual information. **The pattern of the data is the message, and so, labels and other marks on the map are kept to a minimum to avoid clutter and the inevitable problem of overprinting the mapped data itself.**

Yet without some detail, especially in areas unfamiliar to the map reader, the map often can be difficult to interpret. Understanding where highs and lows are, for instance, is useful, but being able to explain them in relation to cities, rural expanses, major industrial areas, and rich or poor neighbourhoods allows a way not only to express spatial patterns but also begin to explain them in relation to places people can relate to.

Labels can be important but they don't always have to be on the map. In printed form, a map might be part of a series, in which case a reference map can be used against which other maps are read. **Layers of labels or other contextual information might be contained in switchable layers on a web map or contained in a pop-up that appears when a place is clicked on the map.** These techniques keep the map clean while giving access to contextual detail when required.

Ideally, reference layers should be created for the mapping task at hand, which includes the basemap and any overlaying information such as boundaries, borders, and labels. But often you might not have ready access to such information, and for web maps, you might choose to use preconfigured data services rather than creating them yourself. If this is the case, be aware that it's highly unlikely they will suit your map perfectly because they are designed for general purposes and to meet a wide range of needs. But it's useful to keep in mind a few basic ways to optimise the use of third-party reference layers for print or web delivery.

The maps in this book are designed predominantly for print, and so they sit on white paper. This works equally well for web maps as you'll see if you visit the gallery of web maps that support this book, at esriurl.com/election2016. But, of course, there are alternatives. You may prefer a dark basemap, and so, colours may be inverted because light colours will appear as more and dark as less.

Poor

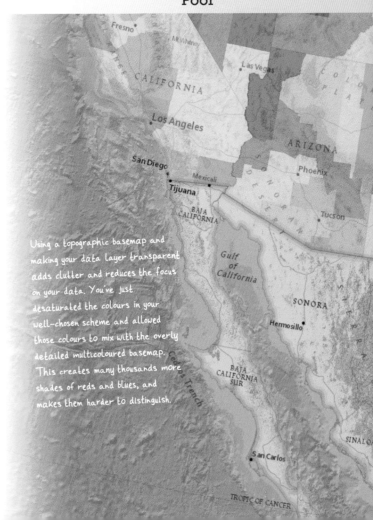

Using a topographic basemap and making your data layer transparent adds clutter and reduces the focus on your data. You've just desaturated the colours in your well-chosen scheme and allowed those colours to mix with the overly detailed multicoloured basemap. This creates many thousands more shades of reds and blues, and makes them harder to distinguish.

Regina

Winnipeg

Designing your own basemap and reference layers has many advantages as you fully control the choices you make.

Maybe the map doesn't need labels at all. A reference map elsewhere in a collection can provide that function. So can a pop-up if the map is digital.

NNESOTA

WISCONSIN

MICHIGAN

IOWA

OHIO

ILLINOIS   INDIANA

UNITED

Denver

Better   **Best for print**   **Best for web**

MISSOURI   KENTUCKY

Oklahoma City   ARKANSAS

TENNESSEE

Dallas

ALABAMA   GEO

MISSISSIPPI

aso

Thematic maps often function on their own without any additional reference detail.

For instance, there are no hard boundaries shown at all on this map. Even boundaries can be removed.

etter choice when using basemap
rvices is to go with neutral
lettes like light grey. Data
ntrasts better against a neutral
ackground and brings it into focus.

Austin

San Antonio   Ho

LOUISIANA

ou might not have much choice in
ypefaces, label position, size, and
ensity. If you can, make sure labels
it on top of the map.

Subtle design of type colour, halos, and position can make a big difference to the map's appearance. Selecting labels to appear only at the appropriate scale helps consistency.

Halos can
be harsh!

Static or dynamic

Going beyond the printed page

The death of paper maps has been greatly exaggerated. Paraphrasing Mark Twain's famous quote, there's no doubt that we're living in a digital age in which maps are a daily staple across our various news feeds. Maps are now portable. They exist in our pockets on our mobile devices. They have become ubiquitous in a way that paper maps never achieved. Yet paper is still an extremely valuable medium for the map. It's high resolution, tactile, and focusses the mind of the mapmaker because of the limitations of the static, printed environment impacting the design process. **The design of a good paper map is often not that different from a digital version of the same data, except that in a digital environment, you can take advantage of dynamic and interactive mechanisms for engagement.**

The properties of a static print map often serve as constraints on the map design process. Their fixed scale and size mean you must cope with possible difficulties occurring from the range of the dataset and its suitability at that scale and page size. Once a map is printed, there's no going back, unless you start again and replace the run. For fast-changing datasets, a static product cannot easily show changes over time and can become outdated rapidly. Hard-copy maps have edges, which often leads to the need for a map series on different sheets. Yet these facts are often useful in helping the mapmaker hone in on what's important to show. Of course, a static map doesn't need to be printed either. Web maps can be static in terms of their content but slippy so readers can pan and zoom.

Web maps offer additional capabilities. Maps can be multi-scale, meaning the level of detail can increase as you zoom in to larger scales. Symbols can be designed to work across these scales in a coherent way. Maps no longer have edges, yet the interface into the map, such as a mobile phone or tablet screen, means there's always an edge, and at larger scales, you can lose sight of context. The ability to hover or click on the map can reveal additional details through pop-ups or labels. These aids can include additional media or links that provide access to additional material.

Interaction is key to designing and using web maps. Animated maps help reveal how things might change over time. Positioning maps on a template to work alongside additional maps helps configure maps in an editorial style to tell specific stories, often called *story maps*. In 3D, you can rotate the map to overcome problems of occlusion.

Many of the maps in this book have a digital twin that accommodates these characteristics. It doesn't make the print maps inferior because they are designed for a static product. The digital versions at esriurl.com/election2016 illustrate many of the ways in which dynamic maps work.

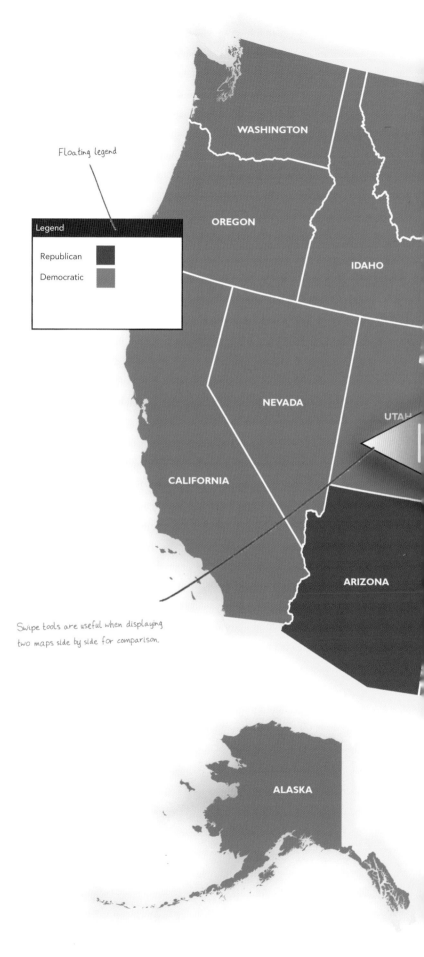

Floating legend

Legend

Republican

Democratic

Swipe tools are useful when displaying two maps side by side for comparison.

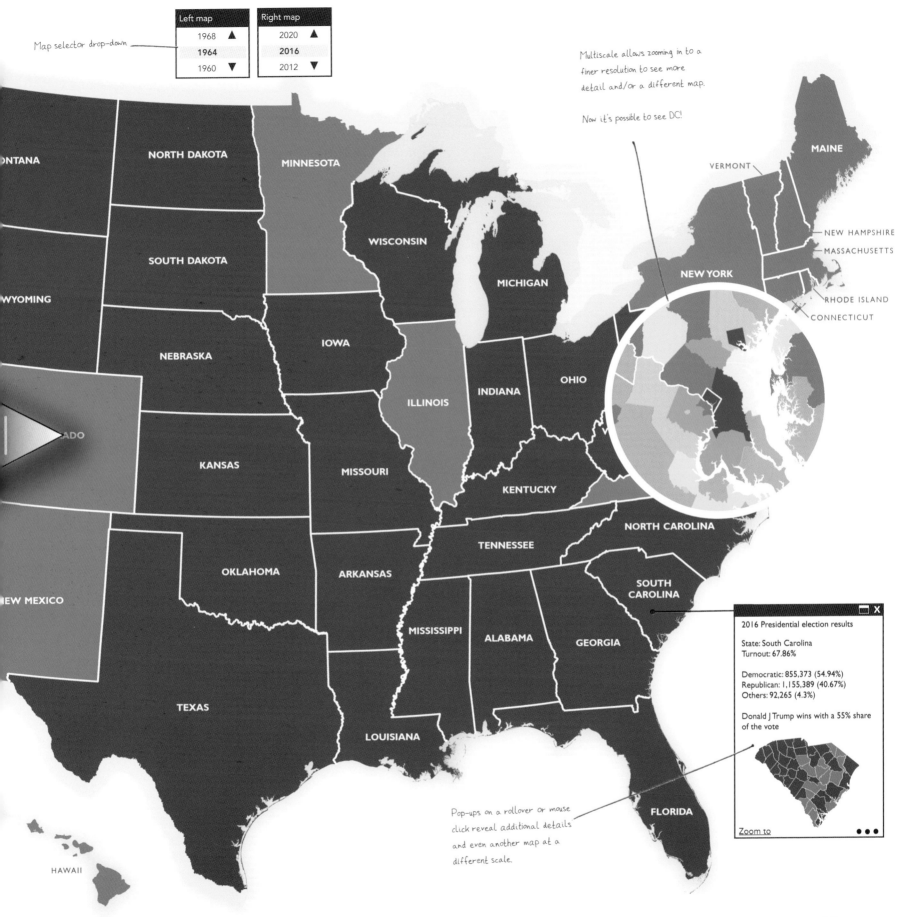

2. Area maps

The path of least resistance in making a thematic map often begins and ends with maps that show data in areas.

Population data usually is reported in aggregate form using administrative areas. Election data is no different because, ultimately, it's the aggregate nature of the counts that has meaning for the eventual result.

Areas can vary in size and scale from very small (precinct or district) to very large (state or even country). It therefore follows that arguably the most common types of maps to make for election data are those that report by these same areas. They are often the easiest to make because you likely will have the data already stored in the same areas, and so data manipulation is often unnecessary or rudimentary.

Even if the eventual map type doesn't use areas, virtually every map in this book is based on data collected as an areal count, although as seen in subsequent chapters, these maps can be manipulated into different geographies, metrics, or feature types.

This chapter explores the range of map types in which the focus is the area and how the decisions you make in processing and representing the data can modify the way the map looks and how it will be read.

Red and blue states

The bottom-line result as unique values

The purpose of a map that shows unique values is to represent how areas differ in type. The bottom line for the election is to show who won, and it doesn't need to be based on numerical information. **The map shows the outcome, categorised by the predominant party on the basis of the number of votes won which got them first past the post.**

Because states are the largest unit of aggregated votes, it's a good place to start the maps in this book. It's the most generalised form of reporting of the results. A candidate takes the state by winning more votes than their opponents in that state. They then capture the Electoral College votes allocated to that state, although it's worth noting that not all states allocate all Electoral College votes to the winner, and instead distribute proportionally.

The data for a unique values map is qualitative and represents the basic differences between features. There are only two categories, namely Republican or Democratic on this map. The two categories are symbolised with red and blue hues to match the political affiliations, but if the theme has no specific colour associations, the symbol would be designed as neutral and not value-laden. In this sense, hue is used as a differentiating visual variable because red is seen as different from blue. There is no quantitative association to them in relation to one another. By keeping value and lightness relatively constant, neither appears more important than the other at a symbolic level.

Importance (who won) comes from our ability to see how much red there is on the map compared with blue. However, as seen in later maps, this can be deceptive because the geography and population density lead to some incongruous effects on the map and the way empirical relationships between data and different geographies are perceived.

Ultimately, a map reader should be able to see efficiently the differences between the categories on a unique values map while not inferring any more importance to one category over another. **On this map, differences are shown between areas which are categorised as one of two alternatives so it's a binary representation of the outcome of the election.**

It's a simple map. It's useful as the headline because an individual's vote counts only in aggregated form. It tells us little of the detail of the results, but if detail is what you're more interested in showing, read on because the rest of the book is for you.

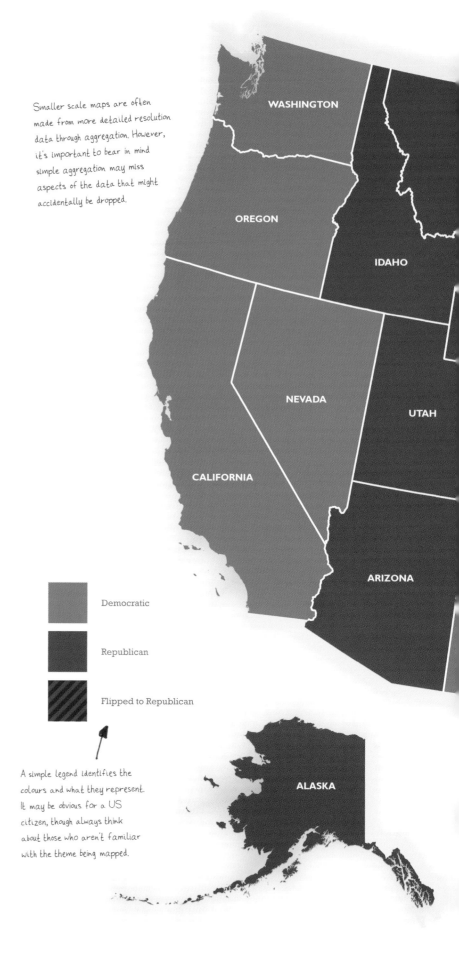

Smaller scale maps are often made from more detailed resolution data through aggregation. However, it's important to bear in mind simple aggregation may miss aspects of the data that might accidentally be dropped.

Democratic

Republican

Flipped to Republican

A simple legend identifies the colours and what they represent. It may be obvious for a US citizen, though always think about those who aren't familiar with the theme being mapped.

Adding a subtle hatched fill of diagonal lines, using a darker complementary shade of red, to some states adds information to the map beyond which are red and which are blue states.

They indicate which states flipped from Democratic to Republican, which is an important aspect of the result.

Labels are hard to position for small areas so putting them off the map and using small leader lines is a good solution. Switch the text colour to work across the lighter background. Abbreviate where necessary.

There's a lot of red on this map compared with blue. In fact, 68% of this map is coloured in red compared with 32% coloured blue. Yes, the Republican Party won but geography visually exaggerates the size of the margin of victory.

MONTANA
NORTH DAKOTA
MINNESOTA
MAINE
WYOMING
SOUTH DAKOTA
VERMONT
NEW HAMPSHIRE
MASSACHUSETTS
WISCONSIN
MICHIGAN
NEW YORK
RHODE ISLAND
CONNECTICUT
NEBRASKA
IOWA
PENNSYLVANIA
OHIO
ILLINOIS
INDIANA
WEST VIRGINIA
DC
NEW JERSEY
DELAWARE
MARYLAND
COLORADO
KANSAS
MISSOURI
KENTUCKY
VIRGINIA
NEW MEXICO
OKLAHOMA
ARKANSAS
TENNESSEE
NORTH CAROLINA
SOUTH CAROLINA
TEXAS
MISSISSIPPI
ALABAMA
GEORGIA
LOUISIANA
FLORIDA
HAWAII

Red and blue counties

Unique values at a finer level of detail

The map to the right uses the same technique as the states map on the previous page but with more detail. **It's a unique values map that shows which candidate won each county based on the number of votes cast in that county.** It's qualitative and shows nothing of the share of the vote or extent of the margin of victory.

The picture it paints of the pattern of voting in the US is, however, different from the state map now that there is more detail. It shows the way in which a state-level map often hides what goes on within a state. States are rarely entirely red or blue, even though it's the state-level aggregation that counts for determining the winner in terms of Electoral College votes.

Not everyone in a state votes the same way, which is what the state map might seem to suggest if you're unaware of how the election process works. Most states contain expanses of one colour over another, pockets of one colour surrounding another, or a patchwork of different voting preferences. **This map lends detail to the understanding of patterns of voting because it shows intrastate variation.**

However, these sorts of summary maps at a fine level of detail can have unintended consequences. They can marginalise those who voted differently. They can overemphasise or exaggerate a victory, perhaps in an area in which a marginal victory was claimed. Whether intentional or otherwise, the map becomes a powerful tool when used in a stark and binary, yes or no, red and blue form. It removes many people from the map who voted a different way from the winning candidate, potentially leaving people feeling disenfranchised.

At this point, it's worth introducing the idea that you often make a map for a particular audience. If you're making a map to show your base the extent of victory, why wouldn't you choose a map type and symbology like this? It's persuasive. It speaks to that narrative. One might even call it *propagandist* if used to support the rhetoric of a convincing, outright, and comprehensive victory. If you're making a map to explain the nuance in patterns of voting to a general audience, or you're on the losing side and don't wish to overplay the extent of a loss that might be closer than the map suggests, maybe this map doesn't suit your purpose.

This map is more detailed than the state version because of the resolution of the data being shown. Whether it gives a better sense of the results is debatable.

State boundaries are in a heavier line weight to distinguish them from the county boundaries, and also to denote their hierarchy as a larger spatial unit of aggregation.

Democratic

Republican

Showing counties that flipped using the same symbology as the state map wouldn't work because they are too small, at this scale, to make the symbology work effectively.

Most states show some intra-state variation, but there are a few states where every county result was the same.

Vermont, Massachusetts, and Hawaii all went Democratic, as did Washington, DC. Oklahoma and West Virginia were entirely Republican at a county level.

MAINE

VERMONT

NEW HAMPSHIRE

MASSACHUSETTS

RHODE ISLAND

CONNECTICUT

NEW JERSEY

DELAWARE

MARYLAND

NTANA

NORTH DAKOTA

MINNESOTA

WISCONSIN

MICHIGAN

NEW YORK

PENNSYLVANIA

SOUTH DAKOTA

WYOMING

IOWA

NEBRASKA

OHIO

INDIANA

ILLINOIS

COLORADO

WEST VIRGINIA

DC

KANSAS

MISSOURI

KENTUCKY

VIRGINIA

NORTH CAROLINA

TENNESSEE

EW MEXICO

OKLAHOMA

ARKANSAS

SOUTH CAROLINA

MISSISSIPPI

ALABAMA

GEORGIA

TEXAS

LOUISIANA

FLORIDA

HAWAII

There's a quirk in Virginia where its cities are also counties. This is a really challenging map problem as the highest density populated areas are tiny on an area map of counties.

County boundaries are represented in a faint light grey line with a bit of transparency thrown in for good measure. This gives the map structure, avoiding the colours simply being a large blob.

Around 82% of this map is coloured red compared with only 18% coloured blue. That's 2,638 red counties compared with 504 blue. This increases the proportion of red compared with blue in comparison with the state map of unique values. The map suggests a much larger win for the Republican Party simply by virtue of the quirks of geography, and population density. On the face of it though, this is quite the victory!

Shades of truth

Share of the vote as a choropleth

Although unique values maps give a binary picture of the bottom-line result, there's often much more nuance in the data that can be explored and communicated in different ways, as the rest of this book explores.

Choropleth maps are a common technique for representing data for areas by variations in shading or pattern. The rationale is easily understood by map readers, which is a large part of its success, though it's often seen as somewhat boring by today's mapping fashions. The main map, to the right, shows the percentage share of the vote gained by the Republican Party at county level.

Data is discrete, with a single value representing each area. **Data is also normalised into percentage rates to account for the underlying heterogeneity of the population and unevenness of the geography.** The idea of a choropleth is to show areas that display similar characteristics. To achieve this, data is classified into groups which have some meaningful class breaks, and areas which are similar are symbolised the same.

This map uses a red hue to match the political affiliation of the Republican Party. **Shades modify the basic red, altering the value of each symbol to create colours that are evenly stepped from light to dark, matched to the low to high data classes, respectively.** This communicates the geographical distribution of the share of the vote across the map. The map reader should be able to recognise efficiently the different classes mapped across the map and to see the areas of high and low. Darker symbols (on a light background) are perceived as meaning 'more'.

The map below shows totals (absolute values) as an example of how not to make a choropleth map. It takes no account of the underlying geography or population distribution, leading to symbols that cannot be compared like for like across the map.

This map only shows the Republican share of the vote. We'd need another map to show the Democratic share of the vote. It's also hard to identify which counties the Republicans actually won unless we had a clear class break at 50%.

Legend must-haves: title, units of measurement, class intervals, colour swatches. Job done!

Republican share of vote (%)

> 76

66 – 76

53 – 65

38 – 52

< 38

Scale is a big factor in what to include on a map. Here, labels and coastal vignette are omitted for clarity.

This map shows total votes by county. Consequently it's a hugely misleading view of exactly the same data as the main map.

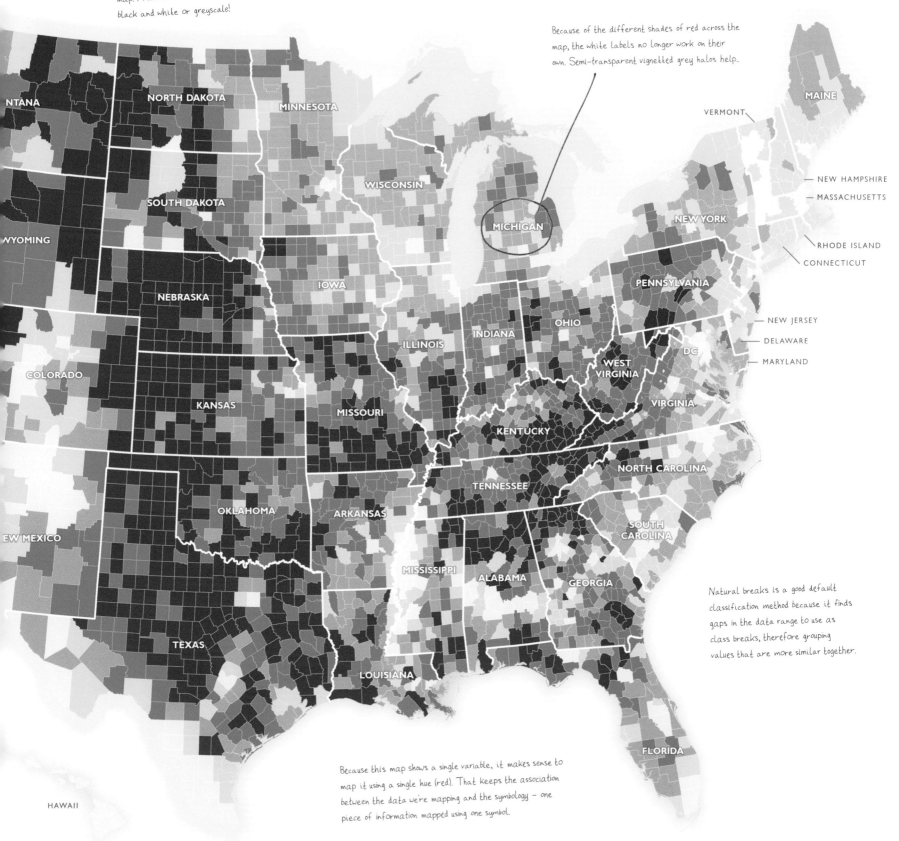

It may be red but this is also a monochrome map. Monochrome doesn't have to refer to black and white or greyscale!

Because of the different shades of red across the map, the white labels no longer work on their own. Semi-transparent vignetted grey halos help..

NTANA

NORTH DAKOTA

MINNESOTA

WISCONSIN

MICHIGAN

MAINE

VERMONT

NEW HAMPSHIRE

MASSACHUSETTS

NEW YORK

RHODE ISLAND

CONNECTICUT

SOUTH DAKOTA

WYOMING

IOWA

NEBRASKA

ILLINOIS

INDIANA

OHIO

PENNSYLVANIA

NEW JERSEY

DELAWARE

MARYLAND

COLORADO

KANSAS

MISSOURI

WEST VIRGINIA

DC

VIRGINIA

KENTUCKY

NORTH CAROLINA

EW MEXICO

OKLAHOMA

ARKANSAS

TENNESSEE

SOUTH CAROLINA

Natural breaks is a good default classification method because it finds gaps in the data range to use as class breaks, therefore grouping values that are more similar together.

TEXAS

MISSISSIPPI

ALABAMA

GEORGIA

LOUISIANA

FLORIDA

HAWAII

Because this map shows a single variable, it makes sense to map it using a single hue (red). That keeps the association between the data we're mapping and the symbology — one piece of information mapped using one symbol.

Seeing things differently

Processing and classifying the data

Classifying data is often fundamental when making thematic maps because you're trying to make a map to show which places are like one another or where the highest or lowest value is. Classification sets up the ability to explore the map through comparison, yet there's no single best method for classifying data. To the right, you see seven methods that add to the natural breaks method used on the previous page. Each map has a different appearance. Some show a lot of midvalues. Some highlight the upper and lower values. **But which is the right classification method? The answer is all of them.** The decision of which technique to use is governed by your map reader's needs. The question is better framed as, what do you want to show and what technique is best suited to the data distribution?

Once the method is chosen, how many classes do you choose? Too few, and you risk overgeneralising your map and showing areas that are quite different from one another using the same symbol. Too many, and your map won't easily support comparison because it'll be difficult to see precisely which shade of colour you're seeing in each area. **The rule of thumb is five to seven classes works well to balance the limits of visual perception while also showing enough detail to make comparisons meaningful.** The histogram below shows how the data distribution works with the methods for this dataset.

Classification of any dataset means you're losing fidelity in the data for the benefit of a visual summary. No one will be able to recover the precise value for an area simply by looking at the map. The best they can do is recognise that it falls between the adjacent class breaks. If you're designing the map for a digital environment, you'll have the bonus of enabling a rollover or pop-up to show the value of the data for each area. **Whatever your choice of classification method, make that choice intentionally.**

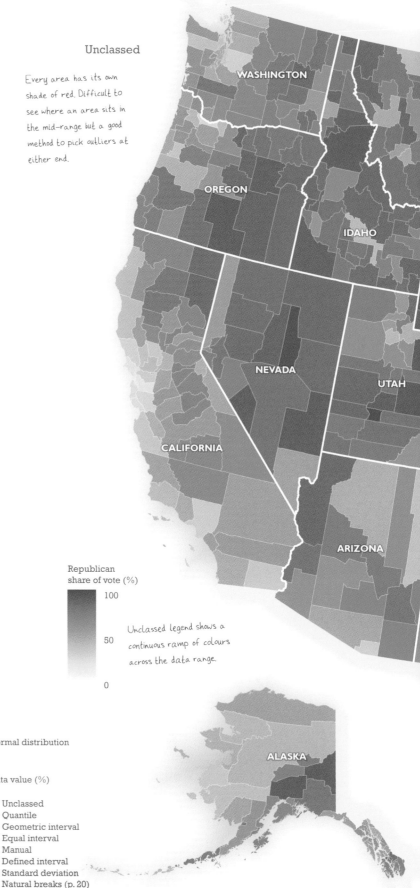

Unclassed

Every area has its own shade of red. Difficult to see where an area sits in the mid-range but a good method to pick outliers at either end.

Republican share of vote (%)

100

50

Unclassed legend shows a continuous ramp of colours across the data range.

0

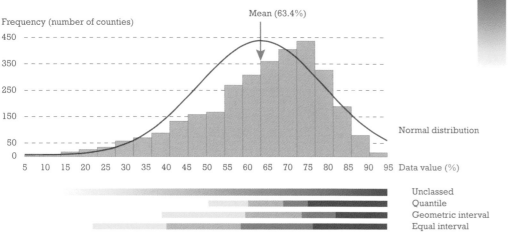

Frequency (number of counties)

Mean (63.4%)

Normal distribution

Data value (%)

Unclassed
Quantile
Geometric interval
Equal interval
Manual
Defined interval
Standard deviation
Natural breaks (p. 20)

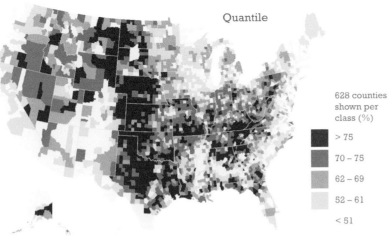

Quantile

628 counties
shown per
class (%)

- ■ > 75
- ■ 70 – 75
- ■ 62 – 69
- ■ 52 – 61
- ■ < 51

Data is ranked and an equal number of data values is placed in each class. Useful for multiple maps when each shares the same scheme to support comparison. Some similar values may be shown in different, adjacent classes or wildly different values may be in the same class.

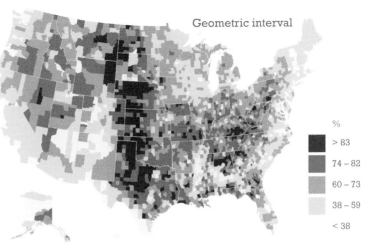

Geometric interval

%

- ■ > 83
- ■ 74 – 82
- ■ 60 – 73
- ■ 38 – 59
- ■ < 38

The class interval has a geometric series and works well for continuous data. It balances highlighting changes in the mid-values and extreme values.

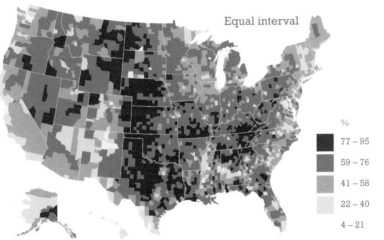

Equal interval

%

- ■ 77 – 95
- ■ 59 – 76
- ■ 41 – 58
- ■ 22 – 40
- ■ 4 – 21

Each class has a fixed interval based on division of the whole range by number of classes. Here, a 17% range of data per class. Often not the best choice because you may end up with empty classes or, conversely, a lot of data values in one class.

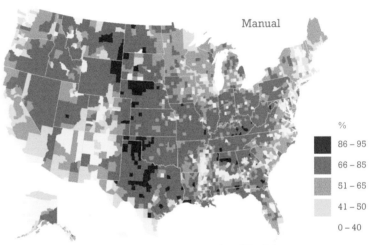

Manual

%

- ■ 86 – 95
- ■ 66 – 85
- ■ 51 – 65
- ■ 41 – 50
- ■ 0 – 40

Requires YOU to look at the distribution of the data range and make decisions about class breaks. It's often used to round equal-interval class breaks to make them less cumbersome.

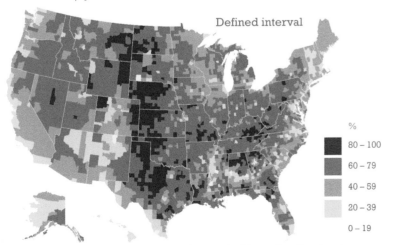

Defined interval

%

- ■ 80 – 100
- ■ 60 – 79
- ■ 40 – 59
- ■ 20 – 39
- ■ 0 – 19

Similar to equal interval but each class is defined by specifying the interval value itself, and the number of classes is calculated automatically. Often leads to a lot of mid-range symbols for data that is normally distributed.

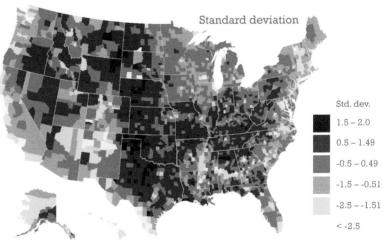

Standard deviation

Std. dev.

- ■ 1.5 – 2.0
- ■ 0.5 – 1.49
- ■ -0.5 – 0.49
- ■ -1.5 – -0.51
- ■ -2.5 – -1.51
- ■ < -2.5

Classifies each data value with respect to the mean of all data values. Can be confusing because SD isn't easily understood.

Side by side

Two choropleths to tell two sides of the story

The two preceding choropleth maps used the Republican share of the vote as the basis for the map. Because those maps are seen on their own, the choice of classification method and the symbolisation schemes need to work only internally. The reading task is to compare one place to another across that single map. Many map-reading tasks require a comparison of one map to another or possibly with many more. When multiple maps are compared visually, the reader shouldn't have to process different classification and symbolisation schemes, which complicates matters considerably. **Multiple maps should be constructed to support visual comparison between the maps and not simply within each map.**

The natural counterpoint to a Republican share of the vote map is the equivalent for the Democratic share of the vote. Any of the classification schemes discussed earlier could be applied to the Democratic map if you were making it in isolation, but for comparison with the Republican map, consistency is important. For instance, if you chose a natural breaks method for each map, they would work independently but not as a pair. That's because your eyes are looking to see whether the patterns are similar or different for areas that interest you. That requires assessment against a scheme of light to dark as the overall hue denotes the party for each map. Natural breaks yields classes containing an irregular number of counties per class, and these classes undoubtedly will differ for each map. It makes the task of comparing maps harder and open to misinterpretation.

A far better approach is to use a quantile or equal-interval classification scheme for both maps. Remember, a quantile scheme puts the range of data values in rank order and allocates an equivalent number of data values to each class, regardless of data value. This means that each map has the same number of counties in the highest class and the next class, and so on. It sets up a consistent basis for visual comparison, and you can state with certainty where a county sits on both maps in terms of its rank among all values.

To the far right, the two maps are imagined as a web map in which a tool or gesture allows swiping between them. This is a common approach for comparing maps in a web browser. It supports the ability to cast an eye on one place and swipe between the maps to reveal the similarity or contrast. To the immediate right is a layout that would be used if the maps were displayed statically. **Placing them in proximity either above and below or side by side helps reduce the friction of distance between them, which might otherwise lead to difficulties in identifying precisely which area is being compared.**

Multiple-map layout

Election maps with only two candidates will inevitably look like the inverse of one another!

For static display, maps positioned next to one another is the only real solution and requires the reader moving their gaze between maps. Boundaries become important visual anchors.

A swipe tool is an effective way to use screen space in a web map for the display and comparison of two maps. Readers can fix their gaze on one place, and swiping back and forth enables processing of the differences between variables on the two maps.

MONTANA
NORTH DAKOTA
MINNESOTA
MAINE
VERMONT
WISCONSIN
NEW HAMPSHIRE
MASSACHUSETTS
SOUTH DAKOTA
MICHIGAN
NEW YORK
WYOMING
RHODE ISLAND
CONNECTICUT
IOWA
PENNSYLVANIA
NEBRASKA
NEW JERSEY
OHIO
DELAWARE
ILLINOIS
INDIANA
DC
MARYLAND
COLORADO
WEST VIRGINIA
KANSAS
MISSOURI
VIRGINIA
KENTUCKY
NORTH CAROLINA
TENNESSEE
OKLAHOMA
ARKANSAS
NEW MEXICO
SOUTH CAROLINA
TEXAS
MISSISSIPPI
ALABAMA
GEORGIA
LOUISIANA
HAWAII
FLORIDA

Share of vote (%)
628 counties per class

Republican	Democratic
> 76	> 44
66 – 76	33 – 43
53 – 65	25 – 32
38 – 52	20 – 24
< 38	< 19

Both these maps have a five-class scheme, meaning 20% of data values in each dataset are allocated to each class, but the class breaks will be different for each given the data distribution. Important to show this.

Two maps in one

Choropleth with a diverging colour scheme

Using two maps to show two datasets on the same page allows you to look at the relationship of the two pieces of information for each area of interest. Yet there's an argument that using two maps can be cumbersome because you're constantly having to visualise two things to process the information and compare one place to another. Additionally, on an election map, it's arguable that each map contains a lot of redundancy because you're likely not too concerned about the areas in which a candidate lost. They lost. Full stop (period).

By accepting that you're interested only in focussing on information about the candidate who won, you can return quickly to the single-map approach. This map to the right effectively shows the winning candidate through a shade of hue as before, red for Republican and blue for Democratic. It also shows the margin of victory from light to dark. **Because you're ignoring the data for the candidate that lost each area, redundancy of data that isn't particularly useful or important in an overview map is avoided.** This, then, is a choropleth map that uses a diverging colour scheme to show the winner for each area and how large their victory was in percentage points.

Diverging choropleths do a good job of showing the relationship for a single theme of information that can be split into two components: the overall qualitative difference in the mapped phenomenon (political affiliation denoted by hue) and the quantitative measure for each. For political maps having potentially different outcomes for each area, this means showing the margin of victory in each area for the winning candidate of that area.

For this map to work, the classification scheme must be consistent for both the Democratic and Republican data. Quantiles would work, though as seen on the previous page, they end up with different class intervals, which can be confusing when seen in a single diverging legend. Equal interval works well because it simplifies a more complex legend. The compromise is that it may not be the best way to represent either of the two sides of data. **Compromise is a recurring theme in this book because cartography is always a compromise, or else there would be a single best way to make the map.**

Of all the maps thus far, this one shows increased levels of detail in the data being mapped. Data manipulation, classification, selection, omission, and symbology help bring nuance to the map. **This map invites greater curiosity and reveals more than the simple red and blue or single-hue choropleths. It begins to answer questions about who won, where, and by how much, all in a single map.**

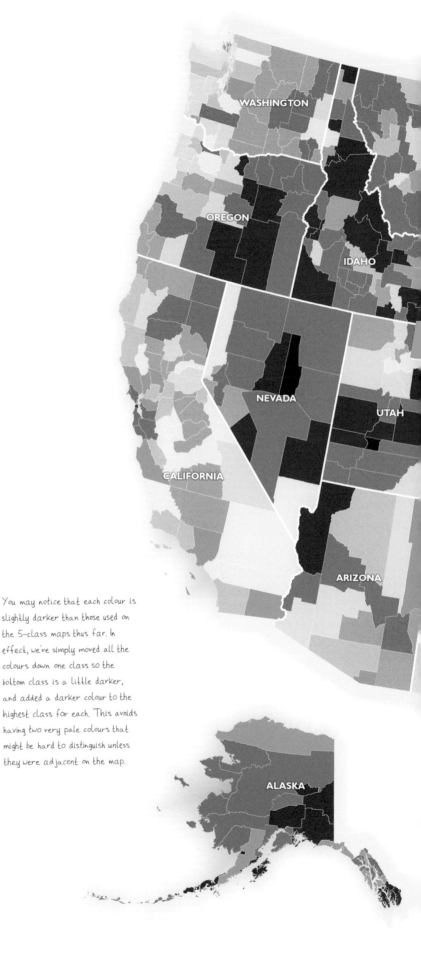

You may notice that each colour is slightly darker than those used on the 5-class maps thus far. In effect, we've simply moved all the colours down one class so the bottom class is a little darker, and added a darker colour to the highest class for each. This avoids having two very pale colours that might be hard to distinguish unless they were adjacent on the map.

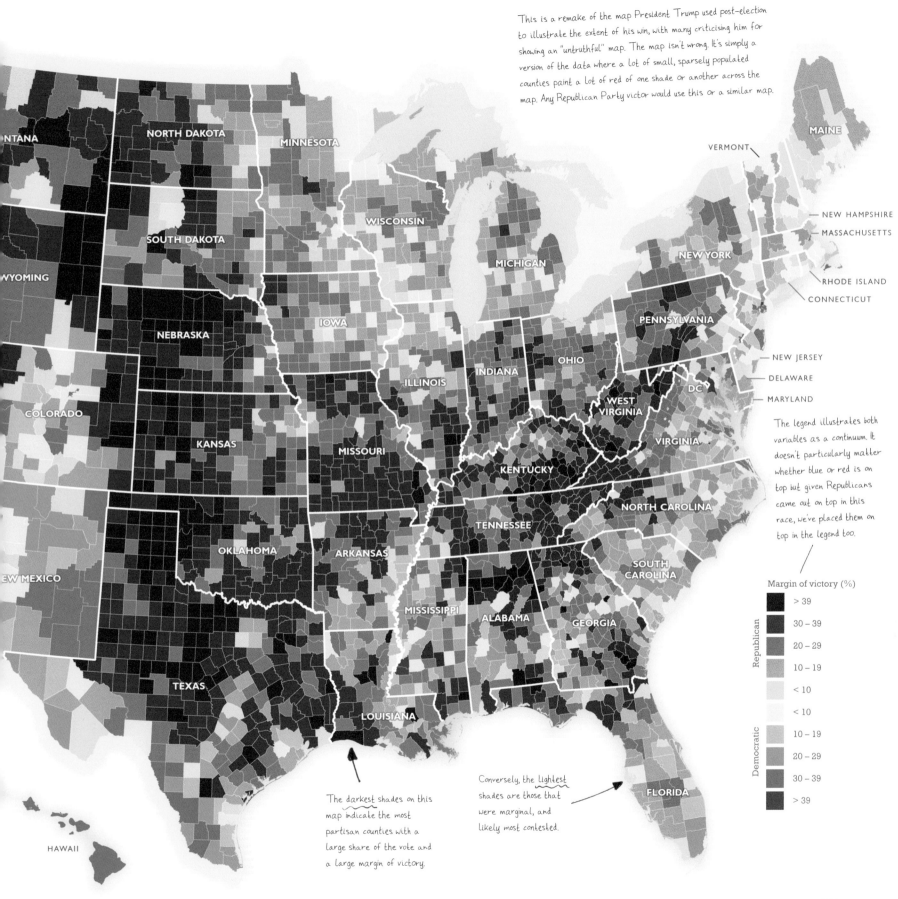

This is a remake of the map President Trump used post-election to illustrate the extent of his win, with many criticising him for showing an "untruthful" map. The map isn't wrong. It's simply a version of the data where a lot of small, sparsely populated counties paint a lot of red of one shade or another across the map. Any Republican Party victor would use this or a similar map.

The legend illustrates both variables as a continuum. It doesn't particularly matter whether blue or red is on top but given Republicans came out on top in this race, we've placed them on top in the legend too.

Margin of victory (%)

Republican

> 39
30 – 39
20 – 29
10 – 19
< 10

Democratic

< 10
10 – 19
20 – 29
30 – 39
> 39

The darkest shades on this map indicate the most partisan counties with a large share of the vote and a large margin of victory.

Conversely, the lightest shades are those that were marginal, and likely most contested.

Red and blue makes purple

Choropleth with a blended-hue colour scheme

To this point, red and blue has been used as the colour scheme to denote the difference between the Republican Party and Democratic Party. The value of the hues to encode some numerical information has been varied, which leads to maps in which light hues represent the areas with a lower share of the vote, or margin of victory, and a more marginal, bipartisan mix of voters. Darker hues show areas of more partisan support. **In some ways, the red and blue maps might paint a picture of a polarised electorate with harsh lines of division at boundaries, and despite marginal results, an area is still red or blue.** By changing the colour scheme from a diverging one to a blended one, a different picture emerges.

Blending red and blue gives purple. It's basic colour mixing, and it leads to a metaphor that purple becomes the dominant colour in areas in which there is a narrower margin of victory. **The use of a blended-hue colour scheme focusses attention on the amount of purple across the map, leading to the belief that, in fact, the electorate may be closer in their voting habits than other maps might suggest.** The winner-takes-all approach to counting votes and declaring a winner often makes the outcome seem more partisan than it is. Making marginal victories a purple midpoint across the scale from strong Republican to strong Democratic majorities helps reveal that picture.

The overall map becomes a series of purple shades. They tend toward a reddish hue as the margin of victory becomes increasingly Republican and toward blue as the margin tends toward the Democratic side. The map perhaps shows an America that is not as divided as diverging hue or unique values maps, especially, seem to show.

A drawback of the approach is that you have a map of colours with similar values and saturation. This similarity presents problems for those with colour deficiency to recognise the subtleties of shifts across a purple spectrum. It's also difficult to identify whether a purple shade that is close to the midpoint tends toward blue or red. It becomes difficult to decipher and is further compromised by adjacent colours. A marginal red county easily can be confused with a marginal blue one, depending on the surrounding areas that trick our eyes into seeing colour as much as a function of its surroundings as the colour itself. With the red and blue unique values maps, all you need to do is discern two distinct hues, whereas here, you see a lot of indistinct purples. You end up trying to process a colour as tending toward some measure of red or blue anyway.

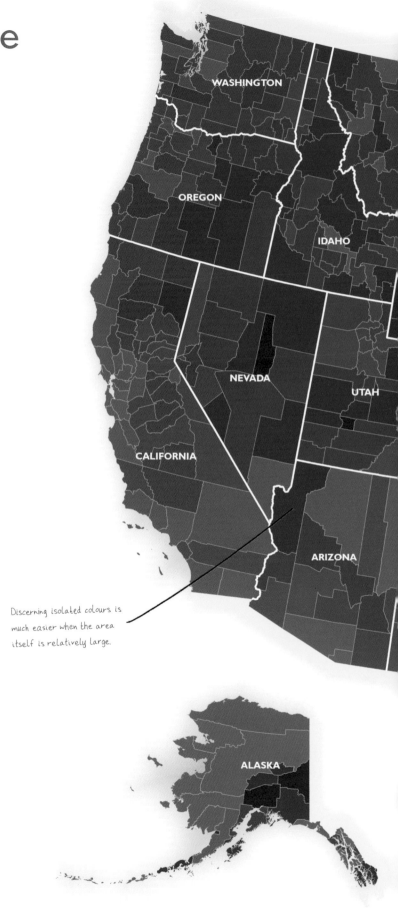

Discerning isolated colours is much easier when the area itself is relatively large.

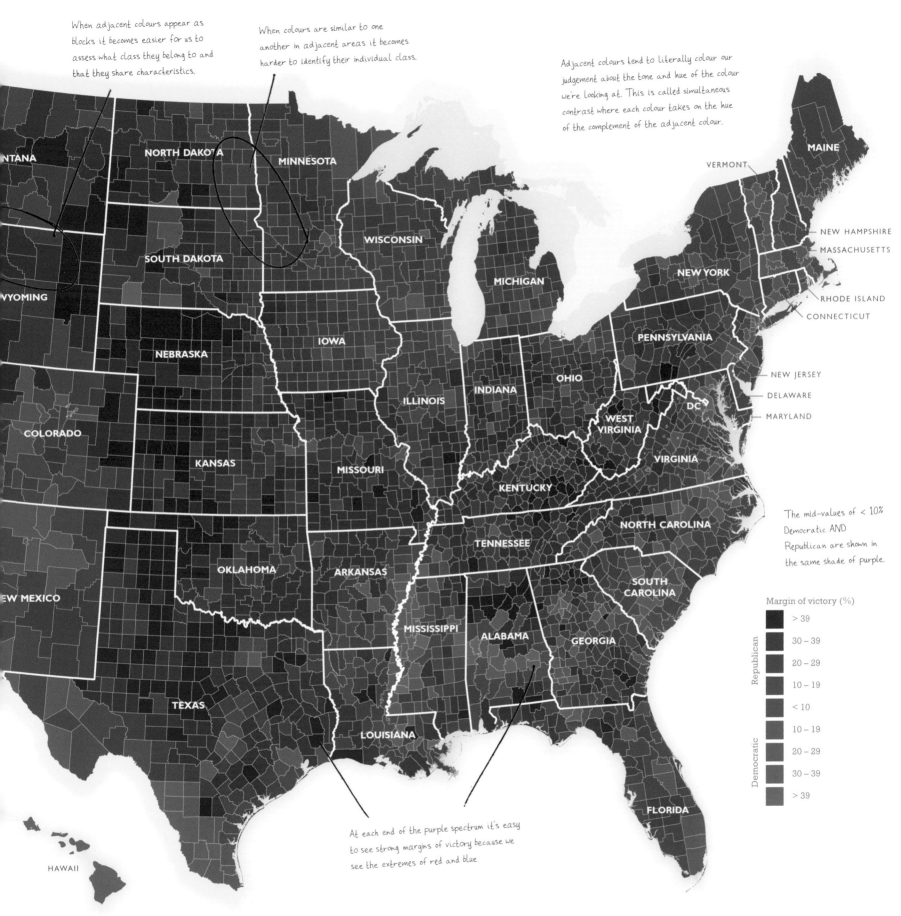

When adjacent colours appear as blocks it becomes easier for us to assess what class they belong to and that they share characteristics.

When colours are similar to one another in adjacent areas it becomes harder to identify their individual class.

Adjacent colours tend to literally colour our judgement about the tone and hue of the colour we're looking at. This is called simultaneous contrast where each colour takes on the hue of the complement of the adjacent colour.

The mid-values of < 10% Democratic AND Republican are shown in the same shade of purple.

At each end of the purple spectrum it's easy to see strong margins of victory because we see the extremes of red and blue

Margin of victory (%)

Republican
> 39
30 – 39
20 – 29
10 – 19
< 10

Democratic
10 – 19
20 – 29
30 – 39
> 39

De-emphasising land mass

Value-by-alpha choropleth

Mapping share of the vote or margin of victory treats every area the same and shows a percentage measurement regardless of the number of votes. No account is taken of whether a county is densely populated or sparsely populated. A county with 10 people in which 7 voted Republican has the same share of the vote as a county of 10,000 in which 7,000 voted Republican. Both are 70 percent. And numbers matter because counties add up to totals at the state level and play a part in the distribution of Electoral College votes. **More people (and votes) do not necessarily show up when you look at the share of the vote or margin of victory maps. Many more people live in small, densely populated places than large, sparsely populated places.**

The value-by-alpha technique adds a secondary layer of data to the map, laid over the main map. Because it is symbolised by varying the alpha channel (transparency), it works by bringing into focus some areas while visually muting others. **The alpha layer represents population density to modify how you see the underlying blended-hue choropleth.** Areas with high population density are shown close to their original colour because the alpha layer tends toward fully transparent. Where there exists a low population density, the alpha layer tends toward opaque. Because the base colour is white, the increasingly opaque overlaid layer makes that area recede toward the same colour as the paper itself, making it less prominent.

This technique introduces a visual correction to the map to accommodate widely varying population (and hence, voting) densities that are a function of administrative boundaries. **It gives visual prominence to the distribution of the results of the voting process by population, and not just the distribution by land surface area.**

Margin of victory (%) as seen on the blended-hue choropleth map

25	
	25
22	
	21

With population density alpha layer (people per sq. mi)

16	
	1
73	
	2

These counties in Utah had very similar margins of victory for the Republican Party, yet have massively different population densities.

The main map (right) uses the blended colour scheme for the underlying margin of victory map. There's still a lot of purple. Virtually all the strong Democratic areas remain visible.

When we use the diverging colour scheme (left) the map still shows the strong Democratic areas but a lot of the dark red has been muted due to a sparse population. That said, there's still more red on the map because many of the marginal share of the vote areas that are purple on the main map revert to a red hue on the diverging map.

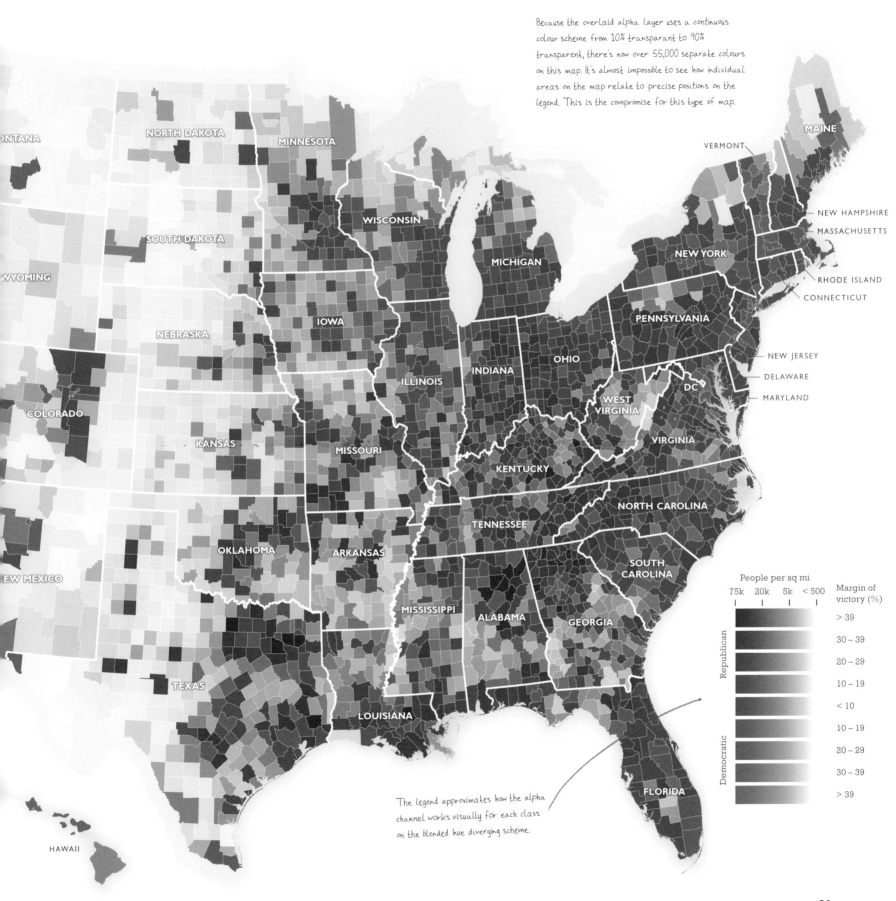

Because the overlaid alpha layer uses a continuous colour scheme from 10% transparent to 90% transparent, there's now over 55,000 separate colours on this map. It's almost impossible to see how individual areas on the map relate to precise positions on the legend. This is the compromise for this type of map.

MONTANA

NORTH DAKOTA

MINNESOTA

MAINE

WYOMING

SOUTH DAKOTA

WISCONSIN

VERMONT

NEW HAMPSHIRE

MASSACHUSETTS

NEW YORK

MICHIGAN

RHODE ISLAND

CONNECTICUT

NEBRASKA

IOWA

PENNSYLVANIA

COLORADO

ILLINOIS

INDIANA

OHIO

NEW JERSEY

DELAWARE

MARYLAND

DC

WEST VIRGINIA

KANSAS

MISSOURI

KENTUCKY

VIRGINIA

NEW MEXICO

OKLAHOMA

ARKANSAS

TENNESSEE

NORTH CAROLINA

SOUTH CAROLINA

TEXAS

MISSISSIPPI

ALABAMA

GEORGIA

LOUISIANA

People per sq mi

75k   20k   5k   < 500

Margin of victory (%)

> 39

30 – 39

20 – 29

10 – 19

< 10

10 – 19

20 – 29

30 – 39

> 39

Republican

Democratic

FLORIDA

The legend approximates how the alpha channel works visually for each class on the blended hue diverging scheme.

HAWAII

Using visual effects for emphasis

Choropleth with hillshading

The idea of using graphical effects to bring focus to certain parts of a map is a natural progression from the principles that lie behind the value-by-alpha approach. **Using ancillary data in combination with an underlying choropleth changes the way you see the distribution of a phenomenon.** If you think of the impact of geography on the overall map, you might suggest population density is an important characteristic to incorporate into the design. Population density already has been used to modify a blended-hue choropleth map but maybe that didn't go far enough. In addition to visually muting areas that have low population densities, you may want to add emphasis to those areas that have high relative population densities. To achieve this, you can apply an additional graphical effect.

On topographic maps, you often see elevation symbolised not only with colours (hypsometric tints), but also other mechanisms to accentuate the impression of relief, such as hillshading. Such effects are designed to emphasise the physical character of a landscape and help you more clearly see the 3D nature of relief—albeit, on a flat, planimetric map. Steeper slopes create longer and darker shadows, and that helps emphasise the more extreme terrain. **This well-established technique from topographic mapping can be applied to thematic mapping.**

Population data can be thought of as a statistical surface, and in that sense, it's no different from a topographical surface that has highs, lows, peaks, and troughs. Each area (county) has a value that is relatively higher or lower than others. If you imagined this in a 3D space, you'd see prisms of different heights (in fact, you'll see prism maps later in the book), and it becomes a 3D landscape to which you can apply a hillshade.

The simplest form of hillshade uses an imaginary light source in the top left, at an azimuth of 315° and at an angle (zenith) of 45°. **Casting light across the surface of the election results with population density represented in the z-dimension produces a range of different shadows.** In fact, the map to the right has 60 different light sources, which have been used to create an almost abstract paint-like quality as light bounces around the statistical landscape. It casts long shadows from small areas with large population densities and creates illuminated plateaus in which adjacent values are broadly the same and there are no shadows. It uses the value-by-alpha map as a base, but the visual effect of the hillshade adds further emphasis to the counties with high population densities that allows them to rise above the landscape below. It's more artistic than analytic but leads to an interesting aesthetic that shifts focus to the places you're emphasising.

There are 60 different light sources that generate the shadows but the top left remains the strongest source so we see heavier, dark shadows to the lower right of those areas that have high population densities. Higher population density areas are also emphasised as sun-lit plateaus.

This map uses three techniques in one. Firstly the blended-hue margin of victory chorpleth map, then the alpha layer modification, and finally the hillshade and shadows modification. Combining cartographic techniques can give rise to new and interesting ways of representing the data.

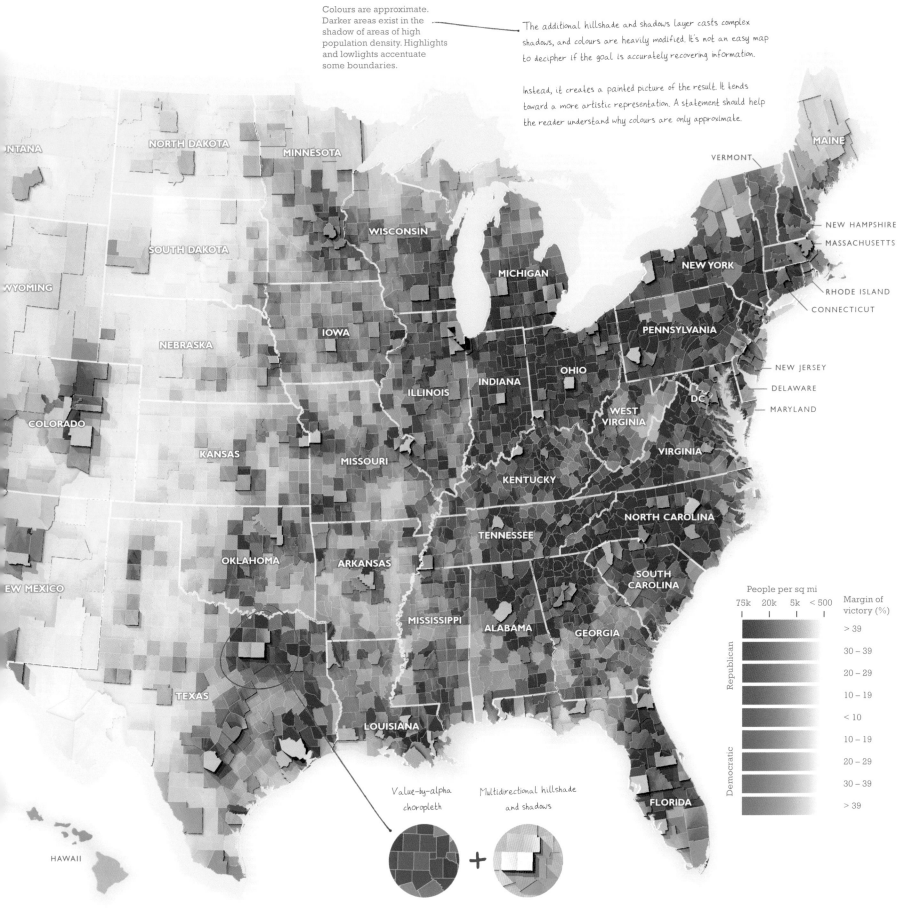

Colours are approximate. Darker areas exist in the shadow of areas of high population density. Highlights and lowlights accentuate some boundaries.

The additional hillshade and shadows layer casts complex shadows, and colours are heavily modified. It's not an easy map to decipher if the goal is accurately recovering information.

Instead, it creates a painted picture of the result. It tends toward a more artistic representation. A statement should help the reader understand why colours are only approximate.

MONTANA
NORTH DAKOTA
MINNESOTA
MAINE
VERMONT
NEW HAMPSHIRE
MASSACHUSETTS
WISCONSIN
SOUTH DAKOTA
WYOMING
MICHIGAN
NEW YORK
RHODE ISLAND
CONNECTICUT
IOWA
PENNSYLVANIA
NEBRASKA
ILLINOIS
INDIANA
OHIO
NEW JERSEY
DELAWARE
DC
MARYLAND
COLORADO
WEST VIRGINIA
KANSAS
MISSOURI
VIRGINIA
KENTUCKY
NORTH CAROLINA
NEW MEXICO
OKLAHOMA
ARKANSAS
TENNESSEE
SOUTH CAROLINA
TEXAS
MISSISSIPPI
ALABAMA
GEORGIA
LOUISIANA
FLORIDA
HAWAII

Value-by-alpha choropleth

Multidirectional hillshade and shadows

People per sq mi
75k   20k   5k   < 500

Margin of victory (%)

Republican
> 39
30 – 39
20 – 29
10 – 19
< 10

Democratic
10 – 19
20 – 29
30 – 39
> 39

Emphasising place, not space

Dasymetric reallocation to new areas

A standard choropleth map uses shades of colour to fill an entire area. Collectively, the entire map is shaded, and that gives you a sense of the distribution of the data from place to place. These are the area-based maps you've seen thus far, except there's a major limitation in all of them that's caused by geography. Put simply, some areas are very large, and some are very small. Some have a low populace which is sparsely distributed. Others have a high populace which is densely distributed. Small areas often are densely populated and large areas sparsely populated. It's a fact that is most inconvenient for cartography. This reality makes choropleth maps give you more a sense of the geography of the arbitrary spatial units used than the distribution of the data itself. **The dasymetric technique is based on the principle that areas used to collect and report population data (e.g. counties) do not have a homogenous population that exhausts space.**

The dasymetric approach makes use of an ancillary dataset that helps reallocate data from one set of areas to another. Most commonly, the ancillary data reflects populated places or some sort of landscape that delineates where people live from where they clearly don't. **This is not simply a case of using a mask to hide unpopulated areas, but instead, using the ancillary dataset to mathematically reallocate all data values to a new set of boundaries.** In the example to the right, the original data is the same county-level data of the election used elsewhere. The ancillary dataset is areas that reflect the impervious surfaces of the physical landscape.

Impervious surface data is a good proxy for reflecting characteristics of where people live. It includes mainly artificial surfaces, impenetrable materials, and rooftops. It effectively shows urbanity. For each county, the area of coverage of the additional data is used to calculate the proportion of data from the original dataset that should be reallocated. The process moves data from one large arbitrary area into many smaller areas that better reflect the distribution of people living in that area. The sum of data for all the small areas still will add up to the same total as the original dataset.

Although this approach does a good job of showing data relative to the distribution of people, not geography, the major downside is that any map that uses this approach will tend to take on the appearance of the structure of the ancillary dataset. It's arguable whether this is worse than mapping entire areas in a solid shade, and by showing the data in relation to the vast swathes of unpopulated land, shown as white space, you better locate where people live and vote.

Boundaries are eliminated altogether to show data as a continuous surface. City labels replace area labels so populated places can more easily be identified and described.

San Bernardino is the largest county in the contiguous US at 20,105 sq mi, and home to over 2 million people. Yet most of the population live in the southwest with the remainder being sparsely populated in mountainous or high desert landscapes. This map reflects both the marginal Democratic victory (51.9% share) as well as geography.

Should roads be included? Possibly not, but we spend an awful lot of time on them so they're often densely populated!

Alaska is the largest state by area but also one of the most sparsely populated. At this scale, only a few small areas in and around Anchorage are visible.

This map uses a technique that reallocates election data from counties into areas that represent the populated places in the United States. Smaller areas represent a proportion of the original data based on its area. They do not necessarily reflect the actual results of the people in those areas.

Any dasymetric map gives the appearance of detail beyond which the original data supports. At relatively large scales there can be a tendency for people to infer a level of detail or precision in the map that is not necessarily true. A statement that at least explains this is often useful to help people interpret what they are seeing.

The same symbology is used on this map as the blended-hue choropleth even though individual areas are hard to see at this scale.

Margin of victory (%)

Republican

> 39
30 – 39
20 – 29
10 – 19
< 10

Democratic

10 – 19
20 – 29
30 – 39
> 39

If you look closely you can see swathes of colour showing predominance within sparsely populated areas. Contrast that with some of the more densely populated cities which show pockets of red, blue, and purple.

Flipped counties

Selective choropleth of key places

There's always a tendency with any dataset to map the data in its entirety. You rarely throw data away because it's natural to think there's always value in showing the whole, not merely some of the parts. **Does a map of every result of every county help us understand the story of an election?** Clearly, if the intent is to provide an overview of every place, then yes, you'd make a map that supports that requirement, and there are plenty of examples in this book that would make a good choice. But there's often a story to tell or a narrative to illustrate, and that rarely demands a map of every piece of information in the dataset. **Omission is a key tenet of cartographic practice.**

Focussing on this election dataset, one question might be to what extent did voting patterns mirror, or differ, from the 2012 election (Barack Obama versus Mitt Romney)? Some areas will reflect a voting pattern which returns the same party, regardless of the candidate. There are also marginal areas, often referred to as *battlegrounds*, in which candidates focus their attempts to sway the electorate to gain valuable votes. **If the story of the election is about a few battleground counties that flipped between election cycles, a map that shows those, and only those, places is useful.**

Rather than show the results for every place, the map to the right shows only those counties which flipped from the 2012 election to the 2016 election, from Democratic to Republican and vice versa. In the same way that white space was used on the dasymetric map effectively to de-symbolise unpopulated places, here white space omits those areas that voted the same in 2012 and 2016. Although not suggesting they are any less important to the result, their voting pattern did not change. They might be considered predictable, or perhaps of less interest, when considering the story of how the winning candidate turned blue counties to red and where.

This map focusses on those marginal places in which enough votes were cast to tip the balance from one party to another, from blue to red and (less so) from red to blue. The map could be as simple as showing a unique value result, but the detail has been extended to show the percentage swing. This clearly demonstrates that many places that supported Obama in 2012 switched their support to Trump in 2016. The main story evidenced in this map is the vast swathes of the population in the upper Midwest who switched from Democratic to Republican, and by quite a swing. Of course, no map tells the complete story, and you might consider adding detail that explains the turnout in these areas. Was it larger than normal with a motivated electorate, for instance? Did a certain type of person, in demographic terms, not go to the polls? **Maps such as these tend to answer one question yet give rise to further questions.**

Some counties flipped Republican to Democratic.

Salt Lake County was won by Romney in 2012 with a margin of over 20% (59% to Obama's 38%). Yet in 2016 Clinton won by a margin of 9% (42% to Trump's 33%). The total swing was greater though, since independent candidate Evan McMullin polled 19% leading to a Republican loss of 38% in total.

Alaska, Hawaii, Idaho, Wyoming, Arizona, Kansas, Oklahoma, Missouri, Massachusetts, Vermont, West Virginia, and Louisiana were the only states which did not contain any counties that flipped between 2012 and 2016.

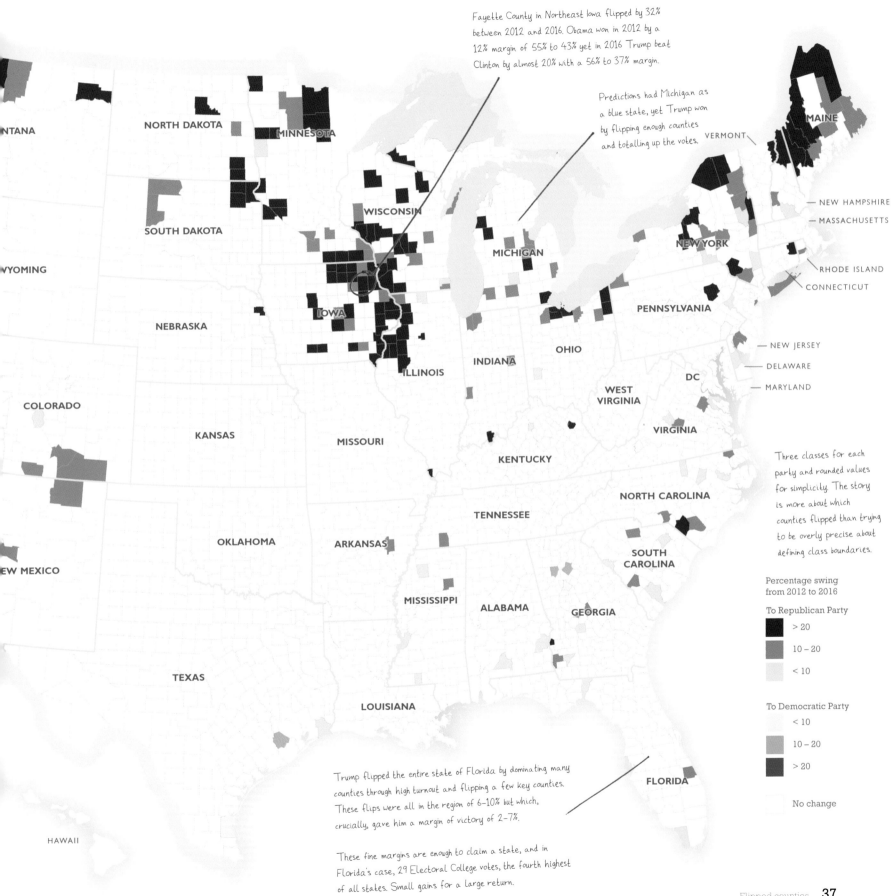

Fayette County in Northeast Iowa flipped by 32% between 2012 and 2016. Obama won in 2012 by a 12% margin of 55% to 43% yet in 2016 Trump beat Clinton by almost 20% with a 56% to 37% margin.

Predictions had Michigan as a blue state, yet Trump won by flipping enough counties and totalling up the votes.

Three classes for each party and rounded values for simplicity. The story is more about which counties flipped than trying to be overly precise about defining class boundaries.

Percentage swing from 2012 to 2016

To Republican Party

> 20

10 – 20

< 10

To Democratic Party

< 10

10 – 20

> 20

No change

Trump flipped the entire state of Florida by dominating many counties through high turnout and flipping a few key counties. These flips were all in the region of 6–10% but which, crucially, gave him a margin of victory of 2–7%.

These fine margins are enough to claim a state, and in Florida's case, 29 Electoral College votes, the fourth highest of all states. Small gains for a large return.

Continuously varying results

Interpolated surfaces

Administrative boundaries can appear too dominant or impact the interpretation more than desired on thematic maps. They are important because they're the containers into which data is poured. Everyone who votes goes into a container by virtue of where the lines are drawn on the map. **People live across a continuous surface, and harsh boundaries rarely separate us.** You can dismiss the boundaries by creating an interpolated surface of the data.

The process is simple. Points are used to represent the values of data for an area, and an interpolator creates a raster grid that exhausts space. There are likely as many ways to interpolate a surface and vary its parameters as there are maps in this book. **The key to a well-balanced surface is to ensure it's neither too detailed nor too smoothed.** The former will end up looking like a lot of pinpricks of high values. The latter will look like the data has little, if any, variation across the map.

Techniques such as inverse distance weighting, spline interpolation, kernel density estimation, k-functions, and kriging all produce different results, and each has different parameters that determine how values are calculated across the map. Exact interpolaters ensure that raster cells that include a sample point get that pixel value. By varying the bandwidth of the search distance around a cell, the number of surrounding data points to include in a calculation, and the weight assigned to the impact of the data points at different distances, the surface can take on many appearances.

Although surfaces avoid the sharp boundaries between the single shades on a typical map that uses administrative boundaries, the drawback is that you're estimating data for every interpolated cell. It's a statistical surface and a best guess at how patterns might occupy intermediate space.

It's worth remembering that votes don't actually vary in the same way as, say, precipitation or temperature. But data that does vary continuously is usually sampled at points and interpolated. We're just applying the same principles to generate a similar map style.

On the dasymetric choropleth map, the abundance of white space showed where people DO NOT live. This map suggests a constantly varying voting population. It's almost the inverse of a dasymetric approach.

Alaska has a population of approximately 734,000 and is the largest state by far at over 665,000 square miles. It also has the lowest population density of any state at 1.2 per sq. mi. Anchorage is home to nearly 300,000 people which makes much of the rest of the state largely wilderness. Yet the map paints a different picture.

It's arguable whether this map type is useful for places like Alaska. The suggestion of a continuously varying population isn't backed up at all by geography.

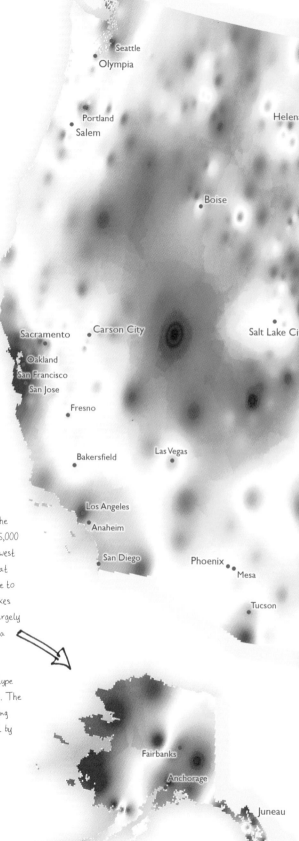

Smoothed surface

The map to the left uses different parameters to create a smoother version for display at a smaller scale.

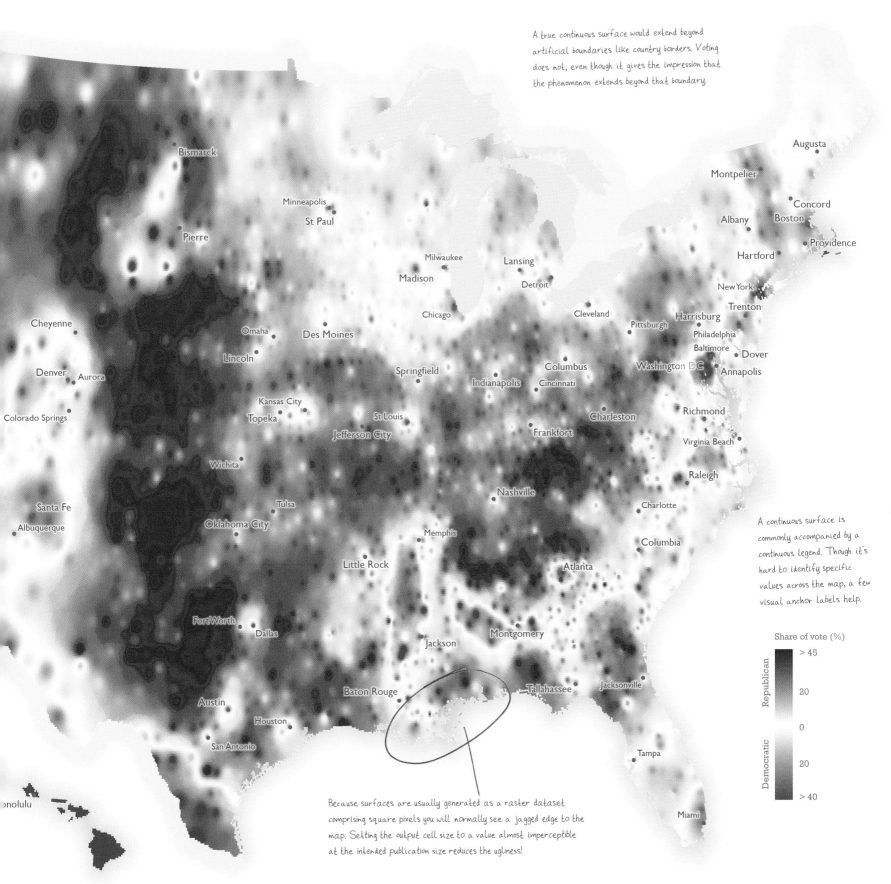

A true continuous surface would extend beyond artificial boundaries like country borders. Voting does not, even though it gives the impression that the phenomenon extends beyond that boundary.

A continuous surface is commonly accompanied by a continuous legend. Though it's hard to identify specific values across the map, a few visual anchor labels help.

Share of vote (%)

Republican

> 45

20

0

20

Democratic

> 40

Because surfaces are usually generated as a raster dataset comprising square pixels you will normally see a jagged edge to the map. Setting the output cell size to a value almost imperceptible at the intended publication size reduces the ugliness!

Looking for relationships

Bivariate choropleth of two variables

Waldo Tobler's First Law of Geography states that **'everything is related to everything else, but near things are more related than distant things'.** In terms of different variables that can characterise the same place, the question arises whether given a high value for one variable, is it necessarily matched by a high value for another variable, and vice versa. To explore this, you can use a bivariate choropleth map which combines two univariate choropleth maps into a single map.

Bivariate maps are particularly useful when you think that two variables might be correlated. The purpose of the map is to find areas of agreement or disagreement between the two variables.

The two univariate choropleths should be limited in the number of classes they use because for a map with *n* classes, the resulting bivariate choropleth will have n² classes. Two three-class schemes yield nine bivariate classes, and two four-class schemes yield 16 classes. **A bivariate map can get unwieldy quickly so limiting class schemes to 3 x 3 to give nine bivariate classes is a good default.** You're looking only to pick up broad patterns so the nuance you get from a univariate choropleth with more classes is unnecessary.

Classify each variable sensibly so there's a good and logical range of each class on the map. You don't want all the data in one class, or you'll end up with a poor range of bivariate symbols. Overlay the maps to mix the colours properly. Complete with a legend that helps demystify the combination. If the colours in the final bivariate legend don't have the correct perceptual steps, go back to the original colour schemes and tweak them. It's the combined colours that are important, not the original ones.

The advice for most thematic maps is to keep things as simple as possible but it's doubly true for a bivariate choropleth because the opportunity for confusion is greater.

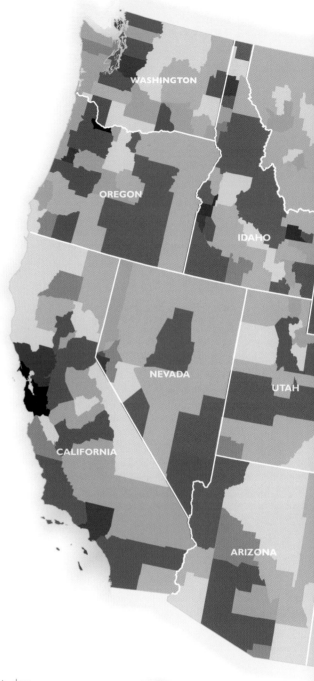

Each of the two variables uses a sequential colour scheme that has a light, neutral colour for low values, becoming darker for higher values.

Complementary colours tend to work best so that when they are mixed they give a good range of identifiable colours but which still retain the appearance of a progression from light to dark on the axis and the diagonal from bottom left to top right.

The original colour schemes are never seen on the map. We only see the combined colour scheme.

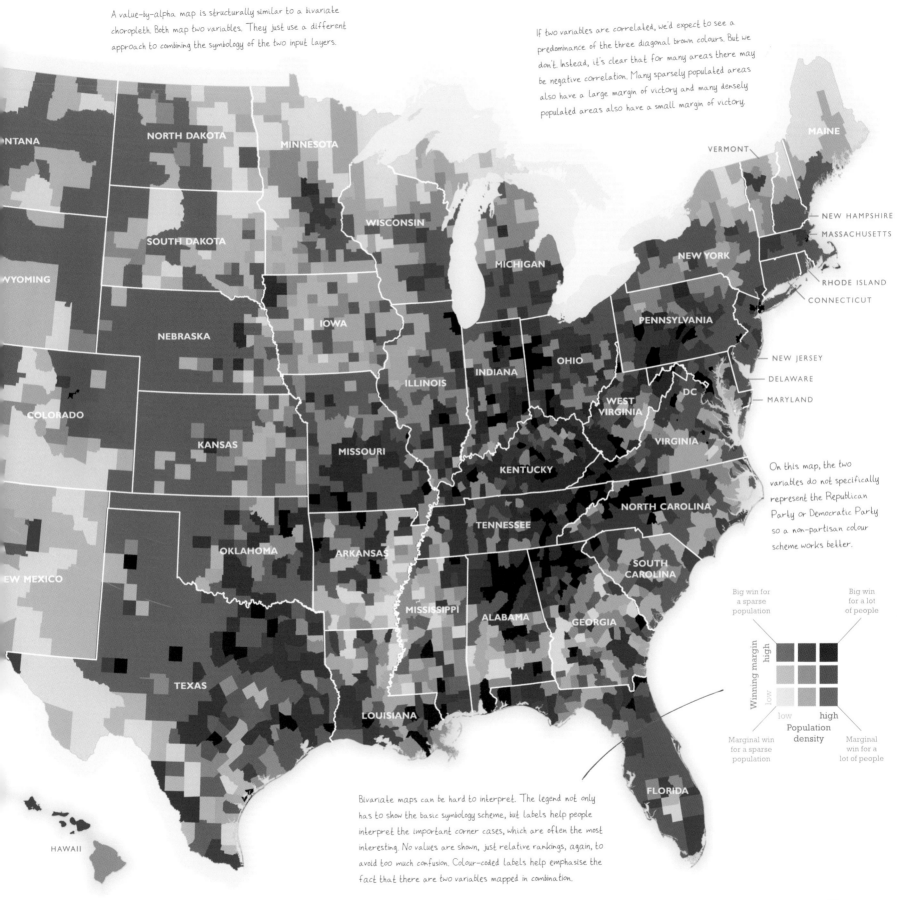

A value-by-alpha map is structurally similar to a bivariate choropleth. Both map two variables. They just use a different approach to combining the symbology of the two input layers.

If two variables are correlated, we'd expect to see a predominance of the three diagonal brown colours. But we don't. Instead, it's clear that for many areas there may be negative correlation. Many sparsely populated areas also have a large margin of victory and many densely populated areas also have a small margin of victory.

On this map, the two variables do not specifically represent the Republican Party or Democratic Party so a non-partisan colour scheme works better.

Big win for a sparse population

Big win for a lot of people

Winning margin
high — low

low — high
Population density

Marginal win for a sparse population

Marginal win for a lot of people

Bivariate maps can be hard to interpret. The legend not only has to show the basic symbology scheme, but labels help people interpret the important corner cases, which are often the most interesting. No values are shown, just relative rankings, again, to avoid too much confusion. Colour-coded labels help emphasise the fact that there are two variables mapped in combination.

All the colours

Fusing three variables into a trivariate choropleth

One glance at the main map to the right, and you'd be forgiven for thinking it's only a random piece of cartographic art with colours painted into areas and a flagrant disregard for the underlying data. **Using a palette of rainbow colours is hardly ever a good idea in cartography.** Perceptually, the array of colours in a conventional red-orange-yellow-green-blue-indigo-violet (ROYGBIV) sequence doesn't fit a numerical scale. Except that this map doesn't encode data in that way.

The trivariate choropleth technique is an extension of the bivariate technique, but it maps three, instead of two, variables. There should be sensible selection of the three variables as you're looking for relationships and possible correlations. Many counties might exhibit high or low levels of all three variables, in which case they would be dark or light, respectively. **Where you see dominant colours from one of the variables, it suggests one variable is dominant in relation to the other two.** And there's a lot of mixing in between.

Colour choice is important, and this sort of map works only if you rely heavily on the principle of subtractive colour, which is the basis of most print images. Subtractive colour is created by using ink (pigments) that partially absorb (subtract) some light wavelengths while reflecting others. Subtractive colour begins as white (the colour of the paper becomes reflected so you see white). As ink is added, it works as a filter, and light takes on the appearance of colour. A layer symbolised using only cyan ink absorbs the red part of the electromagnetic spectrum, reflects the combination of green and blue, and you see cyan. Magenta ink absorbs green and reflects red and blue. Yellow ink absorbs blue and reflects red and green. When you blend these three layers together, you get a full range of colour, which allows you to symbolise each area through the unique combination of different lightnesses of the three inks.

It certainly provides a colourful map but it's difficult to locate precisely each county's colour in the legend. Approximation is as good as you're able to identify. This is largely because of a continuous colour ramp being applied to each of the three variables rather than a classified scheme. As you saw with the bivariate map, classifying data with even the bare minimum of classes (three) gives nine resulting classes. For a trivariate map, it is exacerbated by cubing the number of classes.

A classified version of the legend for this map is shown to the right. With so many colours, it would remain a challenge for the map reader to relate colours from the legend to areas on the map. For this reason, continuous ramps are no worse, even though they come with clear drawbacks for unique identification and data recovery.

Three small inset maps isolate the individual maps to show the individual variables. It's sometimes useful to isolate the components to not only aid interpretation, but to see how patterns exist individually, as much as they do collectively.

Turnout
(% of voting age people who voted)

Margin of victory
(% between winning and second place)

Population
(persons age 18+ per sq km)

Alternative classified legend

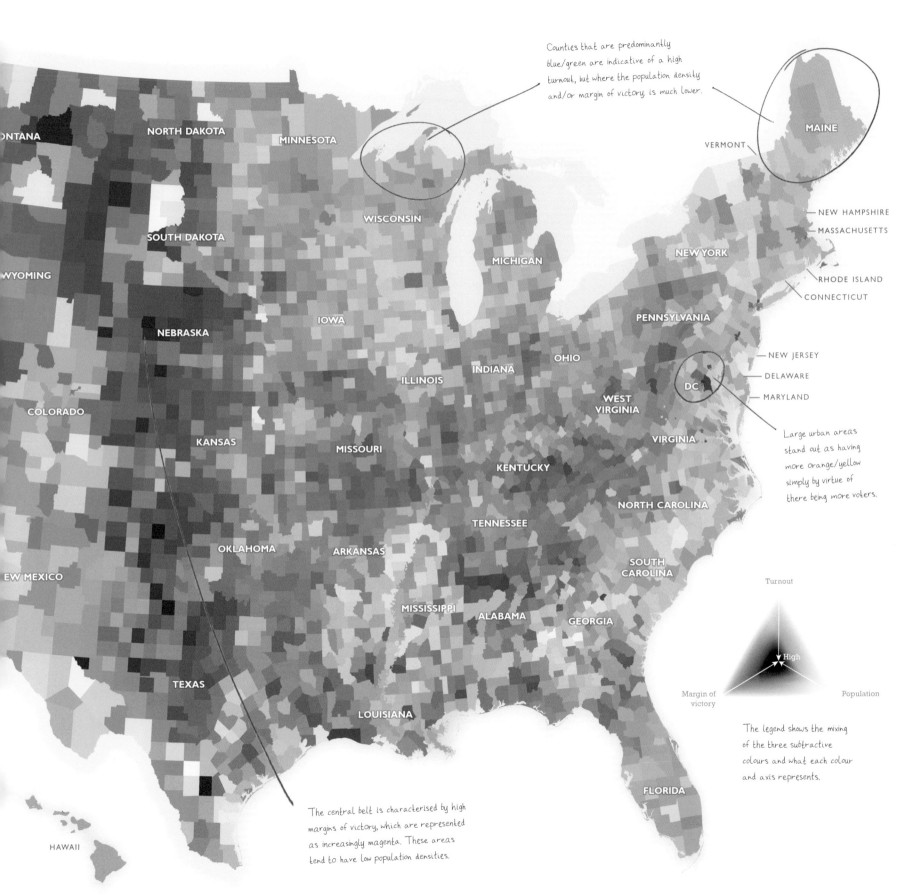

Counties that are predominantly blue/green are indicative of a high turnout, but where the population density and/or margin of victory, is much lower.

MONTANA
NORTH DAKOTA
MINNESOTA
WISCONSIN
SOUTH DAKOTA
WYOMING
MICHIGAN
IOWA
NEBRASKA
COLORADO
KANSAS
MISSOURI
ILLINOIS
INDIANA
OHIO
WEST VIRGINIA
PENNSYLVANIA
NEW YORK
MAINE
VERMONT
NEW HAMPSHIRE
MASSACHUSETTS
RHODE ISLAND
CONNECTICUT
NEW JERSEY
DELAWARE
MARYLAND
DC
VIRGINIA
KENTUCKY
NORTH CAROLINA
TENNESSEE
NEW MEXICO
OKLAHOMA
ARKANSAS
SOUTH CAROLINA
MISSISSIPPI
ALABAMA
GEORGIA
TEXAS
LOUISIANA
FLORIDA
HAWAII

Large urban areas stand out as having more orange/yellow simply by virtue of there being more voters.

The central belt is characterised by high margins of victory, which are represented as increasingly magenta. These areas tend to have low population densities.

Turnout
High
Margin of victory
Population

The legend shows the mixing of the three subtractive colours and what each colour and axis represents.

Don't lose your stripes

Vertical bar fills

There's an almost unlimited number of ways you might design a fill for area maps. **Modern cartography tends to default to solid colour fills, which are modified through changing any combination of hue, value, chroma (saturation), and transparency, but fills can be designed in many ways.** It wasn't long ago that map design was constrained to print technology with pattern fills, usually made from black marks on a white background, as the only real choice. Colour, if used at all, was limited to only a few choices, and registration of complex patterns in offset lithography made the task of combining patterns and colours even more problematic.

This map re-creates the sort of fill one might have designed in, say, the 1980s. It's a pattern fill of sorts, and it's also a proportional fill based on a recurring pattern of vertical bars. There are three colours: a red bar representing the Republican Party, a blue one representing the Democratic Party, and an orange one representing the sum of other candidates. **Each bar is scaled in width to the percentage share of the vote they received per county.** All the Republican bars are justified left in a repeated pattern so there is a consistent baseline across the map. The Democratic Party bars are stacked to the right via offsets to ensure they sit adjacent to the Republican bar and are not obscured. The bar representing the other candidates is similarly offset farther to the right.

The final map is therefore a superimposition of the three variables, one bar for each sector in the dataset, offset from one another to create a complete fill for each area. Where a county is particularly large, the pattern is repeated so the relative proportion of the red, blue, and orange bars still gives a visual mix of colour quantities that reflects the overall share of the vote among the parties.

The value in this map is not necessarily seeing the composition for individual counties, but in the overall pattern that they create in combination. **In an overall sense, you can see where Republicans tended to gain more of the share than the Democrats, or vice versa, and where third-party candidates were relatively successful.**

It's questionable whether this would be many people's first choice for displaying the data for a few reasons. Clearly, technology has moved on, and it's much easier to make solid fills. But more than a technological advancement, in purely visual terms this sort of map is tough on the eyes. It's difficult to recover an answer to the question of the distribution for a given county. It's also hard to compare across the map from one county to another because of the difficulty of retaining complex visual information in memory. **But it's interesting, and it makes you stop and look.**

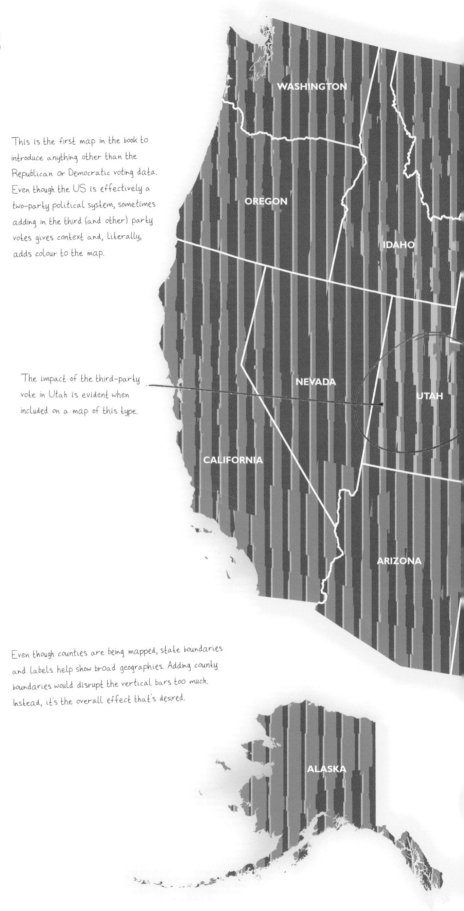

This is the first map in the book to introduce anything other than the Republican or Democratic voting data. Even though the US is effectively a two-party political system, sometimes adding in the third (and other) party votes gives context and, literally, adds colour to the map.

The impact of the third-party vote in Utah is evident when included on a map of this type.

Even though counties are being mapped, state boundaries and labels help show broad geographies. Adding county boundaries would disrupt the vertical bars too much. Instead, it's the overall effect that's desired.

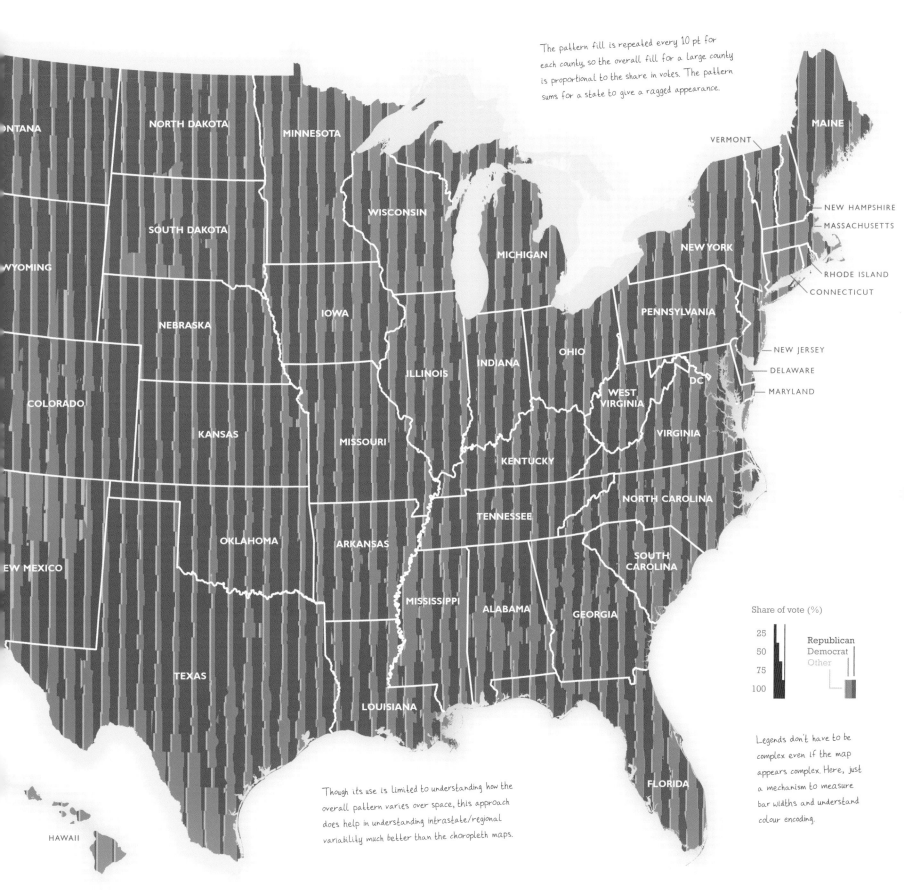

The pattern fill is repeated every 10 pt for each county, so the overall fill for a large county is proportional to the share in votes. The pattern sums for a state to give a ragged appearance.

MONTANA
NORTH DAKOTA
MINNESOTA
MAINE
VERMONT
WISCONSIN
NEW HAMPSHIRE
MASSACHUSETTS
SOUTH DAKOTA
NEW YORK
WYOMING
MICHIGAN
RHODE ISLAND
CONNECTICUT
IOWA
PENNSYLVANIA
NEBRASKA
OHIO
NEW JERSEY
INDIANA
DELAWARE
COLORADO
DC
ILLINOIS
WEST VIRGINIA
MARYLAND
KANSAS
MISSOURI
VIRGINIA
KENTUCKY
NEW MEXICO
NORTH CAROLINA
TENNESSEE
OKLAHOMA
ARKANSAS
SOUTH CAROLINA
MISSISSIPPI
ALABAMA
GEORGIA
TEXAS

Share of vote (%)

25
50
75
100

Republican
Democrat
Other

LOUISIANA

Though its use is limited to understanding how the overall pattern varies over space, this approach does help in understanding intrastate/regional variability much better than the choropleth maps.

FLORIDA

Legends don't have to be complex even if the map appears complex. Here, just a mechanism to measure bar widths and understand colour encoding.

HAWAII

What is a vote worth?

Electoral College votes per capita choropleth

There are many ways to sift the results of an election to determine the winner. And the decisions on how individual votes are tallied and aggregated can have a profound impact on who wins. A first-past-the-post system often will lead to a different result than if the results were based on popular vote or proportional representation.

In the United States, the Electoral College system determines the overall result. It's a system that has as many detractors as it does supporters, and that opinion likely also is based on an individual's political view. **From a political perspective, a voter most likely will prefer the system that gives their candidate the best chance of getting to the Oval Office.**

The Electoral College system means that candidates target states that can yield the most seats, particularly where there's a marginal historical outcome. It also means a candidate who doesn't win the popular vote overall can win more Electoral College votes. Further, turnout also tends to be lower in what might be considered *safe* states because people might not feel their vote counts as much. **The assumption that votes count differently in different places is correct because not all votes are equal.**

The average Electoral College vote represents around 436,000 voting-age citizens, which varies depending on the number of voters over the age of 18 per election cycle. Voters in sparsely populated states such as Wyoming, Vermont, and North Dakota have more voting power because each of the three Electoral College votes for each state represents around 143,000 people. The converse is true in highly populated states such as California, Florida, and New York, which, although represented by far more Electoral College votes, each has a much larger Electoral College vote per capita ratio of one vote per approximately 500,000 voters. A voter in Wyoming has approximately four times more voting power than their fellow citizen in California.

Of course, voting power doesn't accurately describe the voting process because the probability of one individual's vote influencing the result is extremely small. But where it does have an impact is in aggregation. The 2016 election is a good illustration because the candidate who won the popular vote (Clinton with 65,853,514 votes compared with Trump's 62,984,828 votes) did not win the election. Furthermore, because the Electoral College is a function of the number of representatives in Congress, there's a mismatch in legislative power. The 143,000 people in Wyoming have the same number of congressional legislators as 500,000 people in California.

In the US, the president is not elected by popular vote, so each person's vote is not given equal weight. That said, popular vote isn't necessarily a better system either. It's just different.

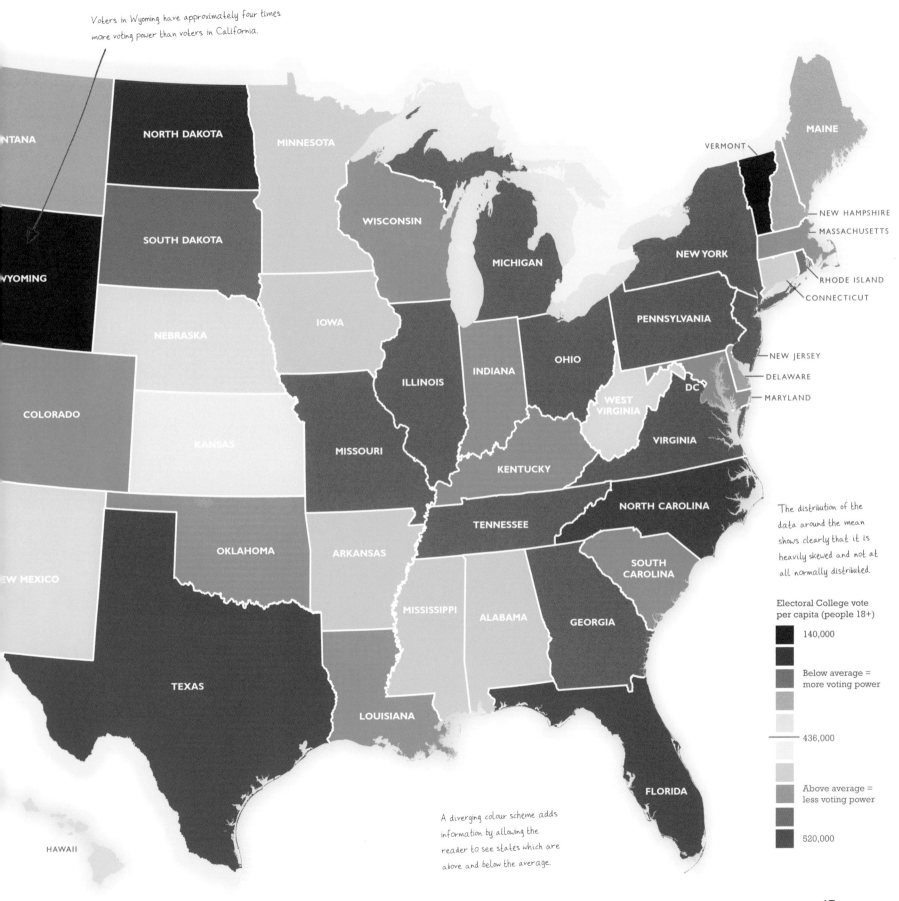

Voters in Wyoming have approximately four times more voting power than voters in California.

The distribution of the data around the mean shows clearly that it is heavily skewed and not at all normally distributed.

A diverging colour scheme adds information by allowing the reader to see states which are above and below the average.

Electoral College vote per capita (people 18+)

140,000

Below average = more voting power

436,000

Above average = less voting power

520,000

One map is not enough

Small multiples over time

One of the first considerations when making a thematic map is deciding whether a single map is enough to show what needs showing. This decision includes an assessment of how much can be fitted onto a single map and whether having two or more maps makes more sense to reduce the clutter on a single map. **Many circumstances merit more than one map**—for instance, when the need is to show maps designed to be compared with one another. This design might be to show different thematic content and leave it up to the map reader to assess how patterns might be similar between variables mapped on different maps. Alternatively, the need may be to show how patterns change over time, and more than one map becomes helpful.

Small multiples are a good way of showing change, either of multiple variables for the same area or of the same theme over time. They allow people to see how numerical measurements may have increased or decreased between dates or as part of longer term trends. For election data or any population-based data, they provide a mechanism to present overall results from one election cycle to the next. This display underpins the ability to see how results have swung from one period to the next.

Because you're using small maps, you inevitably need to generalise them. The tendency is to show summaries for larger enumeration areas rather than show too much detail. The example to the right uses states, not counties. There isn't the space to show county-level detail. **What is lost in detail on each map is gained on the overall map page providing a single image that supports comparisons.** You also can incorporate additional information such as the summary statistics for the candidates and their measures of success (or failure). In this way, the generalised maps are augmented by statements that help the map reader understand the overall strength of victory.

Invisible grids are vital to the small-multiples format. Maps are organised to be read in an understandable way, usually from top left to bottom right, much as you'd read a passage of text one line to the next. The maps have almost no additional detail to ensure they are as uncluttered as possible. This helps their readability. **Simple map types are also more effective as small multiples than those that are visually more complex.** The grid structure, as opposed to an animated map, also overcomes change blindness, a condition in which you can forget rapidly what you've just seen, in favour of what you are currently seeing. Since all the maps are available in a single view, you can scan and trace back and forth much more easily and systematically.

1920   Warren G. Harding | 404 | 60.3%
     James M. Cox | 127 | 34.1%

	Electoral College Votes	Popular Vote
Date	↓	↓
Democratic	303	49.6%
Republican	189	45.1%
Other	39	2.4%

Red states won by Republican Party
Blue states won by Democratic Party
Orange states won by other candidates
Grey states have yet to gain voting rights

Electoral College votes do not always sum to the total due to electors that pledge to vote for other candidates despite the state result.

Percentage of popular vote does not always sum to 100% due to other candidates taking a slim share.

1940   Franklin D. Roosevelt | 449 | 54.7%
     Wendell Willkie | 82 | 44.8%

1960   John F. Kennedy | 303 | 49.7%
     Richard Nixon | 219 | 49.6%
     Harry F. Byrd | 15 | 0.9%

The maps are fairly self-explanatoiry so all that's needed is a statement to help interpret the numbers. This includes the reason why the number of Electoral College votes and percentages don't always add up to the potential maximum.

1980   Ronald Reagan | 489 | 50.7%
     Jimmy Carter | 49 | 41.0%

The alingment of each component on a page of small multiples is critical to guiding the eye. Humans are good at processing repetition, so positioning information in the same place for each map makes processing the detail that much easier.

2000   George W. Bush | 271 | 47.9%
     Al Gore | 266 | 48.4%

Only twice in the last 25 election cycles has the winning candidate not won the popular vote: George W. Bush in 2000, and Donald J. Trump in 2016.

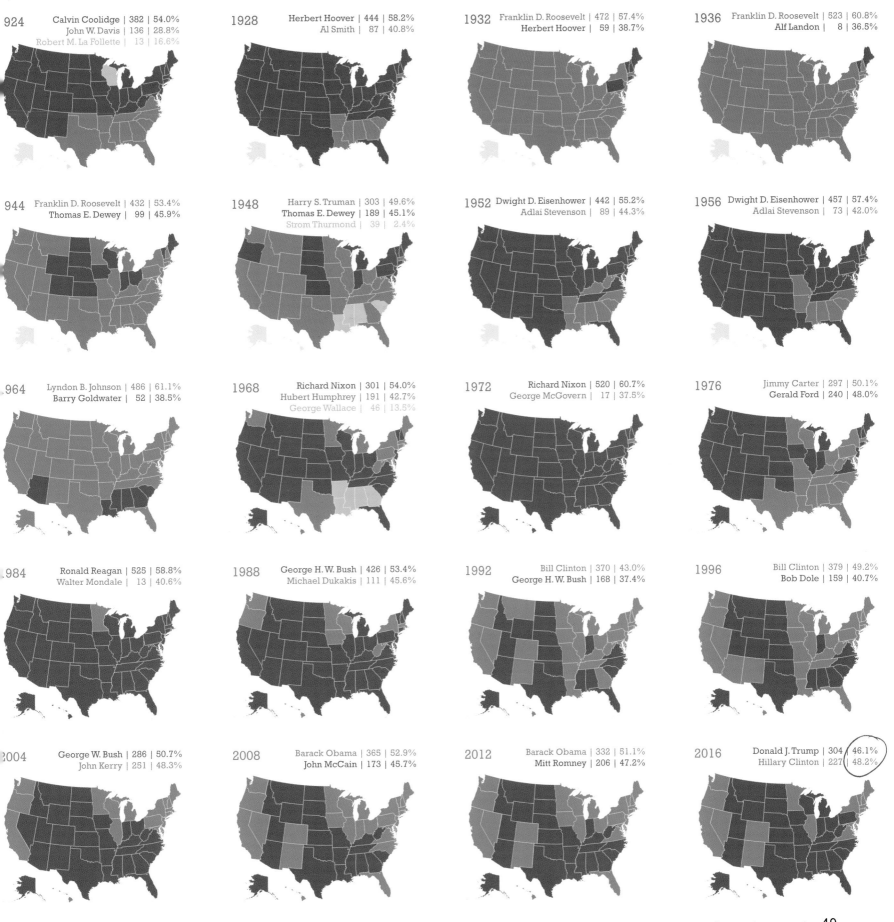

1924 Calvin Coolidge | 382 | 54.0%
John W. Davis | 136 | 28.8%
Robert M. La Follette | 13 | 16.6%

1928 Herbert Hoover | 444 | 58.2%
Al Smith | 87 | 40.8%

1932 Franklin D. Roosevelt | 472 | 57.4%
Herbert Hoover | 59 | 38.7%

1936 Franklin D. Roosevelt | 523 | 60.8%
Alf Landon | 8 | 36.5%

1944 Franklin D. Roosevelt | 432 | 53.4%
Thomas E. Dewey | 99 | 45.9%

1948 Harry S. Truman | 303 | 49.6%
Thomas E. Dewey | 189 | 45.1%
Strom Thurmond | 39 | 2.4%

1952 Dwight D. Eisenhower | 442 | 55.2%
Adlai Stevenson | 89 | 44.3%

1956 Dwight D. Eisenhower | 457 | 57.4%
Adlai Stevenson | 73 | 42.0%

1964 Lyndon B. Johnson | 486 | 61.1%
Barry Goldwater | 52 | 38.5%

1968 Richard Nixon | 301 | 54.0%
Hubert Humphrey | 191 | 42.7%
George Wallace | 46 | 13.5%

1972 Richard Nixon | 520 | 60.7%
George McGovern | 17 | 37.5%

1976 Jimmy Carter | 297 | 50.1%
Gerald Ford | 240 | 48.0%

1984 Ronald Reagan | 525 | 58.8%
Walter Mondale | 13 | 40.6%

1988 George H. W. Bush | 426 | 53.4%
Michael Dukakis | 111 | 45.6%

1992 Bill Clinton | 370 | 43.0%
George H. W. Bush | 168 | 37.4%

1996 Bill Clinton | 379 | 49.2%
Bob Dole | 159 | 40.7%

2004 George W. Bush | 286 | 50.7%
John Kerry | 251 | 48.3%

2008 Barack Obama | 365 | 52.9%
John McCain | 173 | 45.7%

2012 Barack Obama | 332 | 51.1%
Mitt Romney | 206 | 47.2%

2016 Donald J. Trump | 304 | 46.1%
Hillary Clinton | 227 | 48.2%

A landslide for apathy

Non-voters as unique values

Voting in an election is the right of every citizen of voting age in a democratic society. In the United States, being over the age of 18 and a US citizen gives you the privilege of a vote. As reported by the Federal Election Commission, the US has a population of around 330 million with nearly 250 million eligible voters. Combined, there were 136,669,276 votes cast in the 2016 presidential election across all candidates. **The turnout in 2016 was 54.7 percent, which, put another way, is not much more than half the voting population.** Many, for whatever reason, effectively voted 'none of the above' through their decision not to cast a ballot.

Reasons not to vote can be multiple and complex, such as dissatisfaction with the candidates, political campaign fatigue, disinterest, or feeling that an individual vote doesn't count. In systems of proportional representation, individuals may feel more empowered. Yet in the Electoral College system, voters who feel their party is easily outnumbered within their state may not vote from a sense of resignation. Voters who feel their party is assured victory within their state also may decide not to vote because of a potentially misplaced confidence in the outcome.

Getting people out to vote is critical to winning an election. A candidate cannot have supporters sitting at home, which is why any election is preceded by frenetic campaigning. **The map to the right shows all those counties in which more people chose not to vote than the number who voted for the winning candidate.** The result is stark: 'Nobody' won more counties, more states, and more electoral votes than either of the main candidates running for president. If everybody who didn't vote voted 'Nobody', then 'Nobody' would have taken all but 12 states, plus Washington, DC, totalling 447 of the 538 Electoral College votes. Now that's a landslide for apathy.

Despite the majority of counties in Washington voting 'Nobody', and most in Oregon voting Republican, the Democratic Party with fewer counties, but a larger total vote, managed to win both states.

'Nobody' won Arizona (and Hawaii) by a landslide!

The states of apathy

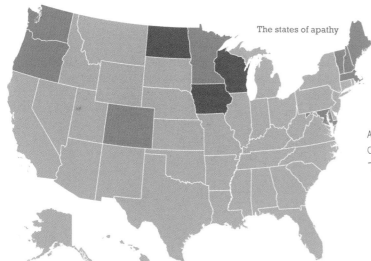

Apathy: 38 states, 447 EC votes
Clinton: 10 states, 72 EC votes
Trump: 3 states, 19 EC votes

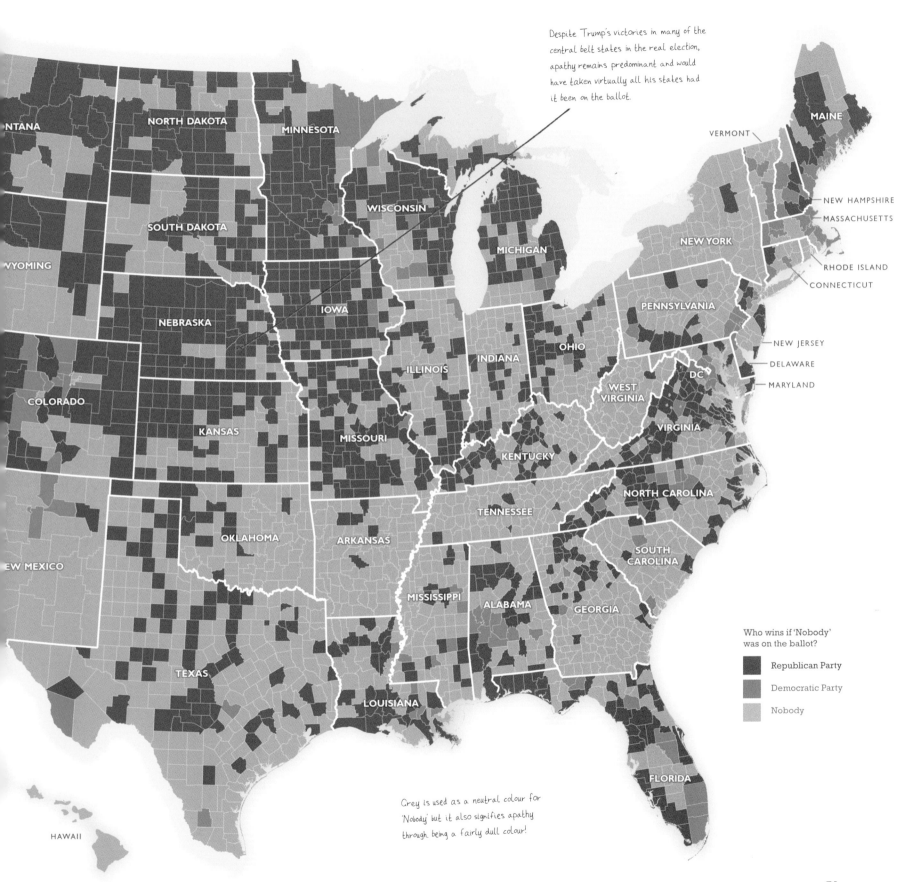

Despite Trump's victories in many of the
central belt states in the real election,
apathy remains predominant and would
have taken virtually all his states had
it been on the ballot.

MAINE

VERMONT

NEW HAMPSHIRE

MASSACHUSETTS

RHODE ISLAND

CONNECTICUT

NEW JERSEY

DELAWARE

MARYLAND

MONTANA

NORTH DAKOTA

MINNESOTA

WISCONSIN

MICHIGAN

NEW YORK

PENNSYLVANIA

WYOMING

SOUTH DAKOTA

IOWA

OHIO

NEBRASKA

ILLINOIS

INDIANA

WEST
VIRGINIA

DC

COLORADO

VIRGINIA

KANSAS

MISSOURI

KENTUCKY

NORTH CAROLINA

NEW MEXICO

OKLAHOMA

ARKANSAS

TENNESSEE

SOUTH
CAROLINA

TEXAS

MISSISSIPPI

ALABAMA

GEORGIA

Who wins if 'Nobody'
was on the ballot?

■ Republican Party

■ Democratic Party

■ Nobody

LOUISIANA

FLORIDA

Grey is used as a neutral colour for
'Nobody' but it also signifies apathy
through being a fairly dull colour!

HAWAII

A landslide for apathy **51**

3. Point maps

A major drawback of many area-based maps is the representation of the entire enumeration area as a solid fill. The implication of totality, regardless of the size of the area, can create a distortion that propagates to the perception of patterns across the map. Geography often gets in the way of the ability to interpret patterns across a map filled to the brim with colours and shades. The lack of white space doesn't give the map's symbology much space to breathe.

Solid fills imply that the reported aggregate value is somehow consistent across the entire reporting area, but that's not the case. It's a single value of data. It's one-dimensional. So instead of representing it in two dimensions as a planar surface that exhausts space, why not represent it using a one-dimensional feature—a point.

The series of maps in this chapter explores options for representing empirical data as a point-based object, or a series of points. You can do a lot with the humble point symbol, and it offers a useful alternative for showing data that represents aggregated counts for areas.

Size matters

Geometric shapes graduated in size

It is clear from the area-based maps that many of the Midwestern and central parts of the US voted Republican. Many of the areas on the maps were represented by some shade of red. So if Donald J. Trump won so many counties, why wasn't his margin of victory so much more than the result? A graduated symbol map begins to explain why.

Geography, in and of itself, is not necessarily a good visual mechanism to interpret the overarching situation. Population totals and densities vary markedly, so comparing like for like between areas is fraught with perceptual and cognitive difficulties. The number of people who voted is crucial to understanding results, yet area-based maps need such data converted into a rate before it can be mapped, therefore potentially losing important context. **Point-based symbol maps can work directly with the raw count data.**

At first glance, the map to the right looks to have reverted back to the simple binary red and blue colour scheme of area-based unique values maps. Yet instead of each area being shaded red or blue, a coloured geometric symbol, a circle, has been placed in each area. **The symbols additionally encode the number of votes by varying their size.** Although area-based maps need normalising (to a rate, ratio, or percentage) to support visual comparison across the map, you can use totals when mapping point symbols because the symbols are scaled consistently between each other. Geography provides only the background for the symbols that sit atop.

On this map, five sizes of the symbol classify the raw data. They quickly illustrate that although many counties are strongly Republican, they also have few voters in comparison with many Democratic counties. **A graduated symbol map can be considered the inverse of the classified choropleth map.** Visually, the choropleth tricks the map reader because the symbols (the shape of counties) exhaust space, and geography dupes you into a false impression because larger areas take visual prominence regardless of their overall impact. Using geometric symbols scaled to a useful variable helps adjust for this disparity and gives a better sense of not only who won, but what the underlying magnitude of the vote was.

The problem with most symbol maps, though, is overlapping symbols. You can get inventive, but ultimately the graphics will win because you tend to map larger quantities in smaller urban areas (that's where people live) and smaller totals in larger, often rural or sparsely populated areas. **You can use transparency or cut-outs in the symbol design, but in a map of so many symbols, it's probably impossible to make a map devoid of overlaps.** It's a compromise when choosing this sort of map type.

Whenever we're placing symbology on top of a base we reduce the amount of space for other detail. Labels are therefore left off this map to avoid them competing for space.

Boundary lines give the map some structure and allow us to see each symbol centred in the area it belongs to.

The western seaboard shows how large populations tend to live in relatively small geographical areas, resulting in numerous large symbols overlapping one another due to their proximity. They are also surrounded by larger areas with smaller populations, and smaller symbols.

The majority of areas in Alaska demonstrate how we often find small data values in large areas.

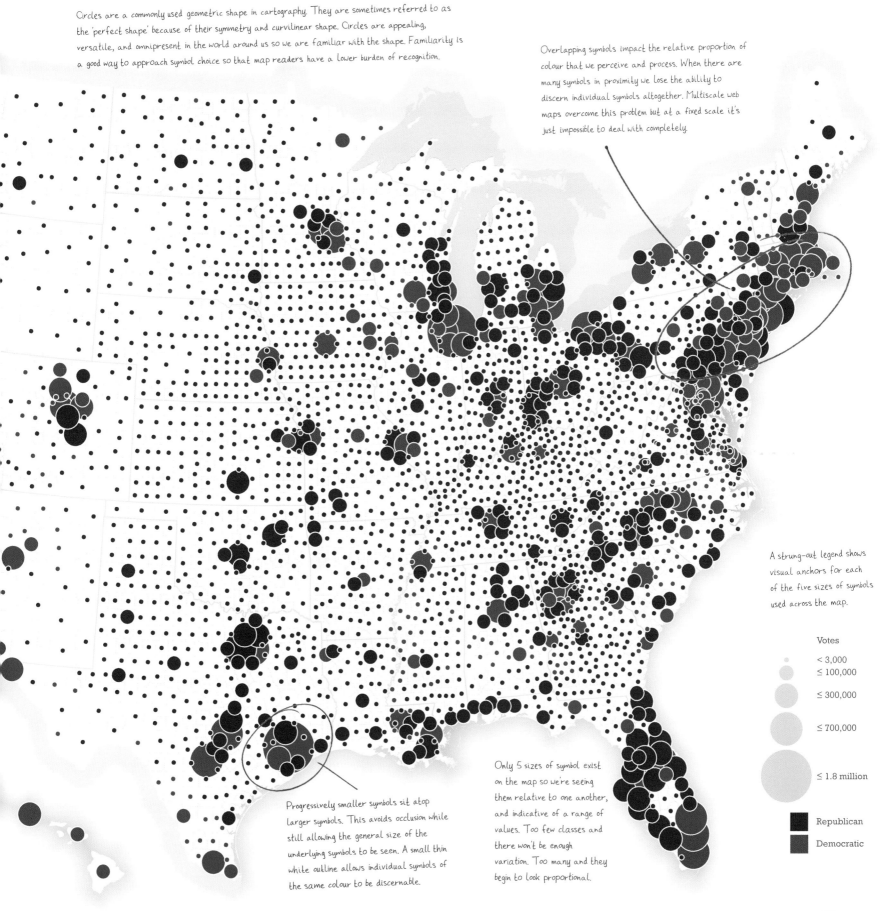

Circles are a commonly used geometric shape in cartography. They are sometimes referred to as the 'perfect shape' because of their symmetry and curvilinear shape. Circles are appealing, versatile, and omnipresent in the world around us so we are familiar with the shape. Familiarity is a good way to approach symbol choice so that map readers have a lower burden of recognition.

Overlapping symbols impact the relative proportion of colour that we perceive and process. When there are many symbols in proximity we lose the ability to discern individual symbols altogether. Multiscale web maps overcome this problem but at a fixed scale it's just impossible to deal with completely.

A strung-out legend shows visual anchors for each of the five sizes of symbols used across the map.

Progressively smaller symbols sit atop larger symbols. This avoids occlusion while still allowing the general size of the underlying symbols to be seen. A small thin white outline allows individual symbols of the same colour to be discernable.

Only 5 sizes of symbol exist on the map so we're seeing them relative to one another, and indicative of a range of values. Too few classes and there won't be enough variation. Too many and they begin to look proportional.

Votes

< 3,000

≤ 100,000

≤ 300,000

≤ 700,000

≤ 1.8 million

Republican

Democratic

The winner's circle

Proportional symbols emphasise the data range

If the graduated symbol map is the counterpoint to a classed choropleth map, a proportional symbol map is the counterpoint to an unclassed choropleth in that each symbol gets its own size. An unclassed choropleth is arguably not that useful as it's hard to see where an individual symbol shade sits along a continuous range of light to dark hues. But you can use map readers' natural perceptual and cognitive skills to good effect with proportional symbols. **If map readers approach a map of symbols thinking they are proportional, you can make them proportional rather than graduated into classes of only a limited number of sizes.**

The size of a proportional symbol is defined by calculating its area. For a circle, it is πr^2, in which the radius of the circle, r, is the data value. A data value of 2,500 gives a circle size of nearly 62 million square units (whatever the map units on the page might be). This quite often creates unwieldy symbol sizes, so data values commonly are scaled using the square root of the value and applying an exponent that scales them all up or down. **Calculating symbol sizes is often a process of trial and error to get a good spread of symbol sizes while not smothering the map.** It's also common practice to use a symbol of consistent size for all values above or below a threshold to avoid outliers impacting the distribution of symbol sizes in the midrange and to make them either visible or not overbearing.

Humans are reasonably good at assessing how much longer one line is relative to another, but we're not so good at assessing relative areas, and we tend to underestimate. In other words, if you're asked to identify a circle, among a range of circles, that is twice the size of another, you likely will pick one that's about 1.8 times the size. And this problem gets worse the larger the symbols. **To accommodate the human limitation of size estimation, you can apply perceptual (or apparent) scaling to symbols.** This is referred to as *Flannery compensation* after the research of James Flannery in the 1950s. This method exaggerates symbol size to compensate underestimation. Flannery compensation is not without controversy because if you measured a perceptually scaled symbol and reverse engineered the value, you'd get the wrong value. Additionally, perceptual scaling is based on the average response to psychophysical studies. Yet none of us see things on average. All of us react differently.

Which method is right? Both are. Data ranges guide which one will be suitable for any given map.

Clinton won many of the major populated areas (large blue symbols). Trump won a large number of counties with relatively small populations (small red symbols). If major cities were spread evenly among states Clinton may very well have won many more states. But they aren't. They are concentrated in a relatively few states.

Transparency is both beneficial and problematic. It makes overlapping symbols visible but in particularly congested areas it's difficult to pick out individual symbols which can be confusing. Darker overlapping areas may suggest 'more data' yet darker hues are purely a function of a graphical artifact of the overlap.

Los Angeles County, California
1,760,729 votes

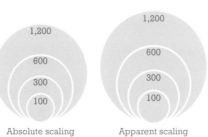

Absolute scaling   Apparent scaling

Many datasets have outliers, values that are quite some distance away from the next lowest or highest value at either end of the range. Cook County in Illinois and Los Angeles County in California are both double the value of the next lowest winning vote total (Harris County in Texas, 706,471 votes).

They are labelled separately to explain their prominence, and give us a way to note the maximum data value on the map. This assists the cognitive process of assessing the relationship between symbol size and data value for these outliers.

Cook County,
Illinois
1,528,582 votes

Nested legends are space efficient, include several indicative symbol sizes, and use rounded numbers for ease of reading.

Votes
1 million

500,000

100,000
50,000

Using transparency always has unintended consequences. On this map it modifies the colours of the boundaries and coastal vignette slightly due to the mixing effect.

Also, colours of smaller symbols are modified slightly due to the predominant larger symbol.

Republican

Democratic

Squaring the circle

Proportional symbol map using squares

Proportional symbol maps are incredibly flexible. They can be constructed using different data types (e.g. numerical or categorical), and they can show data reported at a point or at a centre point that represents an area. For area-based data, they overcome the problem of unequal geography because symbols are scaled to the data magnitude, regardless of geography. **In some ways, proportional symbol maps let the thematic data speak for itself because geography is removed from the equation.** Though they are constrained to the areas they represent, symbols represent data directly.

The map to the right is the same as the previous proportional symbol map that used circles. It displays the number of votes for the winning candidate by county using that data value to scale the symbol. Although circles often are preferred, one drawback is that people sometimes have difficulty assessing values related to the area of a circle. Squares alleviate that difficulty to an extent because area is a^2, where a is the length of one side of the square. It's easier for people to visualise the differences between squares of different sizes and make assessments about the values being mapped.

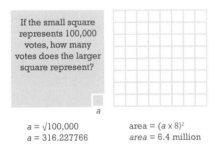

If the small square represents 100,000 votes, how many votes does the larger square represent?

a

$a = \sqrt{100{,}000}$
$a = 316.227766$

$area = (a \times 8)^2$
$area = 6.4$ million

In most cases, each symbol on a proportional symbol map will be positioned either at the point that relates to the data value or aligned vertically and horizontally with the centre of the feature it represents. In some circumstances, this positioning can cause difficulties if the symbol represents an oddly shaped area in which the symbol might appear to fall outside or be awkwardly offset. This can result in some dissociation from its location. An adjustment often can be applied to provide a position that allows people to identify clearly the area the symbol represents. This accommodation avoids overlaps.

Most proportional symbol maps benefit from a good legend, as it provides people with a clear way to assess symbol size and its relationship to the data values. Because the symbols are proportional and square, it's not difficult to assess the data value. There simply aren't many thematic map types that support the ability to recover individual data values from a map's symbols to a reasonably accurate degree.

The illustration, left, shows the basic calculation for square proportional symbols. However, most proportional symbol maps use exponents to scale symbols.

Exponents may force a maximum symbol size for the map, which would then scale all other values. They may also specify minimum sizes or have symbols deliberately stretched across a range of sizes even if many values are similar. The impact makes estimating accurate values difficult since there isn't necessarily a simple linear relationship between the value and symbol size.

This explains why, on this map a value of 1 million votes is roughly the same as the size of the large square, left, which states 6.4 million. The largest value in the dataset is LA County at 1,760,729 votes and this has been set with a maximum symbol size of 150 pt. (a little over 5 cm in width which gives an area of approximately 25 cm² for the largest symbol on the map.)

Los Angeles County, California
1,760,729 votes

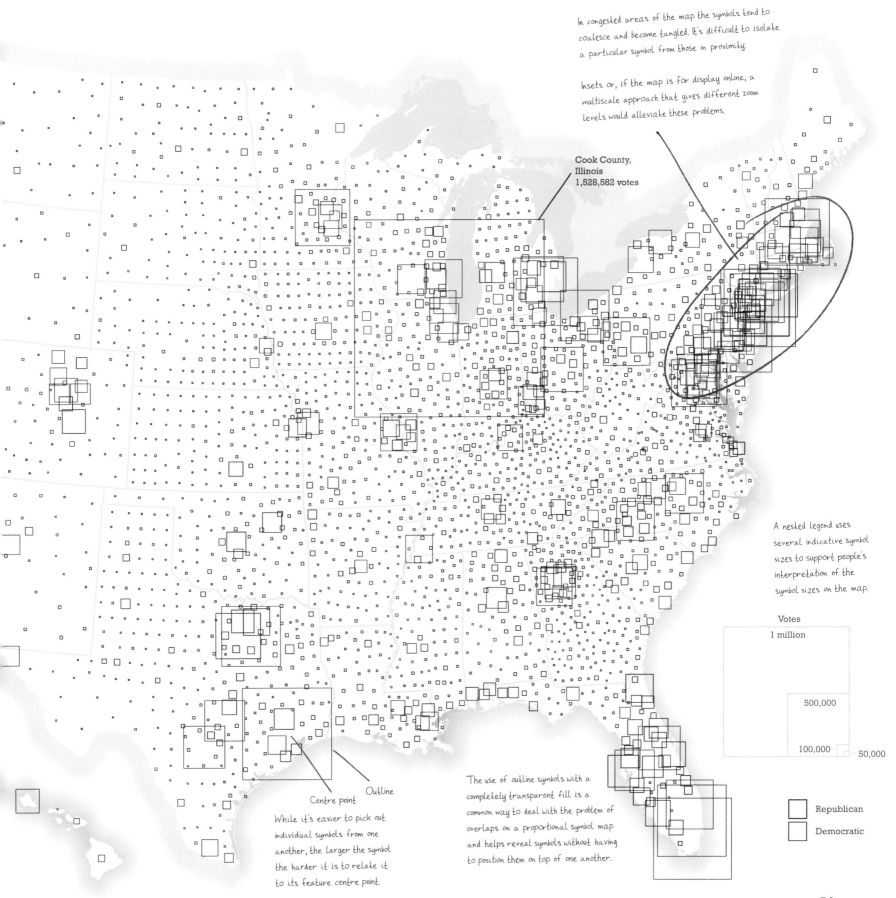

In congested areas of the map the symbols tend to coalesce and become tangled. It's difficult to isolate a particular symbol from those in proximity.

Insets or, if the map is for display online, a multiscale approach that gives different zoom levels would alleviate these problems.

Cook County, Illinois 1,528,582 votes

A nested legend uses several indicative symbol sizes to support people's interpretation of the symbol sizes on the map.

Votes

1 million

500,000

100,000   50,000

Centre point

Outline

While it's easier to pick out individual symbols from one another, the larger the symbol the harder it is to relate it to its feature centre point.

The use of outline symbols with a completely transparent fill is a common way to deal with the problem of overlaps on a proportional symbol map and helps reveal symbols without having to position them on top of one another.

Republican

Democratic

Painting the town red

Proportional symbol map using abstract symbols

Geometric shapes such as circles or squares are normally chosen because they are recognisable and the link between the data values and the property of area of the shape is reasonably understood. So what about other shapes? Although geometric shapes are the conventional choice when making proportional symbol maps, best practices are there to be broken (though it helps if you know what you're doing). **Practically any shape can be used, but the further you veer from simple geometric shapes, the bigger the potential drawbacks.** More complex shapes are harder to process visually in relation to each other.

Take the following symbols of a handprint. What is the relative size of one from the other? And how is it calculated?

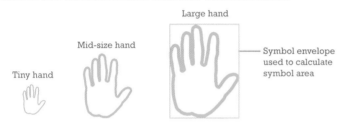

Tiny hand

Mid-size hand

Large hand

Symbol envelope used to calculate symbol area

If area is the property used to size the symbol, what component of the symbol does area relate to? Is it the surface area of the hand, for instance? That's an awkward shape for which it would be almost impossible to calculate the area, even as a rough estimation.

If you use an automated system to generate maps, the area normally is calculated as the envelope of the symbol, which represents the outer bounding box. This is likely a square or rectangle. Although that makes sense in defining a shape on which to base the calculation, it's not as easy for the reader to assimilate what is going on behind the scenes. **Any oddly shaped symbol will have white or empty space around it that isn't seen, and thus isn't processed in attempting to discern the relative or absolute size of the symbol.** And you don't want to add envelope lines to all the symbols. There are enough odd overlaps as it is.

The basic premise of symbol design remains: keep it as simple as possible to limit potential confusion from poor choices in symbol design. The map's context of use may lead to a decision to go for something more artistic. Perhaps a large poster or the cover of a book might be more suited to artistic symbology. Such maps often grab attention more easily than standard geometric symbology because people are attracted to maps that appear different. These maps perhaps are seen as more attractive and engaging relative to the more conventional symbol treatment but they are harder to read.

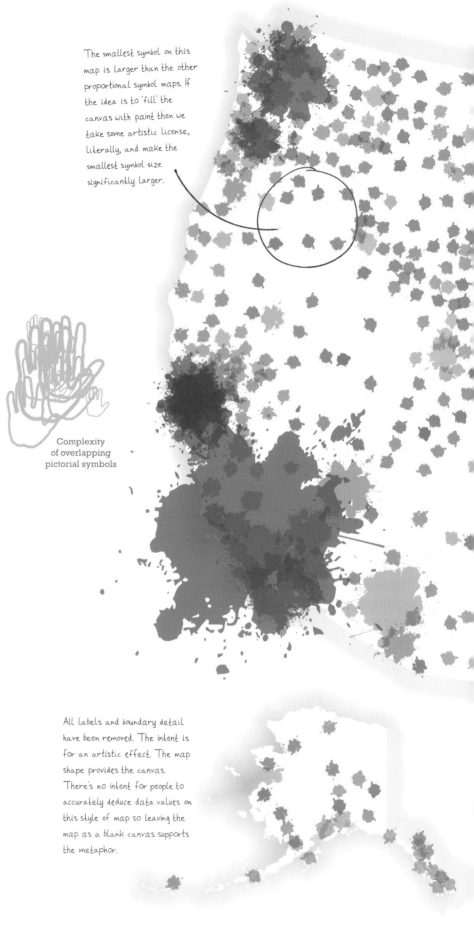

The smallest symbol on this map is larger than the other proportional symbol maps. If the idea is to 'fill' the canvas with paint then we take some artistic license, literally, and make the smallest symbol size significantly larger.

Complexity of overlapping pictorial symbols

All labels and boundary detail have been removed. The intent is for an artistic effect. The map shape provides the canvas. There's no intent for people to accurately deduce data values on this style of map so leaving the map as a blank canvas supports the metaphor.

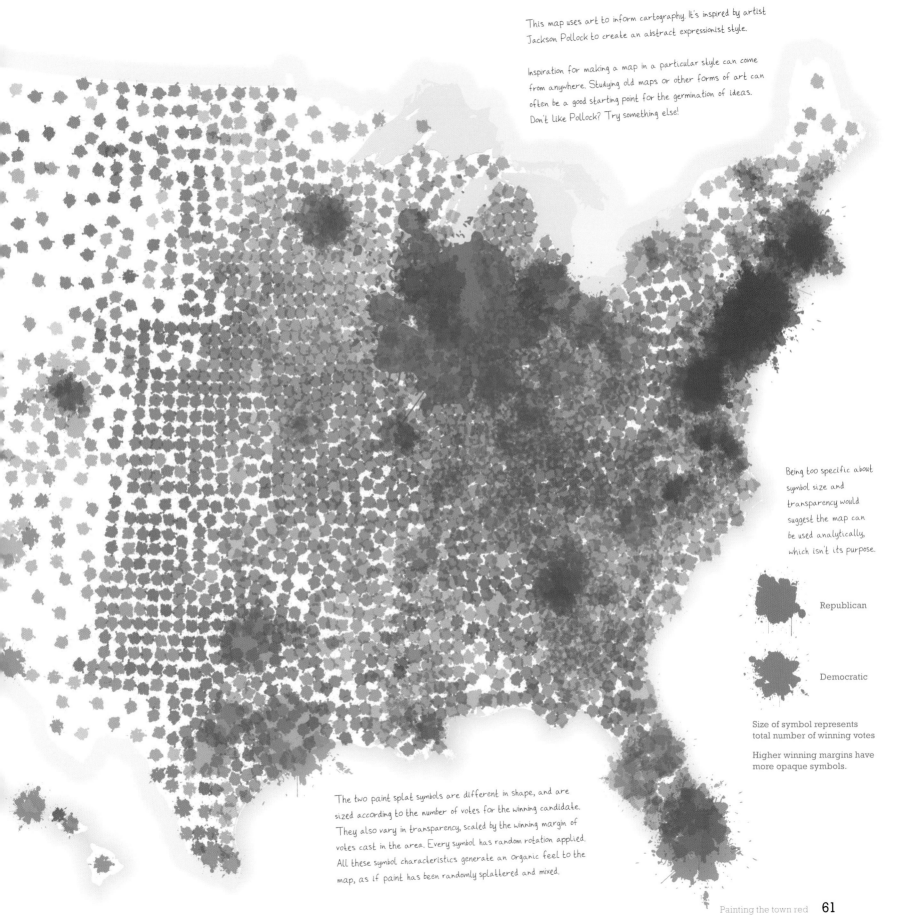

This map uses art to inform cartography. It's inspired by artist Jackson Pollock to create an abstract expressionist style.

Inspiration for making a map in a particular style can come from anywhere. Studying old maps or other forms of art can often be a good starting point for the germination of ideas. Don't like Pollock? Try something else!

Being too specific about symbol size and transparency would suggest the map can be used analytically, which isn't its purpose.

Republican

Democratic

Size of symbol represents total number of winning votes

Higher winning margins have more opaque symbols.

The two paint splat symbols are different in shape, and are sized according to the number of votes for the winning candidate. They also vary in transparency, scaled by the winning margin of votes cast in the area. Every symbol has random rotation applied. All these symbol characteristics generate an organic feel to the map, as if paint has been randomly splattered and mixed.

Donkeys and elephants

Logos as proportional symbols

A map often makes more sense to a reader when it uses colours or symbols that relate to the theme. This book consistently uses red and blue as the base colour scheme because these colours relate to the two parties. People associate red with the Republican Party and blue with the Democratic Party. **Symbols are used as a graphical shorthand to connote meaning, and the use of a stylised donkey for the Democratic Party and elephant for the Republican Party is widely understood.**

These two animals long have been associated with politics in the United States, and their origins can be traced to the 19th century. During the 1828 presidential campaign of Andrew Jackson, opponents often referred to him as a 'jackass" because of his perceived stubbornness. Rather than react to this assumed affront to his character, Jackson instead had pictures of a donkey placed on his promotional posters. Jackson won the election, and by 1870, the left-facing donkey symbol had been popularised by political cartoonist Thomas Nast for the Democratic Party en masse.

The Republican Party elephant first was seen in political cartoons during the Civil War as a reference to the phrase 'seeing the elephant', meaning to experience combat. Nast went on to use it an 1874 cartoon in *Harper's Weekly* as a metaphor for 'the Republican vote', and within a few short years, other cartoonists also were using the right-facing elephant to symbolise the party. Depending on your political perspective, the donkey might be considered hard-working, humble, and dedicated or a symbol of stubbornness. The elephant might be seen to represent intelligence, dignity, and strength, though opponents might prefer to see it as a circus animal. This illustrates clearly that each of us brings bias to interpreting symbology and maps.

The design of the symbols reflects the US flag with standardised red and blue, plus white stars, used in both. **They have become prominent symbols of the two political parties, and maps can make good use of them to provide an immediately recognisable theme.** Interestingly, neither of these animal logos are official logos of the parties, but for a map, the key is to use a graphic that the map reader will identify with, and these symbols do that job much better than the less frequently used official logos.

For logos to be used successfully, maps must be relatively simple and not overloaded with symbols. Overlaps need to be avoided because the complex shape of a logo often causes distractions. **An important aspect of the use of any logo is determining whether you have permission from any copyright holder to use the logo.** These logos are copyright free.

When you use a complex symbol shape or logo, you will often have to make compromises elsewhere. Using the elephant and donkey as proportional symbols for 3,141 counties would be unwieldy. I tried it. Looked awful! The compromise is to revert to states. Less detail, but a readable map.

The logos have been modified to suit their use on a map. The original logos use the same two colours. When you're only looking at one of each animal the differences in shape are obvious, yet on a map the abstract shape of each animal isn't sufficient on its own to reflect the categorical nature of the difference between them. So the Republican elephant is all red and the Democratic donkey is all blue. A simple change but one that helps to visually differentiate them.

Additionally, the elephant is visually more weighty than the donkey at the same size, so the elephant is shrunk a little on the map so that states with the same number of Electoral College votes (e.g. Minnesota and Wisconsin) appear the same weight.

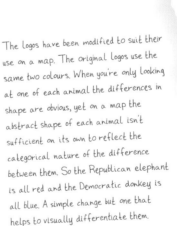 

The use of a halo helps each symbol sit a little above the background map. This helps separate the background and foreground. A compromise of this approach is the halo appears to change the shape of the coastline in places. We could have changed the background colour of US to something darker but then the whole map appears too dark. Decisions, decisions!

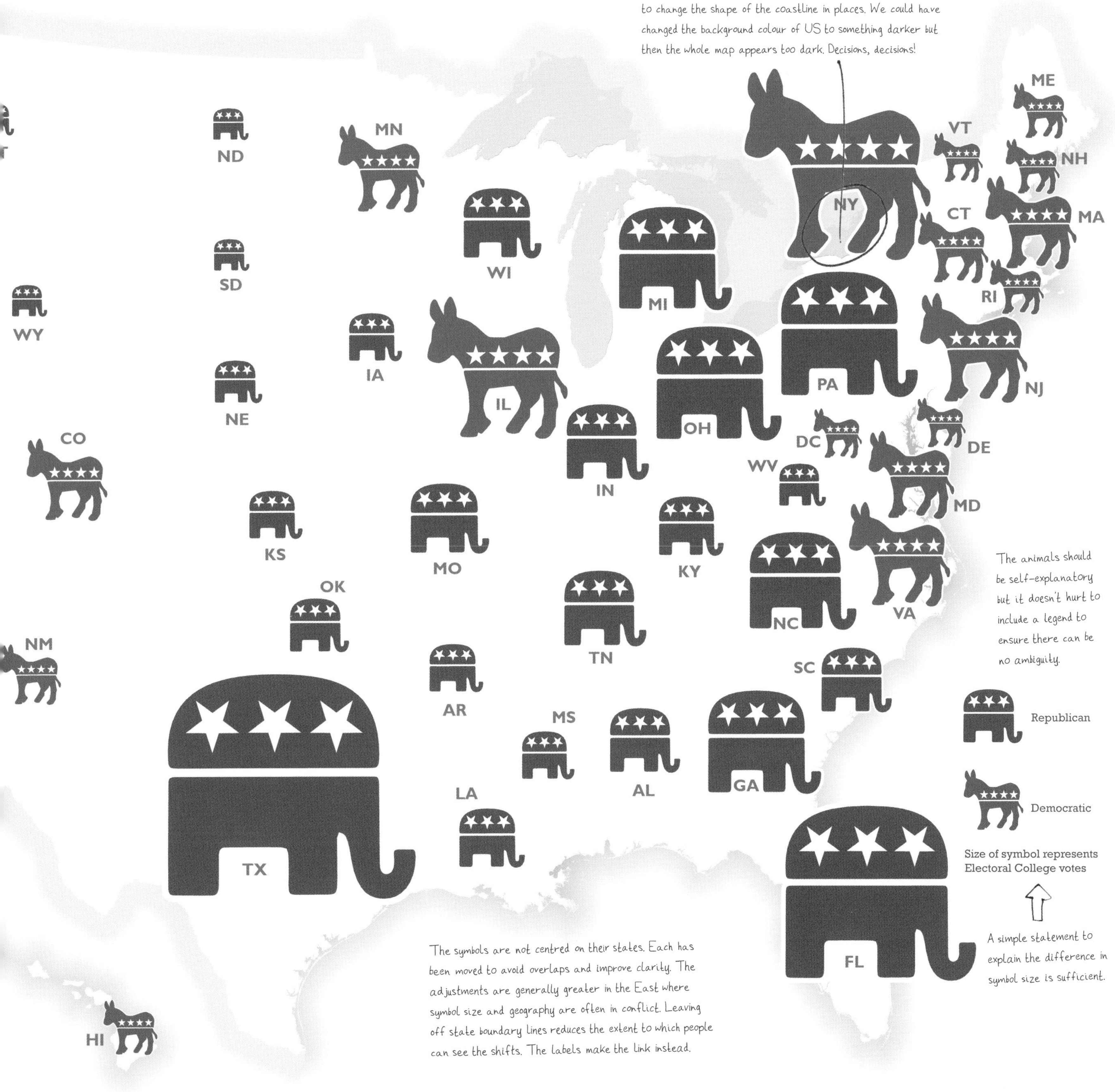

ME

MN

ND

VT

NH

SD

WI

NY

CT

MA

WY

IA

MI

RI

NE

IL

OH

PA

NJ

CO

IN

DC

WV

DE

KS

MO

KY

MD

NM

OK

TN

NC

VA

The animals should be self-explanatory but it doesn't hurt to include a legend to ensure there can be no ambiguity.

AR

MS

SC

Republican

LA

AL

GA

Democratic

Size of symbol represents Electoral College votes

TX

A simple statement to explain the difference in symbol size is sufficient.

FL

The symbols are not centred on their states. Each has been moved to avoid overlaps and improve clarity. The adjustments are generally greater in the East where symbol size and geography are often in conflict. Leaving off state boundary lines reduces the extent to which people can see the shifts. The labels make the link instead.

HI

Donkeys and elephants 63

Hemispheres

Multivariate proportional symbols

Often, the data that sits behind a map contains a wealth of information. The job of any map is to tease out something specific from the data. The process of selection or, conversely, omission is vital to good cartography, but when you want to show more than a single variable (a univariate map), you'll need to consider how to graphically encode multivariate data without reducing the map to visual noise.

The temptation to map more data than the map's symbology physically can hold is always a tension but it's not difficult in practice. There are three questions to answer. **Firstly, is there benefit in showing multiple variables? Secondly, does an individual symbol provide an understandable mechanism to understand the data? Finally, is there clarity when the collection of symbols is positioned on the map?** If the answer is no to any of these questions, chances are that the desire to map multiple variables is compromised by the increased complexity of the graphics employed to do the job.

Keeping things simple remains a key tenet. Restraint in design is vital. Election data includes two main pieces of information for each county: the Republican result and the Democratic result. On most maps in this book, one is shunned to emphasise the other, the winner. But presenting both pieces of information at the same time offers an opportunity to see the differences between the two. Here, to the right, is the simplest geometric shape once more, the circle, used to create proportional semicircles: Democrats on the left and Republicans on the right. Colour encodes the parties, and size encodes the number of votes. **Each symbol therefore contains six pieces of information—namely, location, parties, total number of votes for each party, and relative differences between the total number of votes by virtue of the relative size of each semicircle.**

Symbol overload is problematic if you try to do too much. Look at each of the following symbol designs. They don't work because there's too much going on, even though it's a basic two-segment pie chart. Put 3,141 of any of these on your map, and it'll be one gigantic mess. And 3D symbols have their own unique problems because perspective introduces difficulties in understanding relative amounts.

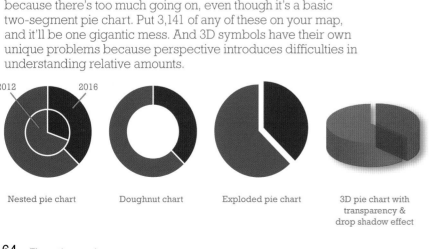

Nested pie chart   Doughnut chart   Exploded pie chart   3D pie chart with transparency & drop shadow effect

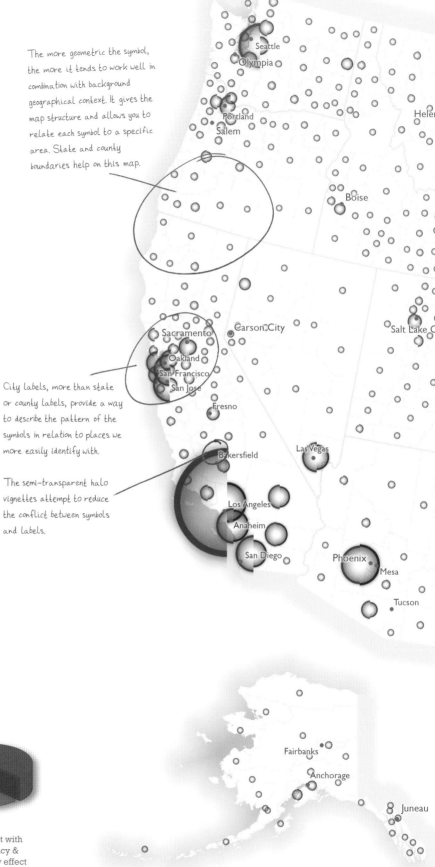

The more geometric the symbol, the more it tends to work well in combination with background geographical context. It gives the map structure and allows you to relate each symbol to a specific area. State and county boundaries help on this map.

City labels, more than state or county labels, provide a way to describe the pattern of the symbols in relation to places we more easily identify with.

The semi-transparent halo vignettes attempt to reduce the conflict between symbols and labels.

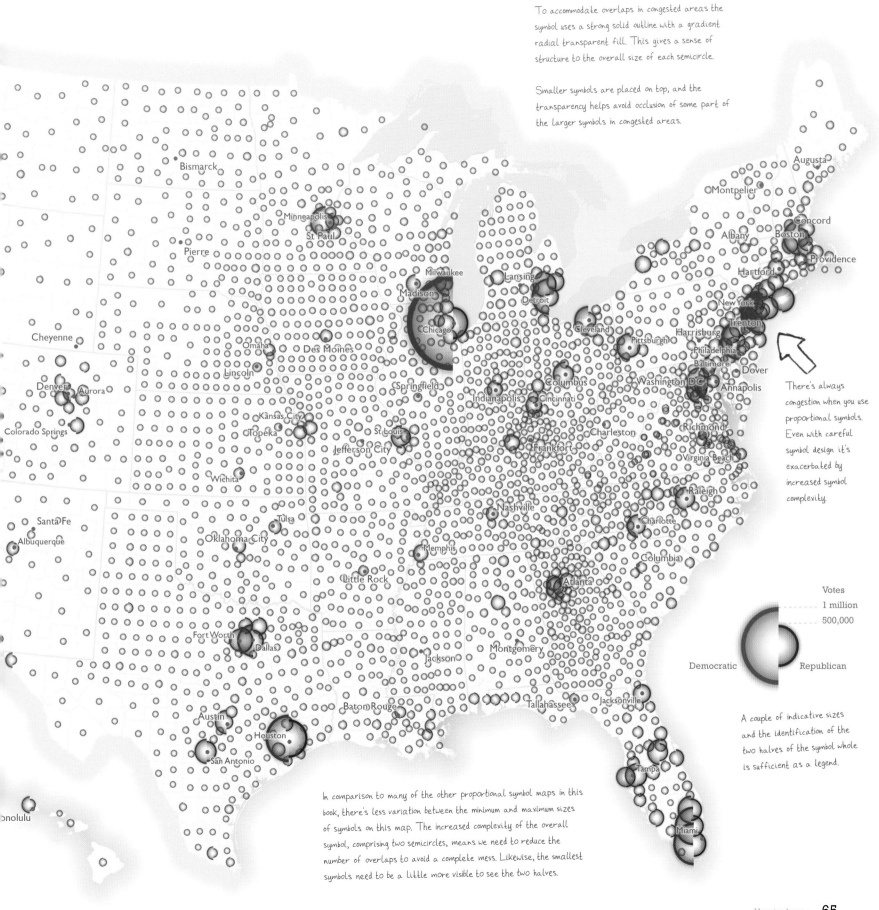

To accommodate overlaps in congested areas the symbol uses a strong solid outline with a gradient radial transparent fill. This gives a sense of structure to the overall size of each semicircle.

Smaller symbols are placed on top, and the transparency helps avoid occlusion of some part of the larger symbols in congested areas.

There's always congestion when you use proportional symbols. Even with careful symbol design it's exacerbated by increased symbol complexity.

Votes
1 million
500,000

Democratic   Republican

A couple of indicative sizes and the identification of the two halves of the symbol whole is sufficient as a legend.

In comparison to many of the other proportional symbol maps in this book, there's less variation between the minimum and maximum sizes of symbols on this map. The increased complexity of the overall symbol, comprising two semicircles, means we need to reduce the number of overlaps to avoid a complete mess. Likewise, the smallest symbols need to be a little more visible to see the two halves.

Purple haze

Mixing red and blue dots

One of the drawbacks of area-based maps is their tendency to poorly reflect different population densities using solid area fills. **The dot density map could be considered an area map, but areas are seeded with dots instead of a block of solid colour.** Potentially, map readers could count the dots to recover an actual data value but, in practice, this is an unreasonable objective. What the dot density map does well is naturally encode population density (or, in this case, the density of voters) to offset the impact of different densities in differently sized geographical areas. Dense areas of data get a lot of dots. Sparse areas get fewer, and so the visual weight of the mass of dots leads you to see more and less accordingly.

The dot value and size are crucial to create a coherent overall map. **Each dot on the map represents a data value. At smaller scales, you'll need to simplify the overall appearance by classifying one dot to represent multiple values—say, 100 votes**. The size of the dot is selected in combination with deciding the data value to give a distribution of dots that meets two main objectives. Dots should not coalesce too much in the areas of highest density, or else it's impossible to determine how much the dots are overlaid. In sparse areas, you must try not to leave the impression of too few dots. **As with most cartographic design objectives, you're looking to strike a balance and not make a decision to solve one problem that causes an even bigger problem elsewhere.**

For this map of election data to the right, dots represent the total Democratic and Republican votes per area, which are mapped at the same time using blue or red hues to differentiate them. By using the same data classification and symbol size for each set of data, you end up with a map full of red and blue dots. **The result of all these mingled red and blue dots is the impression of an array of purples, mauves, and violets that reflect a blended political landscape.** In some respects, it's akin to a blended choropleth map but one which accounts for differences in voter density at the same time. It's also clear where the more partisan areas are, although some of them clearly have sparse population densities.

A drawback of this technique is the potential to assume a dot implies location. It doesn't. **Dots are randomly placed point symbols and do not imply a unique location.** For this reason, it's advisable to avoid maps that classify each dot as representing a single vote unless you're obfuscating location somehow. Compounding this issue, as map scale increases, there's inevitably a temptation to look at individual neighbourhoods and assume the dots represent the vote of a specific household or individual. For these reasons, dot density maps are discouraged at larger scales.

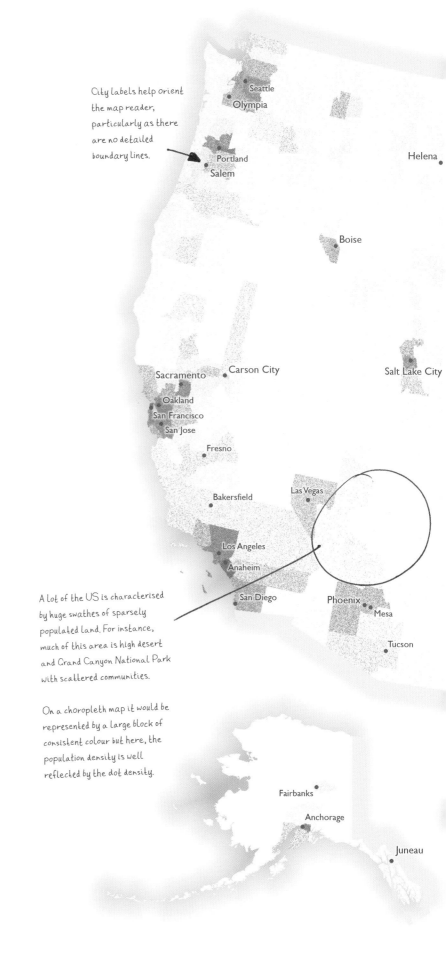

City labels help orient the map reader, particularly as there are no detailed boundary lines.

A lot of the US is characterised by huge swathes of sparsely populated land. For instance, much of this area is high desert and Grand Canyon National Park with scattered communities.

On a choropleth map it would be represented by a large block of consistent colour but here, the population density is well reflected by the dot density.

In order not to oversaturate the map with dots in the densest areas there's a consequence for the most sparsely populated areas where it's hard to see the hue of individual dots. It's unlikely that any dataset can be accommodated at both ends of the data range, particularly at small scale.

In the densest areas it's impossible to see individual dots but there's enough mixing of red and blue to give the impression of different shades of purple and, so, a mixed vote.

The underlying structure of the counties can still be seen from the sharp boundaries that exist between densely populated areas and neighbouring sparsely populated areas. There's therefore no real need for boundary lines.

The crucial detail is to explain the value attributed to each dot. Useful to show different indicative density patterns so some assessment can be made of the densities seen on the map.

100% Democratic   50:50   100% Republican

1,000,000
500,000
100,000
50,000
10,000
1,000

1 dot = 100 votes

vote density

Purple haze   **67**

One dot per vote

Maximum detail with dasymetric dot density

Despite addressing the problems of unequal geographies and population densities, the dot density map still suffers from the same problem as a choropleth map in that it exhausts space. Each area acts as a container within which dots are randomly positioned regardless of varying population densities. This situation can be remedied by employing the dasymetric technique.

By using a second, ancillary dataset to define areas of populated space, you can reallocate the data (votes) into the revised areas and use the dot density technique to distribute random points. **This solution arguably provides a more realistic way of seeing the spatial pattern of voting as it combines the advantages of the dot density and dasymetric techniques, allowing maximum data resolution.**

The downside to this mapping technique again relates to the impression of what each dot represents. It's perhaps even exaggerated when compared with an area-based dot density map because you are trying specifically to ensure dots are located in populated areas. **When maps are at large scale (e.g. street level) and show one dot representing one vote, the potential for incorrect inferences is inevitable.** Falsely implying a more detailed level of precision in maps made from aggregated data is always something to be aware of when mathematically and spatially reallocating numbers across a map.

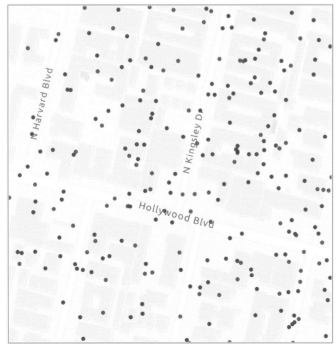

1 dot = 1 vote

Each dot is 04 pt in size and 99% transparent so it's impossible to see every unique dot at this scale.

Dots overlap and coalesce, to show density and a mix of colour in urban areas and less so in rural areas.

If the dots were solid colour there would be no differentiation between a sparsely populated area or a densely populated area.

At very large scales such as the inset left, each dot implies a precision and location far beyond what the data source actually supports. This can easily lead to incorrect assumptions on the part of the map reader. This impression is compounded by the fact that the dot value is now equal to a single vote.

In terms of its technical construction, the map is 'correct' yet it misrepresents the relationship between the data and geography.

If your house were on this map you'd be pretty upset at how the map 'reveals' your voting preference. It doesn't. It just looks that way. At small scales, such as the main map, the individual dots cannot possibly be seen as a unique locator.

What this map shows us that other maps don't is the pattern of distribution of voting between major metropolitan areas and rural communities. There's a clear differentiation in the distribution of voters with Democrats gaining a greater share in urban centres and Republicans gaining the smaller towns and rural areas in between.

There are **62,984,828 red Republican dots**, and **65,853,514 blue Democratic dots** on this map: one dot for each of the 128,838,342 people who voted for Donald J. Trump or Hillary Clinton.

Areas that Republicans won tend toward a red hue. Areas that Democrats won tend toward blue. Many areas appear purple, which shows places that had a relatively even share of the vote.

Why is most of the map apparently empty? Because it is! Most of the US is unpopulated space. This is why any dasymetric map will show the difference between where people live, and the unpopulated spaces in between.

Large dots and/or small dot values fill more of the map. Small dots and/or large dot values leave a much sparser appearance. Maps like this require experimentation to get to a pleasing result.

Putting people on the map

Minimum detail with isotype symbology

Most maps attempt to clarify the spatial component in data. That's their primary job in translating data to something meaningful, and why most of the maps you'll see of election data use boundaries or specific locations as a framework for classifying and symbolising the underlying data. Yet location can be expressed in many ways, from areas that aggregate data and exhaust space to dots that show precise locations of individual features. **What about the middle ground in which the map uses approximate locations to show approximations of data?**

Rather than trying to create the most precise map, you can use a map that deliberately overgeneralises the data for a useful impression of the outcome of the election. Some 136,669,276 votes were cast in the election. The previous dasymetric dot density map had almost the same number of dots (minus the 7.8 million votes for candidates other than Donald J. Trump or Hillary Clinton). The map was useful in showing the distribution of people across the US as opposed to showing the same data binned into counties which map geography, not people. But it's not necessary to have 136 million dots on a map to convey a similar story. **Here, to the right, each dot represents 250,000 votes. The map shows the dasymetrically distributed data aggregated and classified.** Each symbol then takes on a colour and shape indicative of the candidate.

The total number of symbols on the map sums to approximately 136 million, and the number of symbols for each candidate approximates their total popular vote. The map has a few more Clintons than Trumps because Clinton won the popular vote. The others are included on this map, shown by 31 symbols. The symbols are distributed in a semi-random fashion but approximate the distribution of population density in the US, in much the same way as the dasymetric dot density map. **The product is a heavily generalised map. It's accurate but also imprecise.** The data aggregation is coarse but you might argue that larger symbols distributed in the same way as the dasymetric dot density map give a clearer impression of the pattern of voting. Certainly, if an argument against a one dot equals one vote map is the false precision it might imply, this sort of treatment overcomes the problem.

The symbols are designed to be repeated. Without geography, they conform to the International System Of TYpographic Picture Education (ISOTYPE) pattern. An example of pure isotype is given later in the book. The map to the right uses the idea of simple pictures and combines them with a rudimentary geography of population distribution.

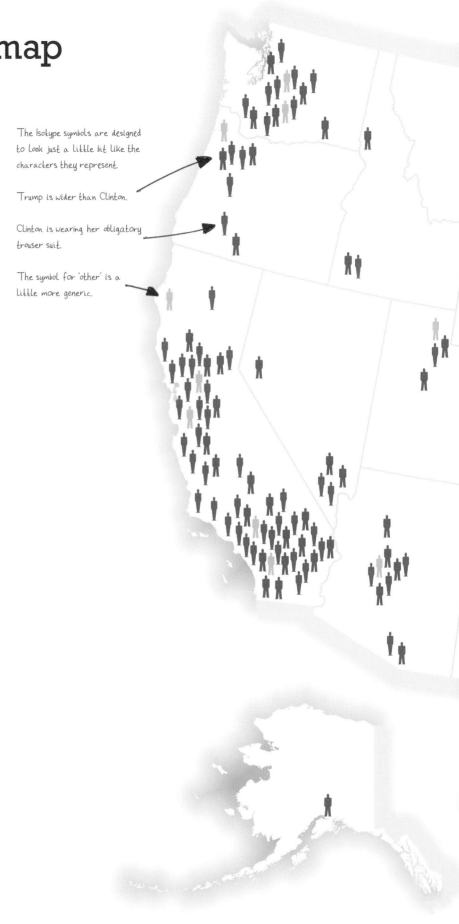

The Isotype symbols are designed to look just a little bit like the characters they represent.

Trump is wider than Clinton.

Clinton is wearing her obligatory trouser suit.

The symbol for 'other' is a little more generic.

There are 547 symbols on the map. At 250,000 per symbol the map represents 136,750,000 votes which approximates the real number of 136,669,276 votes. There are 264 Clinton symbols, 252 Trump symbols, and 31 Other symbols, which reflects the distribution of the popular vote.

The map should be fairly self-explanatory but the number of votes encoded into each symbol and the placement strategy is important to note in the legend.

Each figure = 250,000 votes

Trump

Clinton

Other

Votes are distributed to approximate population density.

The distribution of the popular vote is much more evenly spread between the two main candidates than many maps suggest. There are blue and red symbols in relatively even quantities throughout the populated areas of the US which is likely to reflect reality.

Making the point

A hybrid proportional symbol dot density approach

Proportional symbol maps have benefits and disadvantages. So do dot density maps. **But there's no reason to use only one mapping technique per map.** In fact, every map technique has its positives and negatives, and compromise is often the main outcome of the mapmaking process. So why not combine techniques?

With such a large variation in both voter density and geographic distribution, many of the point-based symbol maps struggle to accommodate both ends of the data range and geographical continuum. So, many large areas have low numbers, and fewer yet significant small areas have larger numbers. With a dot density technique, you often must accept increased coalescence of individual dots in the small, high data density areas to ensure you see some dots and variation in larger, less densely populated areas. With a proportional symbol map, you often end up with small symbols at the lower end of the data array, and a lot of them, in lieu of ensuring you don't end up with a few extra-large symbols at the upper end.

The map to the right provides a hybrid solution. **It uses a generalised dot density technique for areas that have fewer than 70,000 total votes.** Each dot represents 1,000 votes, and they are both larger than on the other dot density maps in this book and incorporate transparency so there's a little mixing of the colours in areas of overlap. **Proportional symbols are used for areas over 70,000 votes.** The symbol design is a little brave because the symbols are designed to appear 3D.

Using 3D symbols is generally inadvisable. **Humans are not good at estimating differences in the relative areas of different geometric symbols and even worse at estimating the relative volumes of 3D shapes.** An entire map of 3D symbols would be too much, yet when used judiciously (to bend best practice), 3D symbols can provide an interesting option. Instead of 3,141 3D symbols, this map has only 239. So with less than 10 percent of the map's symbols being proportional, it's increased the size of the smallest symbol and made them independently visible.

A map like this is designed to use the best bits of two different techniques. When you design the symbols in sympathy with one another, it becomes less obvious that two techniques are used. If you get the balance right, they work in harmony and allow you to innovate with your symbol design. Seeing relatively larger and smaller quantities is the main purpose of the map. Being able to use it analytically is less so.

Ensuring the large proportional symbols are seen as distinct from the dot density symbols is important and so creating a very visible difference helps cement that idea.

Why a globe? Why not? It could have been a two-dimensional circle, or a smooth 3D sphere but adding detail, particularly a graticule that adds to the illusion of it being a 3D object, helps give the symbol some structure.

Adding a shadow on the underside and a very faint drop shadow accentuates the 3D relationship of the symbol sitting atop the rest of the map.

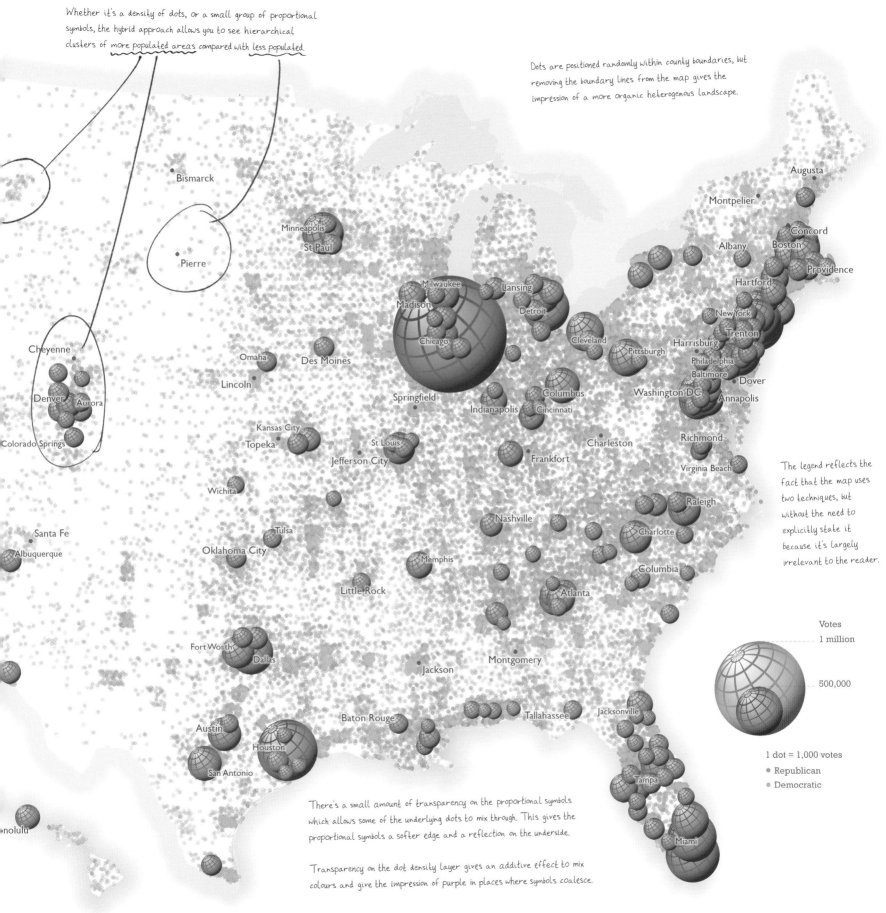

Whether it's a density of dots, or a small group of proportional symbols, the hybrid approach allows you to see hierarchical clusters of more populated areas compared with less populated.

Dots are positioned randomly within county boundaries, but removing the boundary lines from the map gives the impression of a more organic heterogenous landscape.

The legend reflects the fact that the map uses two techniques, but without the need to explicitly state it because it's largely irrelevant to the reader.

Votes
1 million

500,000

1 dot = 1,000 votes
• Republican
• Democratic

There's a small amount of transparency on the proportional symbols which allows some of the underlying dots to mix through. This gives the proportional symbols a softer edge and a reflection on the underside.

Transparency on the dot density layer gives an additive effect to mix colours and give the impression of purple in places where symbols coalesce.

When is an area a point?

Combining a choropleth with coloured dots

Geography often gets in the way of a map's intended message. This is especially evident when mapping a large country such as the United States, which has vastly different-sized areas and population densities. Although you can modify the way you symbolise areas to mitigate the problems, you're fighting a losing battle. **Large areas always seem more important on the map compared with small areas.**

Mapping data as point-based symbols provides one solution but these also come with difficulties such as overlapping symbols in the most densely populated areas. So why not use the best of both worlds and create a map that uses both areas and points in harmony? **This might be termed a hybrid 'pointillist' approach, borrowing terminology from the artistic technique of pointillism.**

Areas that fall below a certain threshold in areal extent are retained as areas and shown as a choropleth, and those that exceed the threshold are shown using points of the same size. Areas and points are symbolised using the same colour scheme. There's no automatic way to determine what value of area is going to work as the threshold but the objective is to create a well-balanced map. This could be defined as not so few areas that the map looks more like a dot density map with poorly scaled symbols. **Ensure the size of the point-based symbols is broadly the same as an area that meets the threshold value.** In this way, you'll get a map on which the vast majority of symbols, whether area- or point-based, are similarly sized. This recalibrates the overall visual impact so each enumeration area is broadly visually equivalent to another.

A drawback can be that some small areas still appear lost on small-scale maps but this is a minimal cost for the benefit of correcting for the extremes of visually dominant large geographical areas.

On the map to the right, the threshold is set at 2,000 sq km, and population density is incorporated into the symbol design by applying the value-by-alpha technique from a nearly fully opaque (white) overlay to reduce the visual importance of low population density areas, to transparent for high population density areas. Circular symbols are used because they look better. Other geometric shapes might be useful, but for the US, squares may be seen as areas rather than points. The distinction is important.

One might argue that the map appears empty but, in truth, it is. People live in small areas among vast swathes of uninhabited geography. **Giving false importance to geography is a major cause of misinterpreted population maps which this technique addresses.**

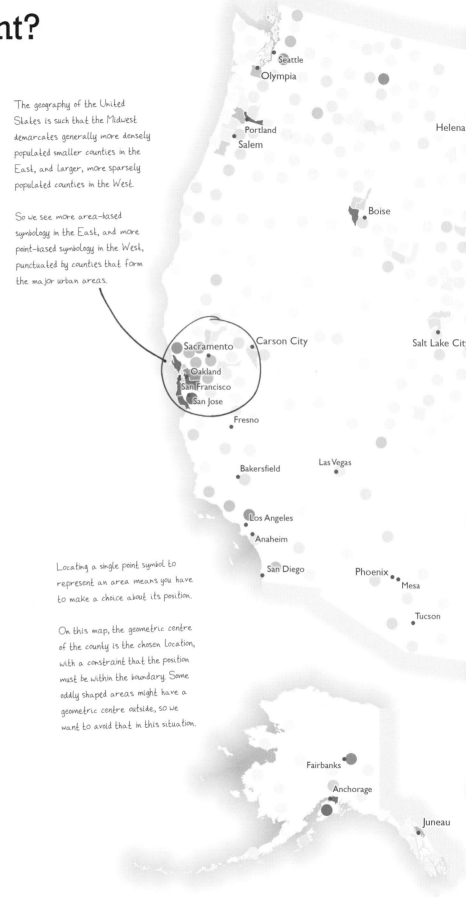

The geography of the United States is such that the Midwest demarcates generally more densely populated smaller counties in the East, and larger, more sparsely populated counties in the West.

So we see more area-based symbology in the East, and more point-based symbology in the West, punctuated by counties that form the major urban areas.

Locating a single point symbol to represent an area means you have to make a choice about its position.

On this map, the geometric centre of the county is the chosen location, with a constraint that the position must be within the boundary. Some oddly shaped areas might have a geometric centre outside, so we want to avoid that in this situation.

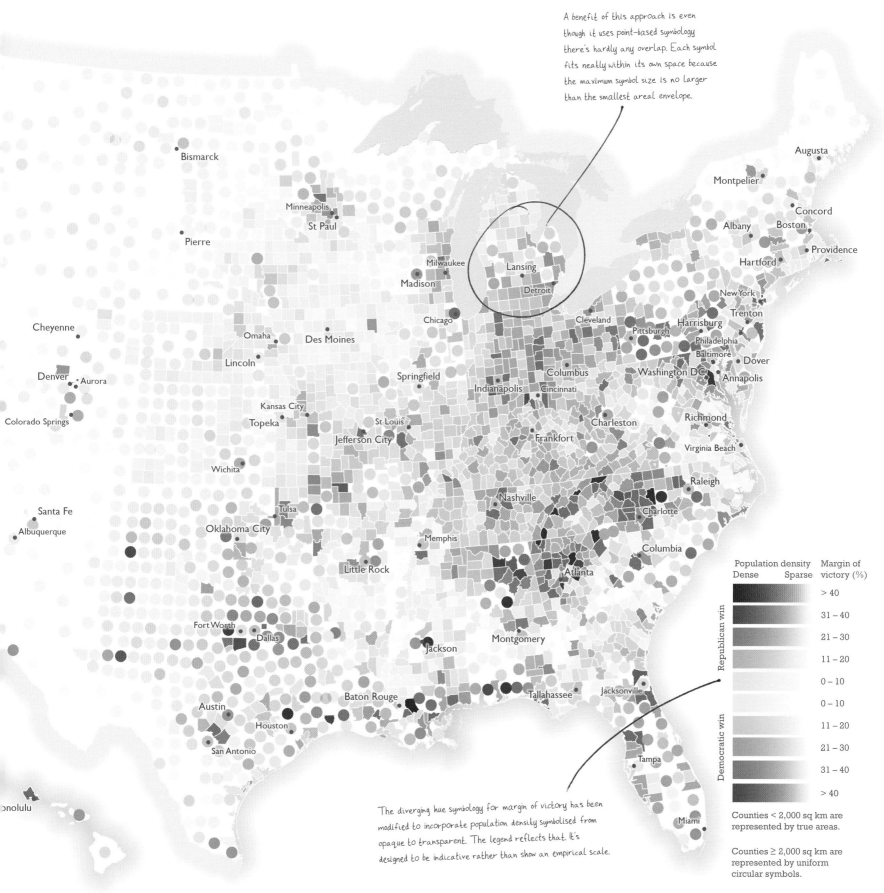

A benefit of this approach is even though it uses point-based symbology there's hardly any overlap. Each symbol fits neatly within its own space because the maximum symbol size is no larger than the smallest areal envelope.

The diverging hue symbology for margin of victory has been modified to incorporate population density symbolised from opaque to transparent. The legend reflects that. It's designed to be indicative rather than show an empirical scale.

Population density
Dense — Sparse

Margin of victory (%)

Republican win

> 40
31 – 40
21 – 30
11 – 20
0 – 10

Democratic win

0 – 10
11 – 20
21 – 30
31 – 40
> 40

Counties < 2,000 sq km are represented by true areas.

Counties ≥ 2,000 sq km are represented by uniform circular symbols.

Crop circles

Equal-area binned, gridded symbols

You'll often see abstract representations of data in map form. These often are designed to provide a headline graphic rather than recover information. **The rise of infographics as a form of graphic design has spurred the manipulation of maps into a range of alternative visual representations of data.** Infographics tend to be striking and engaging to provide a means to communicate information quickly and clearly.

To the right, geography is eschewed in favour of a regular grid of uniform symbols—circles. They cover the mapped area, and therefore take on the approximate shape of the United States. **The symbols are reporting data systematically sampled at that point across the map.** In that sense, an uncertain amount of variation and detail is likely to have been lost in converting data from administrative areas to a surface and, finally, to a set of regular points. On top of this uncertainty is the issue of interpolation because the underlying data that's been sampled already has been manipulated onto a surface rather than representing the county data in its original form.

You can imagine this sort of map style being used alongside other visuals such as headline numbers, a few graphs, and minimal text, yet at its heart, it's just a grid of multivariate map symbols. Each symbol encodes several pieces of data using different variables (hue, value, and size), which are varied within the constraints of the symbol size. There's little space for variation, which has the impact of being almost impossible to decipher it in an empirical sense.

The style does a good job of showing density of data values and how they vary relatively across the map. The map reader can clearly see more and less. That's the takeaway.

Aggregating data into cells of equal area fundamentally depends on using an equal-area projection. Here, the Albers Equal Area Conic projection suits the technique well.

Because it equalises area, you can map percentages, ratios, medians, means, and counts equally well and maintain visual consistency across the map.

Shapes have meaning and circles, with soft gentle curves rather than sharp angles, are easy on the eye. They convey a message of wholeness and mirror many familiar objects such as planets, wheels, and balls. In fact, on a map of the United States they mimic the natural patterns you often see across large swathes of the country that have centre pivot irrigation and circular crop patterns.

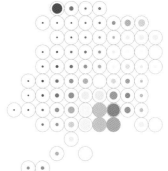

The symbol can take three variables, namely majority winner, share of the vote, and a measure of population (total votes) shown by varying the hue, value, and size of the symbol.

Limiting the map to two variables, and removing the total number of votes, would give the less complex, but arguably less interesting map to the left.

Interpolation and binning of data into new shapes can have unintended consequences. The cells that represent Minnesota and Colorado appear to show a predominance of red, when in fact both states were Democratic wins.

The symbol size used on this map is based on a sampling grid of 50 x 50 km, creating a 2,500 km² area in which each circular symbol is situated. This gives a good balance between symbols that are too large and lack spatial variation across the map, and too small that they become indecipherable.

The legend is in grid form to show the bivariate colour scheme along with the symbol size. It's simple and indicative, rather than something quantifiable and measurable.

Share of vote (%)

Increasingly Republican   Increasingly Democratic

Relative number of voters   High   Low

The map errs more toward a graphic representation with its regular gridded pattern. There's very little room for additional detail like labels. Boundaries are superfluous. Using a faint grey outline for each circular cell helps give the overall map some structure, and ensure many of the smaller symbols don't appear to be floating in empty space.

Map box of doughnuts

Equal-area binned, gridded symbols with a hole

At first glance, the map to the right looks like the one on the preceding page (gridded crop circles), and the idea is essentially the same: to create a regular grid of symbols that represent different components of the data. Both maps began as an interpolated surface of the county-level data. They then were created using a grid to sample the values of the surface.

But the approach to the symbology of the two maps is different. On the gridded crop circles map, each symbol is a single entity, coloured and sized according to different variables. Here, there are two symbols as two separate layers, with one sitting atop the other. **This design creates a multivariate symbol combination in which it's the border of the symbol that encodes the combination of the variables being mapped, and not the fill of the symbol.** It's effectively a set of gridded doughnut symbols.

The use of white space in cartography is a valuable tool. The basic underlying symbol here is a circle of constant size, shaded using a diverging colour scheme to indicate percentage share of the vote. Across the top is another set of circular symbols, all coloured white and sized according to the total number of votes. This symbol literally punches through the other symbol to create the hole in the doughnut.

The increased white space in areas of low numbers of voters then forces symbols to recede in the visual plane. The converse of this is that areas with high numbers of voters (the most populated areas) appear as either almost whole filled doughnuts or ones with a small hole. **Doughnuts with small holes have more dough and so come into focus to a greater extent.**

The intriguing combination of the two main variables, share of the vote and total number of voters, provides many combinations, and so the map is only useful as a way to assess different places relatively, not empirically.

One consideration for any multivariate symbol is information overload. **Maps such as these easily can tip into the realm of being totally unusable by packing too much information into the same symbol.** The impact makes individual symbols difficult to interpret, and the overall map display becomes harder to explore and patterns much more difficult to see and decipher. This map veers toward an abstract representation of the data but abstract can be useful for building visual interest in a story.

Mmm ... doughnuts!

Low population, and high share of the vote

You could use pretty much any shape as a basic symbol design (squares and hexagons are used elsewhere in the book to give a different effect) but circles have both reflection and rotation symmetry around the centre for every angle so they look good when positioned in a regular grid.

A drawback, however, is the perception that there are many small missing pieces in between adjacent circles. While that's not the case for the analytical aspect of calculating values, it's an unavoidable consequence of choosing a symbol type that doesn't exhaust space and tessellate.

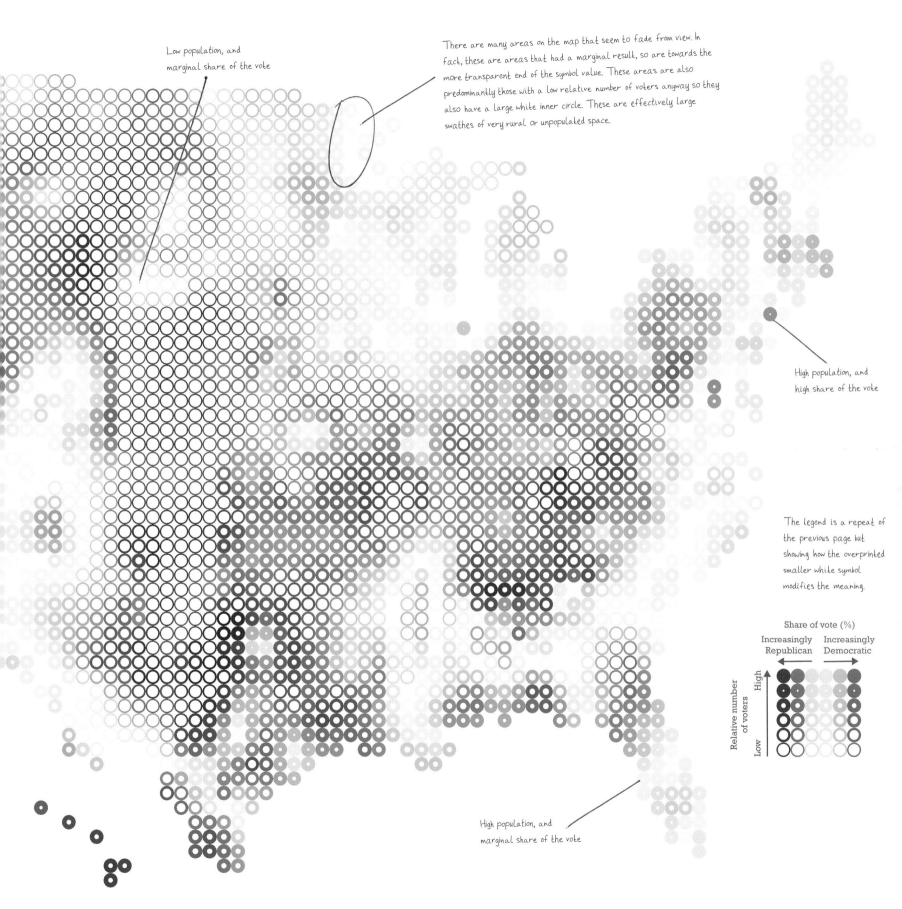

Low population, and
marginal share of the vote

There are many areas on the map that seem to fade from view. In
fact, these are areas that had a marginal result, so are towards the
more transparent end of the symbol value. These areas are also
predominantly those with a low relative number of voters anyway so they
also have a large white inner circle. These are effectively large
swathes of very rural or unpopulated space.

High population, and
high share of the vote

The legend is a repeat of
the previous page but
showing how the overprinted
smaller white symbol
modifies the meaning.

Share of vote (%)

Increasingly   Increasingly
Republican   Democratic

←   →

High

Relative number
of voters

Low

High population, and
marginal share of the vote

Bipartisan patterns

Purple gridded squares

Grids often are used in cartography to provide a framework for mapping a phenomenon. Perhaps the most obvious is also a data type, a *raster*. A raster exhausts space and encodes each individual cell (pixel) with a value denoting some quality of the landscape over which it sits. Each cell usually captures the average data value for the area it covers, although other characteristics of data can be captured easily. This map isn't a raster grid, but it borrows the same ideas.

A grid is placed across the data to be sampled. **Each grid cell is then encoded with the mean value of the election result for the area it covers.** The grid squares then are symbolised to tell a story. The map is more graphical than cartographic, and there's inherent uncertainty in what each grid cell represents because it's the function of a fair bit of data processing, but the result reveals information that can be interpreted relatively. **Rather than showing share of the vote or margin of victory, the map to the right is symbolised in shades of purple to identify broad patterns of bipartisanship.** It's a simple metric extracted from the data that creates its own distinct patterns.

Grids are used more often in other tessellated forms to create cartograms in which each grid cell represents a unit such as a state or county. Here, they represent only geographical space with each grid cell being 50 × 50 km and occupying 2,500 km². The impact is that in places with highly variable data, the final map may lose richness of detail. In other areas in which there is either little variation or you're sampling from large counties, you may be implying a greater variation than exists. Does this matter? It depends on what the map is used for. With a vague but acceptable legend and used as an overview, the map gives general patterns. But to find out the result in a specific county, it's not much use.

Intertwining valleys of marginal areas, which might be ripe for targeting a swing vote from one party to another in an election campaign.

The main map deliberately mutes the symbology for those areas with an increasingly partisan share of the vote to allow us to see the marginal, bipartisan areas.

Compare that to the map to the left which has the unadulterated shading scheme. It's actually very difficult to see the marginal areas in isolation. Using symbology to tease out a particular view of a dataset is part of the art of cartography.

Most regular grids have uniform individual grid squares. They all align. Just to add some interest, this one orients each individual grid square to the curved graticule of the underlying Albers Equal Area Conic projection. It's for no other reason than a little added visual interest, but which doesn't detract from the overall function of the map.

A simple single hue transition that allows you to interpret where the marginal areas are more prevalent.

Margin of victory
Small: marginal

Large: partisan

Large swathes of areas which had partisan support for either the Republican or Democratic candidate. These are areas that are likely not a good bet for an opposition candidate to flip.

Let the data speak for itself

Proportionally sized labels as literal symbols

Most maps use symbols to encode meaning and communicate some characteristic of the data. **In fact, much of the art and science of cartography is applying knowledge of visual and graphical systems and human perception and cognition to design a visual representation of data through symbols.** This isn't a task to make a map look pretty but rather a way to carefully encode meaning into how symbols reflect data.

The symbols act as signs to direct understanding, but at the same time, you're tacitly accepting that something will be lost in the translation from the data value to the graphical version. Or you could put the data values on the map and save a lot of effort on the part of the cartographer and the reader.

To the right, the percentage share of the vote is used as a label, so every result is shown explicitly on the map. Nothing is hidden or manipulated through classifying or symbolising data in a certain way. **There's no hiding of the labels because they are being used as a literal symbol, and every piece of data is on the map.**

Placing labels in a uniform size and colour, though, creates an extremely monotone map with no visual hierarchy to help the map reader distinguish patterns. **The form of the labels has been manipulated to give emphasis and support the map reader's need to differentiate between highs and lows without having to read every number.** A diverging colour scheme is used to encode the larger share of vote values with a more saturated hue. This is the same diverging shading scheme used elsewhere but here applied to the text rather than a geometric symbol or fill. Additionally, the labels are changed in size so areas with a larger number of total votes are larger. This helps emphasise the larger, more densely populated cities and de-emphasise sparsely populated rural areas. Finally, transparency is applied to add variation between large share of the vote labels and small share of the vote labels. These are only graphical effects applied to the labels to build visual hierarchy into the design.

Clearly on a map like this, space is an issue, so the geography and scale of the mapped area determines label size and positioning. The density and size of the labels works for the United States at this scale. In a multiscale environment, you might show only the most important (largest?) symbols at smaller scales and reveal all the data labels at larger scales as you zoom in. In this way, you'd capture some of the key data values as headlines at small scales before revealing detail.

There's no space for other labels on a map like this but it shows that you can accommodate 3,141 separate pieces of data on a small-scale map without having to use symbols.

Because there's no space for boundaries and labels the background to the map is an important consideration. One approach would be to go entirely graphic and remove all other detail, just leaving the labels as the only features. Instead, a very faint diverging hue choropleth at county level is used as a background. It helps support the relationship between map labels and counties, particularly in areas with large, oddly shaped boundaries.

Geography has a habit of throwing up difficult cartographic conundrums and you'll notice on this map some labels seep into the oceans and lakes. Many larger cities border water bodies, and are more tightly arranged. But labels don't have to be positioned with absolute accuracy. In fact, not having them all constrained within land makes for a more organic look and feel.

The map is pretty self-explanatory, and that's the point. The legend can afford to be brief and descriptive.

Label indicates percentage share of the vote

Democratic win in blue

Republican win in red

Larger number of votes shown by **larger labels**

Larger share of the vote shown as more **opaque**

Smaller share of the vote shown as more transparent

The key feature of the map is to ensure you can fit all the labels on without overlaps. It's often a matter of trial and error (and good labelling software!). Work from the smallest area with the largest label and build the labels out from that.

A map stack

Stacking one map on top of another

Keeping things simple in cartography is always a good maxim, and one of the simplest ways is to stick with a familiar technique done well. Except that this process would reduce most people's cartographic palette to maybe half a dozen basic techniques. **Sometimes, it's important to reveal a little more in the data than a single map can show, so instead, use two.** Except rather than make two maps that sit side by side, position one on top of another.

The example to the right takes two of the variables of the election data: margin of victory and turnout, juxtaposed. Margin of victory is presented as a choropleth and turnout as a proportional symbol. One map sits on top of the other, and so the relationship between two variables for each area on the map can be seen. It's a quick technique to apply and avoids some of the more convoluted data processing you might need to perform to create a value-by-alpha or bivariate choropleth map. It's two alternative but complementary techniques but which demand a higher cognitive load because the map reader must visually combine data for two variables for each area.

The key to making this sort of map work is in focussing on not only the design of the choropleth underlay and the proportional symbol overlay but how the two work together. This comes down to ensuring that whatever decisions you make in designing one map, it doesn't overawe the other. For instance, you don't want to obscure too much of the area under large proportional symbols. On the other hand, there's little point in having too many little points across the map. It's a question of tweaking the symbol design of the two maps until a happy balance is achieved.

Designing the two maps in sympathy helps. Using proportional symbols with transparent fills is often a good fallback. Using fill colours that match the underlying choropleth colour scheme not only reinforces the relationship between variables but helps graphically marry the two maps together. Scaling the size of proportional symbols doesn't have to be linear either. This map uses a logarithmic scale to bring all the symbols into a reasonable spread of sizes with enough variation across the map but not nearly as much as you'd use on a single-themed proportional symbol map.

The drawbacks normally are seen in the typically congested areas of the map and as a function of the large-data/small-area problem that routinely creates difficulties. Using a smaller range of proportional symbol size than you might use on a single-themed map can help. Overall, you're leaving the task of comparing two variables to the reader. Your map is trying to present them both side by side, even though in reality, one is on top of the other.

This map, perhaps more than many, emphasises how difficult it is to balance design when you're working with a set of boundaries that differ so markedly. The left side of the map works better than the right. If the largest halo symbols are any smaller, the right side might improve a little, but the left would be worse off. It's that compromise again!

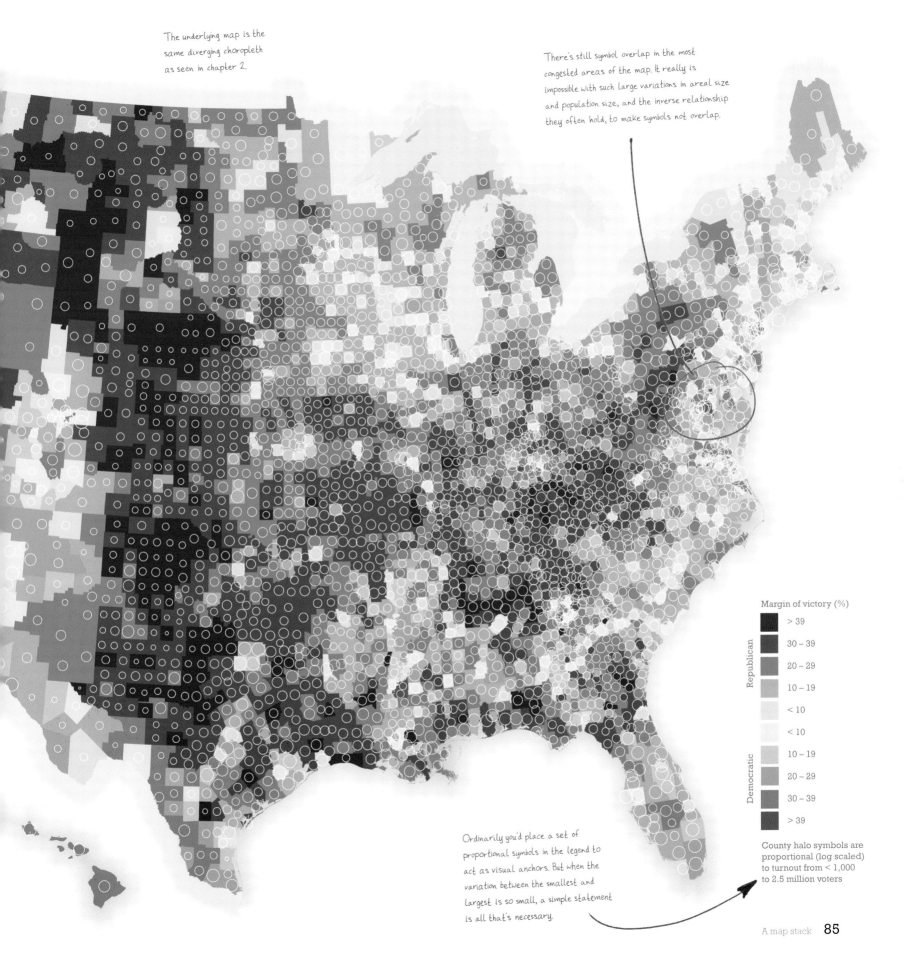

The underlying map is the same diverging choropleth as seen in chapter 2.

There's still symbol overlap in the most congested areas of the map. It really is impossible with such large variations in areal size and population size, and the inverse relationship they often hold, to make symbols not overlap.

Ordinarily you'd place a set of proportional symbols in the legend to act as visual anchors. But when the variation between the smallest and largest is so small, a simple statement is all that's necessary.

Margin of victory (%)

Republican
> 39
30 – 39
20 – 29
10 – 19
< 10

Democratic
< 10
10 – 19
20 – 29
30 – 39
> 39

County halo symbols are proportional (log scaled) to turnout from < 1,000 to 2.5 million voters

4. Line maps

Maps comprise many types of lines. In many population-based thematic maps, lines are used as a graphical way of encoding the boundary between neighbouring places. This might be between counties or county-equivalents in the case of US election data or some other administrative unit used as a container for the collection and reporting of the data being mapped.

Lines can be used in many ways though, and although they're not necessarily the first choice for thematic maps that report population-based data, a range of techniques can bring a different perspective.

In fact, line-based thematic maps are a familiar sight for some thematic data such as temperature or atmospheric pressure, yet they rarely are used for other data. That's about to change as you read further.

The maps in this chapter show how linear features can be used in interesting ways or how new lines can be derived from the data to represent it in engaging and innovative ways.

Data as a fluid surface

Humble contour

People are distributed across space much more fluidly than discrete, hard, arbitrary boundaries such as counties suggest. **Population distribution can be thought of as a continuous surface rather than containerised.** And for other data represented as surfaces, such as temperature or precipitation, it's common to use isolines (lines of equal value) or contours to represent a phenomenon.

A contour line joins a set of equal values and threads itself through values represented by a surface. Contours are nested, so the areas between adjacent contours are between the values of the two contour lines. They allow the imagination of a surface that is continuously varying.

Contours are generated by assessing adjacent values of data points. Imagine each area having a point at which the value (margin of victory) is known, and then you can thread contours through the surface. But where do you position the point? Ordinarily, an area's geometric centre is used, which would be fine in many instances. **Except many counties are sparsely populated across vast areas and the population is more tightly packed into a few areas, so use a population-weighted point to locate the data points.** To an extent, this strategy helps draw contours in more geographically accurate positions.

Labelling contour lines is crucial for people to interpret them. Use them sparsely, nested in a ladder where possible and aligned either to the page or in an uphill fashion. On many topographic maps, labels are the same colour as the contour line itself. This wouldn't work for lines that are light to dark, such as the map to the right, as the lighter labels would be too hard to read. Instead, light grey allows labels to recede a little but be equally legible.

Scale of the final map will help determine the contour interval. For this map, it's possible to use an interval of 5 percent between adjacent contours. For a smaller scale map, the contour interval might be 10 percent and the amount of generalisation (smoothing) applied to the lines would increase. **Contour interval and amount of generalisation are a matter of experimentation and balance to end up with a good amount of detail and with contours that neither have sharp changes in direction nor appear overly smoothed.**

Any map of empirical data can be considered a surface. Points provide the known values, and a surface of values in between can be interpolated. Assumptions are inevitable about the relationship between distance and adjacent values in any model of interpolation but the result provides a fascinating way of releasing containerised data and mapping it with more fluidity.

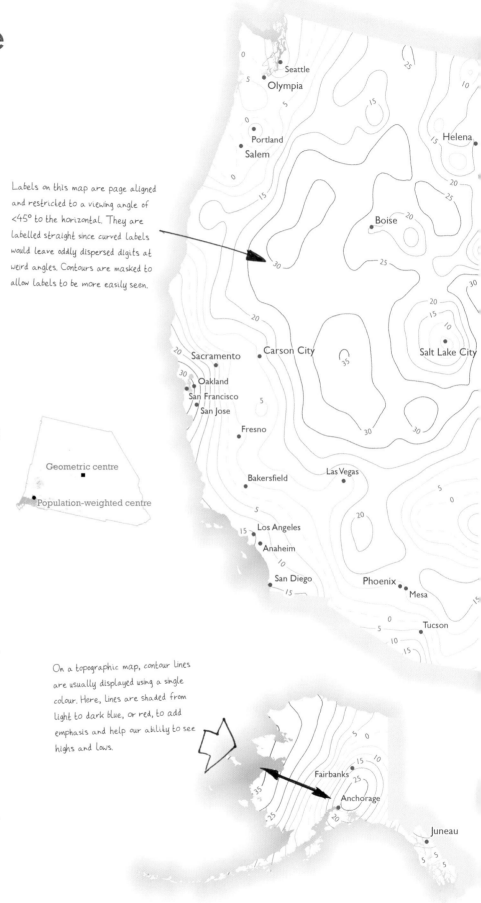

Labels on this map are page aligned and restricted to a viewing angle of <45° to the horizontal. They are labelled straight since curved labels would leave oddly dispersed digits at weird angles. Contours are masked to allow labels to be more easily seen.

Geometric centre

Population-weighted centre

On a topographic map, contour lines are usually displayed using a single colour. Here, lines are shaded from light to dark blue, or red, to add emphasis and help our ability to see highs and lows.

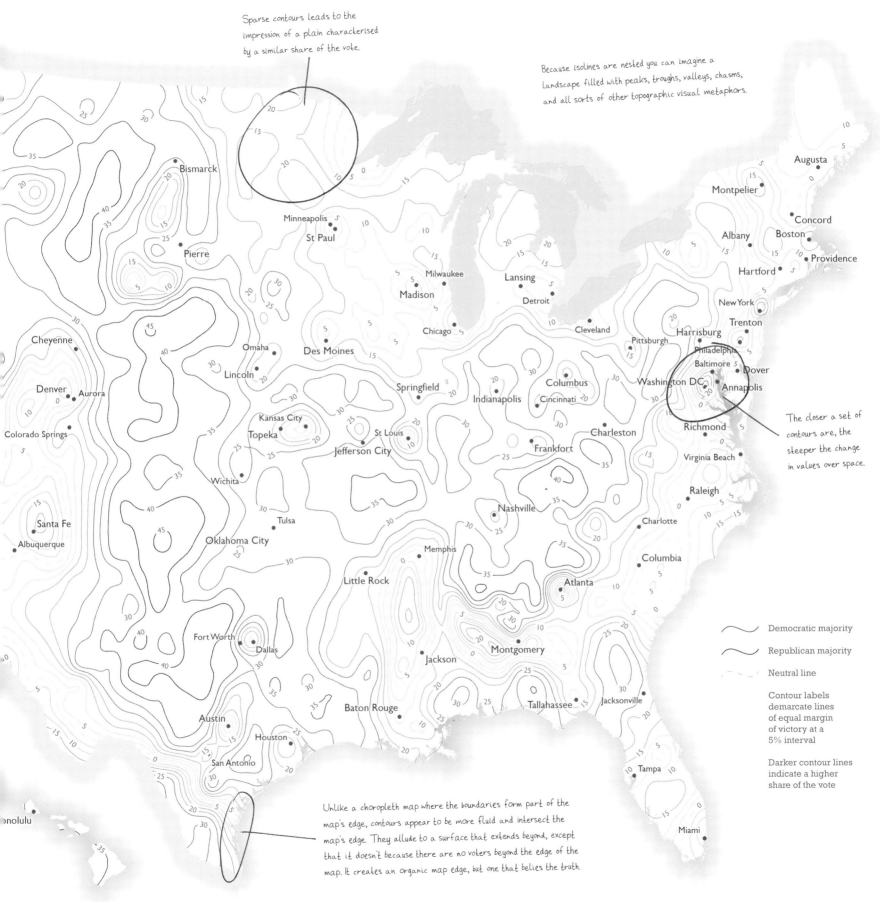

Sparse contours leads to the impression of a plain characterised by a similar share of the vote.

Because isolines are nested you can imagine a landscape filled with peaks, troughs, valleys, chasms, and all sorts of other topographic visual metaphors.

The closer a set of contours are, the steeper the change in values over space.

Unlike a choropleth map where the boundaries form part of the map's edge, contours appear to be more fluid and intersect the map's edge. They allude to a surface that extends beyond, except that it doesn't because there are no voters beyond the edge of the map. It creates an organic map edge, but one that belies the truth.

Democratic majority

Republican majority

Neutral line

Contour labels demarcate lines of equal margin of victory at a 5% interval

Darker contour lines indicate a higher share of the vote

Colouring between the lines

Isopleth (filled contours)

Choropleth maps exhaust space by presenting population data bounded by the administrative units used to collect and report the same. Their inherent drawback is that the structure of the administrative units fundamentally shapes the view of the data. If boundaries are drawn in a different place, the map would take on a different appearance, not only in the organisation of space but in the way in which data is computed. As noted in chapter 1, this is referred to as the *modifiable areal unit problem* (MAUP). **Whatever subdivision of space you have on a choropleth map, the map suffers from the MAUP.** It's inherent in all choropleth maps because summary values are influenced heavily by the position of boundaries and the shape and scale of the unit of aggregation. **An isopleth map is one way to escape the often brutal boundaries of a choropleth.**

Technically, an isopleth refers to a thematic map in which lines connect points of equal value and intervening spaces are shaded as a single class. The example to the right is the same as the contour map on the preceding page, which itself was based on an interpolation of the data values represented by population-weighted centroids. The space between the contour lines of equal value has been shaded using a choropleth scheme. **The overall impact is to create a map that has hallmarks of the choropleth but also a much more organic, fluid appearance that does away with the county boundaries.** It's a manipulation of the visual containers used to represent space though the introduction of error and uncertainty through the original interpolation, and subsequent classification shouldn't be underestimated.

Sometimes referred to as a *filled contour* technique, it's essentially a hybrid of a contour map and a choropleth map. The ability to process the fill of the intervening spaces rather than only the contour lines is helped by the way in which colours nest. Assuming a useful colour scheme that goes from light to dark (or, in this case, a diverging colour scheme for the two parties) is used, the filling in of the in-between space creates an impression of a continuous landscape of voting. Ridges of red, peaks of blue, and valleys of small margins become evident.

This map moves away from the rigid structure of the choropleth-equivalent map of percentage margin of victory. **It's worth noting that darker colours don't mean more voters but a greater margin of victory, which is also a potential misinterpretation of the choropleth version.** Additionally, the inference of the surface extending beyond the physical and administrative boundaries is a drawback. The data stops at a coastline or national border, and this is an awkward artifact of the method. A further drawback is that it's hard for readers to reverse engineer what they're seeing back to a specific county value.

While an isopleth is based on the plotting of contours, it's really the equivalent of a choropleth map, but with contoured boundaries instead of the more normal administrative boundaries.

There's no real need to label the contours themselves as it's the area between the contours that we see and process, not the boundary in between.

This map uses a diverging hue shading scheme with very marginal areas shown in the lightest shades. You could equally use a blended-hue scheme from red to blue through a mixed purple middle colour which would show the many intervening valleys as purple.

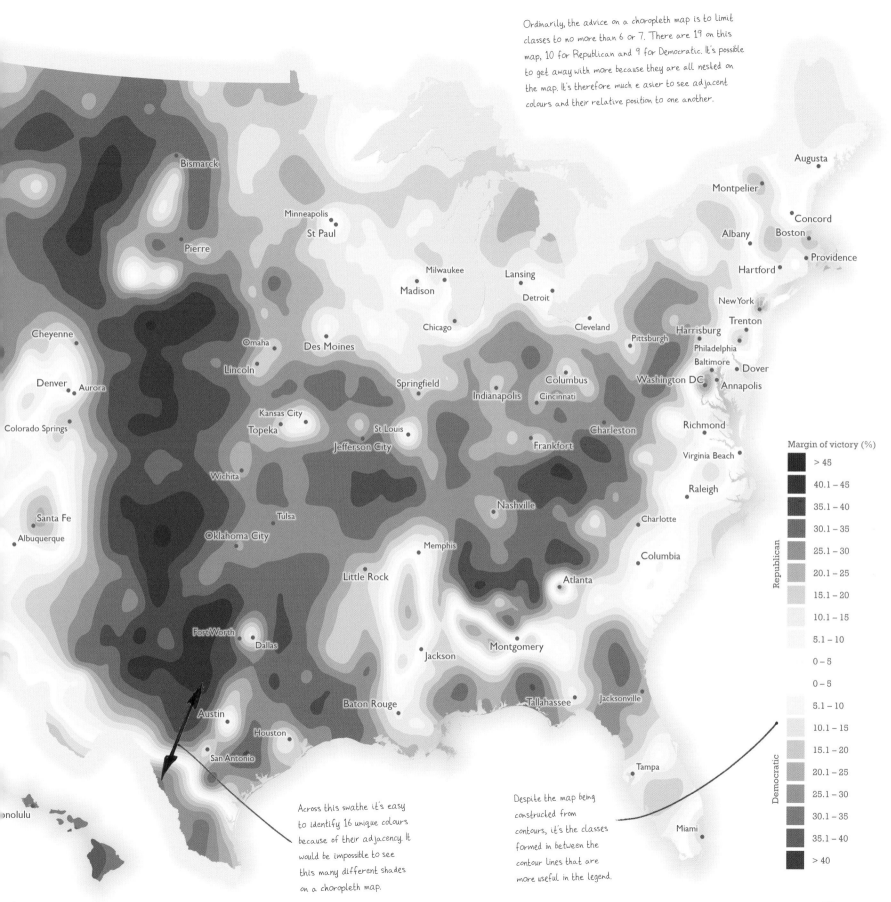

Ordinarily, the advice on a choropleth map is to limit classes to no more than 6 or 7. There are 19 on this map, 10 for Republican and 9 for Democratic. It's possible to get away with more because they are all nested on the map. It's therefore much e asier to see adjacent colours and their relative position to one another.

Across this swathe it's easy to identify 16 unique colours because of their adjacency. It would be impossible to see this many different shades on a choropleth map.

Despite the map being constructed from contours, it's the classes formed in between the contour lines that are more useful in the legend.

Margin of victory (%)

Republican

> 45
40.1 – 45
35.1 – 40
30.1 – 35
25.1 – 30
20.1 – 25
15.1 – 20
10.1 – 15
5.1 – 10
0 – 5

Democratic

0 – 5
5.1 – 10
10.1 – 15
15.1 – 20
20.1 – 25
25.1 – 30
30.1 – 35
35.1 – 40
> 40

Throwing shade

Illuminated contours

Contours sometimes are criticised for difficulty in interpretation because they rely on a reader's ability to work out when a series of contour lines is leading downhill as opposed to uphill. When used for mapping empirical data, the same is true, except the search is for where data points are higher compared with lower. **To the trained eye, it's not such a difficult process, but appreciating how proximity of contour lines indicates steepness or how spacing indicates a concave or convex slope is not innate to every map reader.**

One way to address this limitation is to modify the symbology of the contour line to emphasise the pattern among localised contours. **Introducing tonal variation along the contour line, borrowing ideas from relief shading, helps emphasise the pattern of relief in the empirical surface.** When creating the most basic form of shaded relief, an imaginary light source is cast from the upper-left corner of the map. Those faces with an aspect toward the light source will be illuminated. Those faces with an aspect away from the light source will be in darkness. And those faces which have some illumination will be in a midtone.

The technique is referred to as *illuminated contours* and was popularised by the Japanese cartographer Kitiro Tanaka in the mid-1900s. In some versions of this method, varying line widths approximate the limited way in which light might strike some faces in a similar orientation to the light source, compared with the use of thicker lines for faces that are perpendicular to the light source.

The nested nature of smoothed contours and the tonal shift from light in the upper left to dark in the lower right all combine to give the map a 3D quality. **Illuminated contours accentuate the peaks and troughs across the landscape and help the map reader see where slopes are represented by lines that are clearly climbing to a peak or descending to a trough.**

Illuminated contours are a great example of a technique that goes beyond most software defaults. Contour lines are a well-established way to show lines of equal value cutting through a surface of data points—in this case, the percentage margin of victory. But by modifying their form and combining them with other techniques (such as filled contours with a blended-hue colour scheme in the example to the right), you can make a map that has an altogether different look and feel. **Its utility is in helping people see the shape of the empirical landscape more than would a single technique alone.** The peaks are promontories above the landscape. The pseudo-3D approach assists with making them easier to see.

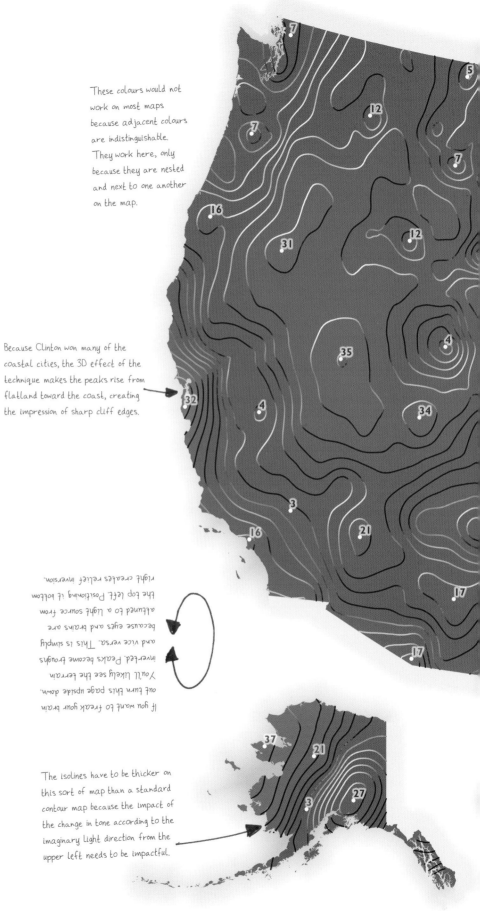

These colours would not work on most maps because adjacent colours are indistinguishable. They work here, only because they are nested and next to one another on the map.

Because Clinton won many of the coastal cities, the 3D effect of the technique makes the peaks rise from flatland toward the coast, creating the impression of sharp cliff edges.

If you want to freak your brain out turn this page upside down. You'll likely see the terrain inverted. Peaks become troughs and vice versa. This is simply because eyes and brains are attuned to a light source from the top left. Positioning it bottom right creates relief inversion.

The isolines have to be thicker on this sort of map than a standard contour map because the impact of the change in tone according to the imaginary light direction from the upper left needs to be impactful.

The areas with a high percentage margin of victory for both Republican or Democratic are shown as peaks, tending toward red or blue accordingly.

The pits of bipartisanship appear as flatland weaving between the peaks.

Compared with the basic contour and the isopleth (filled contour) maps, this version is perhaps more artistic.

Any map that is different from the norm automatically attracts attention. But keeping that attention may be better served with a different map.

Adding spot heights, colour coded by political party provides a visual anchor for people to reference key highs and lows across the map without having to interpret specific colours.

Spot heights = percentage margin of victory at that point.

Red labels = Republican majority.

Blue labels = Democratic majority.

Margin of victory (%)

Republican

> 45
40.1 – 45
35.1 – 40
30.1 – 35
25.1 – 30
20.1 – 25
15.1 – 20
10.1 – 15
5.1 – 10
0 – 5
0 – 5
5.1 – 10
10.1 – 15
15.1 – 20
20.1 – 25
25.1 – 30
30.1 – 35
35.1 – 40
> 40

Democratic

Over the garden fence

Alpha-blended boundary lines

The purpose of many thematic maps is to identify areas that display similar characteristics. There's a natural tendency for map readers to view a map in this way, and the map itself is often set up to support that process. A choropleth map, for instance, places data values into classes, and readers look for where the fill of the symbol is the same across the map. Administrative boundaries on the map are softened so they become less visible. They're often just visual noise because they represent only the container for data visualisation.

But what if the focus is on the boundaries themselves? Instead of trying to see places as similar and build maps to reveal that characteristic, the map might be constructed to reveal how neighbouring places are dissimilar.

On the map to the right, the shared boundary between adjacent counties is seeded with the data from the two counties. This allows calculating the margin of voting difference between adjacent counties. The difference value is shown by encoding the line hue as red for neighbouring Republican wins, blue for neighbouring Democratic wins, and black for neighbouring wins by opposing parties. **The lines are modified by varying the transparency, with opaque showing larger margins and near full transparency for small differences.** Varying the line widths also helps emphasise the differences.

The symbology is deliberately provocative, and the metaphor of leaning over your neighbour's fence for a chat about politics is shown by permeable walls for which either Democratic neighbours are adjacent to Democrats or Republicans neighbour Republicans. Where neighbours voted the opposing way, a black line suggests a barrier to conversation, though lighter greys suggest less of a political divide.

By focussing on the county boundaries and not the fill of the areas in between, a completely different picture emerges of the patterns of neighbouring similarities and dissimilarities.

In some places, it's possible to see how even state boundaries sometimes mask the more contiguous voting preferences of neighbouring communities such as here on the California/Nevada border.

In a 3D approach, such as this map of Florida to the left, height of the boundary lines could be used to further the metaphor. The higher the wall, the greater the dissimilarity between neighbours, creating walled enclaves.

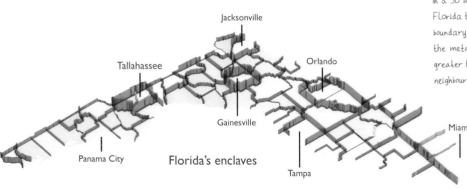

Florida's enclaves

Native Indian reservations show very clearly on this sort of map as very different among surrounding counties. Here, the Standing Rock, Cheyenne River, Pine Ridge, and Rosebud Indian reservations in South Dakota are clearly demarcated.

The symbology treatment plays to the notion of hard boundaries. An impenetrable almost black solid line is used to show boundaries where neighbours voted very differently. Very transparent thin lines gives an impression of a permeable boundary where neighbours share values, in terms of the election outcome at least.

The legend has a series of statements and graphics to clarify the symbology. It's not really feasible to measure the width of the lines but an indicative set of different widths helps assess approximate values.

Democratic neighbours

Republican neighbours

Opposing neighbours

Width of line is proportional to difference in the % share of the vote

<10   >80

The more transparent the line, the less the difference between neighbours

Background tint shows red territory or blue territory

There's a very faint tint of blue or red that sits behind the main symbology for the county boundaries. This helps emphasise the red and blue territories.

Gauging the situation

Directional symbology

Multivariate symbology encodes pieces of information that support one another and help create a map that invites closer scrutiny, and which perhaps provides greater depth. These sorts of maps often allow map readers to go beyond the basic questions of 'where is more?' or 'how much exists in this place compared to that place?'. **It's the interaction between the different pieces of information in the symbol that helps people explain more complex patterns than univariate symbols can show on their own.** Multivariate symbols can be designed in many ways but often use familiar visual metaphors as a framework for their design. For elections, a common visual metaphor is the dial, gauge, or what sometimes is called a *swing-o-meter*. Why? Because election results are said to swing one way or another between election cycles. The idea of a needle that moves is a simple way to understand a more complex symbology.

In the example to the right, a gauge is a useful graphical device and acts as a metaphor for the electoral swing between 2012 and 2016. **The 2016 presidential election result gives only an indication of the contemporaneous election cycle but knowing how it differs from the past can provide a more interesting story.** Understanding how the result has changed from the previous election gives the reader more detail and context. In this sense, the map shows the difference between the 2016 election result and the 2012 election result contested by Barack Obama (Democrat) and Mitt Romney (Republican). This earlier dataset gives context to how the result has changed.

There are several ways the symbols encode information. The symbol itself is an arrow, which implies direction. In politics, shifts in voting patterns move toward the right (increasingly toward Republicans) or to the left (increasingly toward Democrats). **So the arrows immediately reference shifts to the right or left depending on the change in percentage share of the vote between 2012 and 2016.** Showing the arrows in a red or blue hue indicates whether the county remained Republican or Democratic. And a dark-grey arrow shows those counties that flipped. These are also some of the places with the largest swing.

The dial can be designed in many ways. Here, it swings from a fulcrum centred in the county. Alternatively, it could sit upright and appear as an odometer or temperature gauge with an arrow moving left and right. Because the symbology is an odd shape, care must be taken to avoid too much overlap and clutter. It's inevitable on a small-scale map but the overall effect is a good payoff for some congestion.

Red arrow with a large swing to the left indicates a Republican hold, but with a much reduced share of the vote. Here, in Utah the strength of opposition to the Republican candidate is evident.

Blue arrow with a swing to the right indicates a Democratic hold, but with a reduced share of the vote.

The symbology is quite complex, and leads to inevitable congestion in some areas. There's very little additional detail you could add to the map but a background of faint county boundaries helps associate the symbol with its geography.

Because the swing-o-meter is effectively a pendulum, the fulcrum is positioned at the geographical centre of the county. Curiously, this also helps limit overlaps compared with the symbol being centred in the county. That's one of those happy accidents that occurs when trying to find an optimal solution.

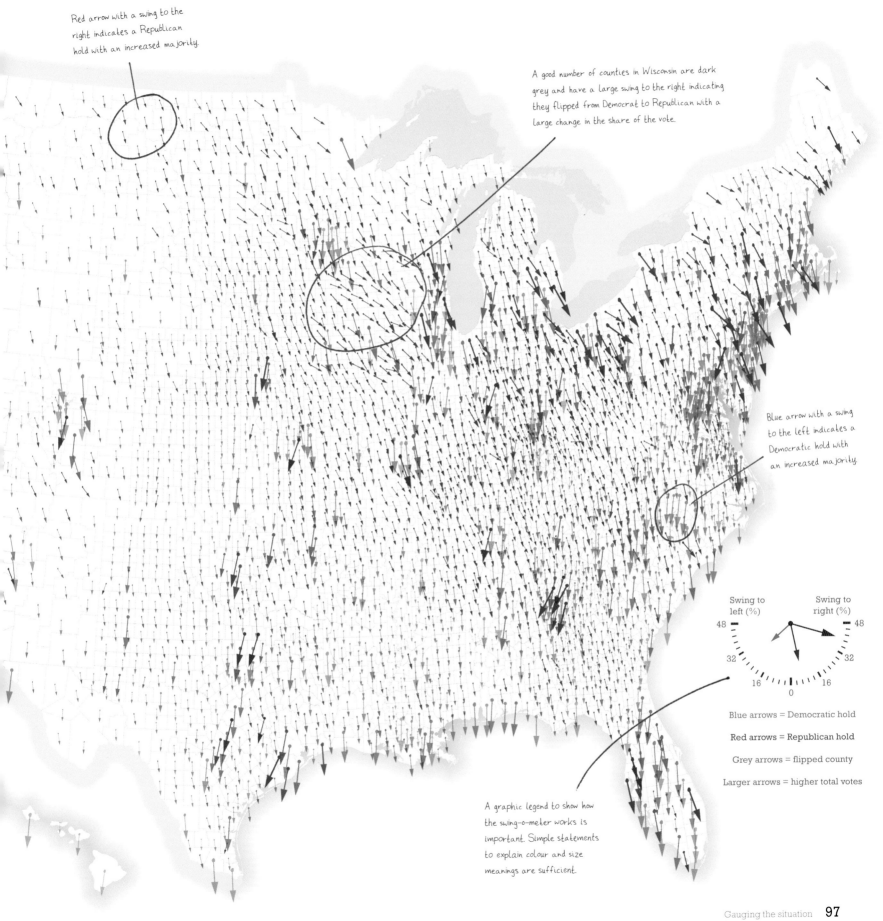

Red arrow with a swing to the right indicates a Republican hold with an increased majority.

A good number of counties in Wisconsin are dark grey and have a large swing to the right indicating they flipped from Democrat to Republican with a large change in the share of the vote.

Blue arrow with a swing to the left indicates a Democratic hold with an increased majority.

Swing to left (%)   Swing to right (%)

48   48

32   32

16   16

0

Blue arrows = Democratic hold

Red arrows = Republican hold

Grey arrows = flipped county

Larger arrows = higher total votes

A graphic legend to show how the swing-o-meter works is important. Simple statements to explain colour and size meanings are sufficient.

Winds of change

Trends over time using directional arrows

Although the swing-o-meter map style on the previous page attempts to encode several pieces of information into a single symbol by varying hue, size, and orientation, it can end up getting congested in places. That's the cost of trying to treat every data point equally and applying consistent symbology rules. Instead, you could make choices deliberately to focus on some data points at the cost of others.

This map is like the swing-o-meter version except it replaces the gauge with a left- or right-pointing arrow to show the direction of swing. On the swing-o-meter map, the percentage difference between 2012 and 2016 was encoded by the orientation of the arrow. The size of the arrow indicated total votes. On this version, the orientation is fixed as either left or right, so total votes are no longer shown, and instead the size of the arrow reflects the difference in percentage share of the vote between 2012 and 2016. So although it might look like the swing-o-meter map, different pieces of information are encoded using a different approach to assigning data variables to the symbol's visual variables.

On the swing-o-meter map, the flipped counties are shown in dark grey to help the map reader see them as qualitatively different from the red or blue arrows. On this map, there are no grey arrows. Instead, there are two shades of red and two shades of blue. **A dark-red symbol shows a swing to the right that resulted in a flipped county, whereas light red indicates a county that didn't flip. The direction of the arrow showing whether there was a trend right to improve the Republican stake in that county or left to erode the margin of victory is compared with the 2012 result.** The same is true of the blue arrows, where darker ones indicate a flipped swing to the left and lighter ones indicate a hold but with a swing to increase or decrease the margin of victory.

The impact of this approach to the symbology is to distinguish the important flipped counties from the congested symbology of all counties. **The map therefore assists the reader by deciding what is of most importance and filtering it through symbology choices.** It leads the reader to see the flipped counties more easily and, through receding the other symbols, suggests they're not worth exploring in detail. It brings focus to key pieces of information through the haze of the whole. The downside is that this map isn't going to be useful if you're interested in some of the non-flipped counties but that's the compromise on this occasion.

It would be possible to show all symbols in the same blue or red but that reduces legibility and forces you to work harder to tease out details.

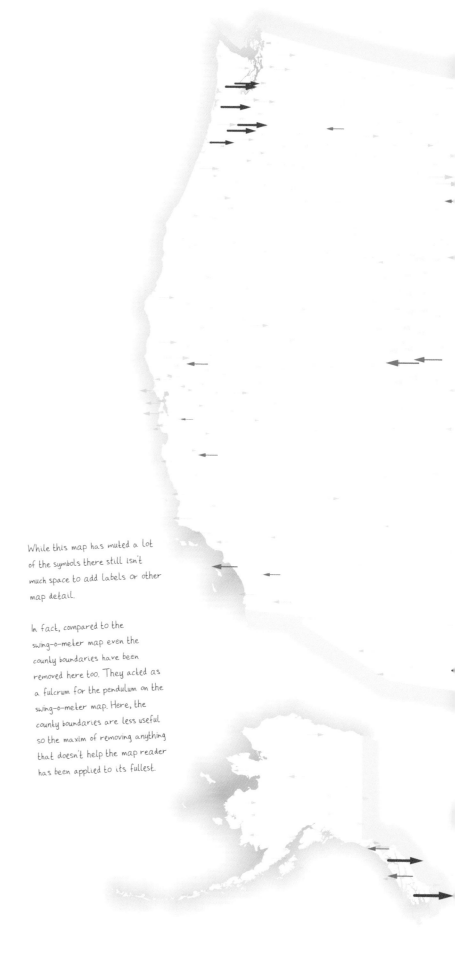

While this map has muted a lot of the symbols there still isn't much space to add labels or other map detail.

In fact, compared to the swing-o-meter map even the county boundaries have been removed here too. They acted as a fulcrum for the pendulum on the swing-o-meter map. Here, the county boundaries are less useful so the maxim of removing anything that doesn't help the map reader has been applied to its fullest.

This map clearly teases out and highlights regional patterns, particularly here in Wisconsin where underlying social and economic conditions set the tone for such a strong shift to the right, flipping many counties from 2012.

Comparing the northeast region on this map with the swing-o-meter map shows how the two create a very different aesthetic. On the swing-o-meter map the congestion creates an overall wave effect as symbols coalesce and overlap. Here, we've isolated the flipped counties. There's a stark, yet cleaner impression of which counties moved right, and voted for a different party from 2012.

The Atlanta, GA, metropolitan area shifted strongly Democratic even though the state was won by the Republicans.

%	Swing to left	Swing to right
10		
30		
60		

Dark blue = flip to Democratic
Light blue = Democratic hold

Dark red = flip to Republican
Light red = Republican hold

Not all left swings will be blue arrows, and not all right swings result in red arrows so a neutral colour in the legend illustrates the way the difference in percentage share is encoded into the symbol size.

Value (symbol lightness) is usually used to show some sort of empirical difference between symbols of the same type. Here, it's used to show a qualitative difference between counties that have been held, and counties that have flipped. It's used to change the visual heirarchy of symbols, to differentiate them, and make some seem more important on the map than others.

Path to victory

Flow map from candidacy to the finish line

Virtually every map in this book deals with the representation of the results of the 2016 presidential election. But other aspects of the election process might call for a map, which allows you to explore a technique that isn't suited to a meaningful delivery of the results themselves. **Flow maps show movement of goods, ideas, or some other tangible or non-tangible phenomenon.** For elections, this may be the campaign trail or the flow of donations to the respective candidate's campaign fund.

Once candidates are given the nomination by their respective parties, campaigning goes into overdrive. The candidates spend their time on the road (or in the air) visiting towns and cities for rallies, visits to establishments, photo opportunities, and to meet people, all with the intent of gaining their vote. The path to victory can be mapped as a flow.

Some flow maps show a network (e.g. roads, pipelines) which have a structure and, perhaps, quantity of a flow measured between places. Radial flow maps show links between paired places, sometimes known as *origin–destination* or *desire line* flow maps. They might show intangible connections (e.g. cell phone calls) or a generalised pairing of places. Radial flow maps that encode empirical information can become distributive when they show movement from a single origin to multiple destinations. **The map to the right is of the origin–destination type, which shows connectivity and direction of flow between two places.** There's a starting point for each of the two candidates and a series of lines that take them on a journey across the United States through time, until they finish on election night.

Flow line maps are hard to design because of the complexity of overlapping lines that is almost inevitable. Minimising the amount of data being mapped is the first task to consider. With nearly 200 lines on this map, it's at the limit. Some might consider that it's gone beyond being practical, but seeing congestion tells its own story about the campaign trail.

Attempts to minimise overlaps and using curves to allow long lines to have a more aesthetically pleasing appearance across the map are important. **Because the map is showing movement in a single direction, each line has an arrowhead.** There's also a buffer around each location so lines begin and end a short distance away from their true origin or destination. This ensures there's a way to see the locations without further congestion. Conversely, there's also a lot of empty space on the map. More than half the states didn't receive a single visit from either candidate.

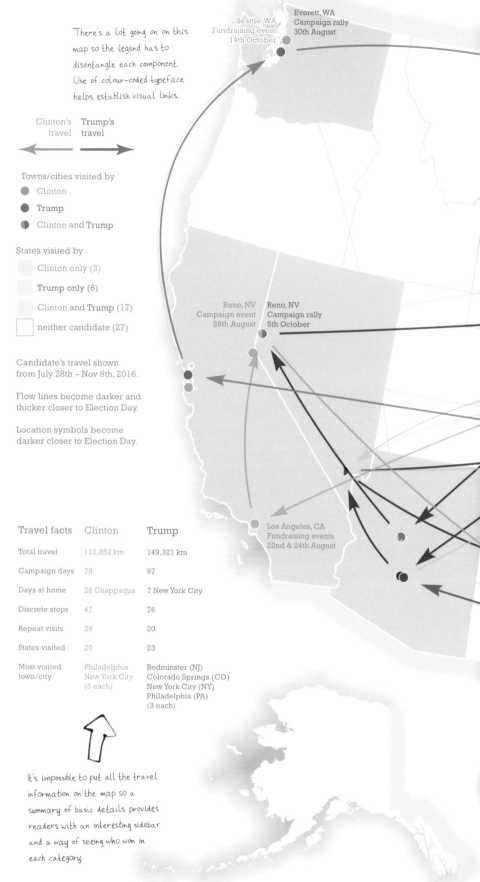

There's a lot going on on this map so the legend has to disentangle each component. Use of colour-coded typeface helps establish visual links.

Clinton's travel   Trump's travel

Towns/cities visited by
- ● Clinton
- ● Trump
- ◐ Clinton and Trump

States visited by
- Clinton only (3)
- Trump only (6)
- Clinton and Trump (17)
- neither candidate (27)

Candidate's travel shown from July 28th – Nov 8th, 2016.

Flow lines become darker and thicker closer to Election Day.

Location symbols become darker closer to Election Day.

Seattle, WA
Fundraising event
14th October

Everett, WA
Campaign rally
30th August

Reno, NV
Campaign event
26th August

Reno, NV
Campaign rally
5th October

Los Angeles, CA
Fundraising events
22nd & 24th August

Travel facts	Clinton	Trump
Total travel	112,852 km	149,321 km
Campaign days	78	97
Days at home	26 Chappaqua	7 New York City
Discrete stops	47	76
Repeat visits	29	20
States visited	20	23
Most visited town/city	Philadelphia New York City (6 each)	Bedminster (NJ) Colorado Springs (CO) New York City (NY) Philadelphia (PA) (3 each)

It's impossible to put all the travel information on the map so a summary of basic details provides readers with an interesting sidebar and a way of seeing who won in each category.

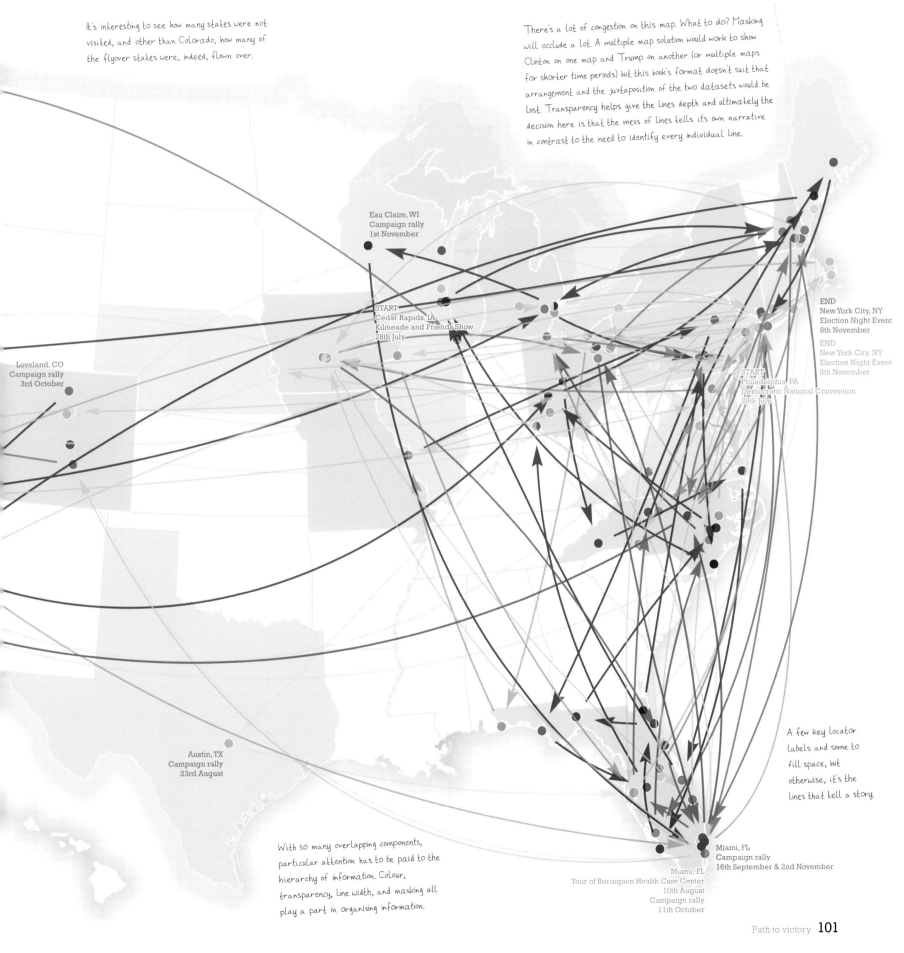

It's interesting to see how many states were not visited, and other than Colorado, how many of the flyover states were, indeed, flown over.

There's a lot of congestion on this map. What to do? Masking will occlude a lot. A multiple map solution would work to show Clinton on one map and Trump on another (or multiple maps for shorter time periods) but this book's format doesn't suit that arrangement and the juxtaposition of the two datasets would be lost. Transparency helps give the lines depth and ultimately the decision here is that the mess of lines tells its own narrative in contrast to the need to identify every individual line.

Eau Claire, WI
Campaign rally
1st November

START
Cedar Rapids, IA
Kilmeade and Friends Show
28th July

END
New York City, NY
Election Night Event
8th November

END
New York City, NY
Election Night Event
8th November

Loveland, CO
Campaign rally
3rd October

START
Philadelphia, PA
Democratic National Convention
28th July

Austin, TX
Campaign rally
23rd August

A few key locator labels and some to fill space, but otherwise, it's the lines that tell a story.

With so many overlapping components, particular attention has to be paid to the hierarchy of information. Colour, transparency, line width, and masking all play a part in organising information.

Miami, FL
Campaign rally
16th September & 2nd November

Miami, FL
Tour of Borinquen Health Care Center
10th August
Campaign rally
11th October

Road to nowhere

Strip map for a political trip across America

This map shows a transect through the data along a path, specifically the roads that make up Route 66 from Santa Monica, California, to Chicago, Illinois. Route 66, also known as the Mother Road, was established in 1926 as part of the original highways in the US National Highway System. It's been immortalised in song, film, and literature. It was a route taken by many who migrated west in the 1930s, and communities grew up along its route. **The map is designed to show how politics play out along the route, even though it is officially no longer part of the National Highway System and more commonly is referred to as Historic Route 66.**

And what better way to map a route or a journey than in the style of the strip maps from John Ogilby's 1675 road atlas *Britannia*. **Ogilby used an innovative scroll technique to split his routes into segments, each of which illustrated a thin extract of the overall route map.** The larger journey could be plotted by reading the strips and their relationship with neighbouring strips. The orientation of the strips is always vertical in a heads-up fashion, and the routes are plotted vertically within each strip. This being the case, each strip rotates the map to fit the strip, which is why each contains a small compass rose to ensure the reader can identify direction.

Strip maps are a form of schematic map. In the example to the right, the highway has been intersected by the counties it crosses and the data from the election results has been used to symbolise the line itself. Route 66 joins two large urban areas which traditionally vote Democratic and did so in 2016, yet it crosses the country through very differently sized counties with fundamentally different political results.

ROUTE 66

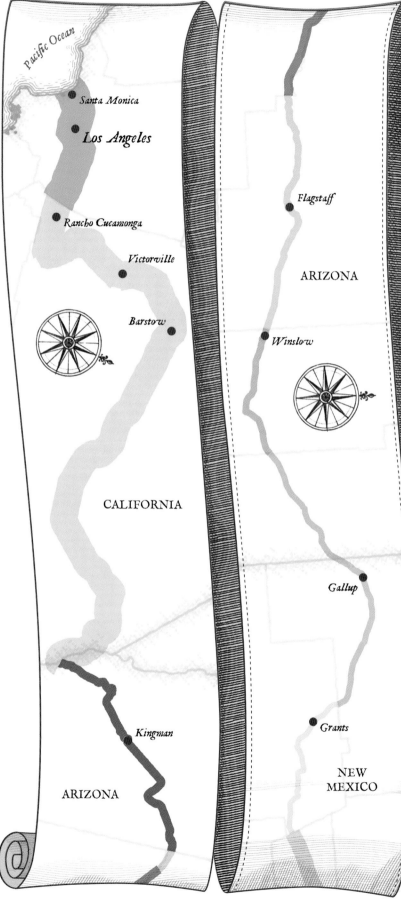

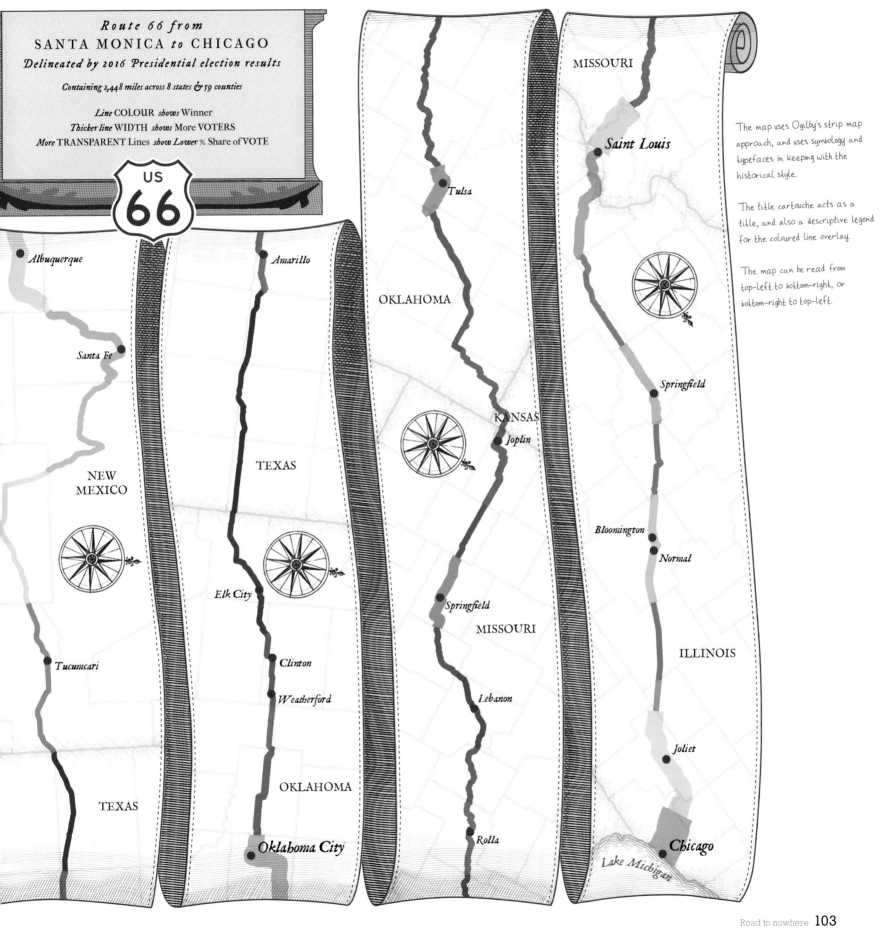

Stars and stripes

Exposing cross-sections across the map

A common technique in cartographic representation is the use of transects, otherwise known as *profiles* or *cross sections*. **They are constructed by drawing a line across the planimetric map and using the values of data that intersect the line to construct a vertical graph.** This then is plotted as an elevation, or a side view of the data as if the terrain has been sliced open to reveal its vertical structure. The most common form of vertical section uses a straight line, as in the example shown to the right. Several horizontal lines are drawn across the map of the United States, which provide a series of horizontal planes and vertical graphs at equal intervals.

The vertical graphs that rise above each line give an impression of the volume of data underneath—in this case, the share of the vote won by the winning candidate. Vertical sections could be shaded in many ways. **A single colour might be used for a single variable being explored, but here, the profile lines cross counties in which the winner is one of two candidates, so the vertical section is shaded categorically according to the winner.** Republican share of the vote is shown above a centre line and the Democratic share of the vote below. This gives a sense of the quantitative aspect of the results across the line. The more red, the greater the Republican share; the more blue, the greater the Democratic share.

Vertical sections provide an interesting look at the data and offer a different visual aesthetic from many maps but, of course, they have one major drawback in that they lose a lot of data from the final display. **The lines drawn across the map miss many counties (and cities), and so the final vertical sections show only a small proportion of the data.** Is it enough though? Sometimes these sorts of cartographic approaches are enough to give an overall sense of the pattern, particularly, as in this case, in which there are clear, albeit generalised, patterns from east to west that go blue, red, blue.

To orient map readers to the graphs, the legend includes a vertical scale so readers can interpret the vertical magnitudes. This is critical because the vertical dimension includes some vertical exaggeration to ensure a good variation in elevation across each profile. **The choice of precisely how many lines, and where to draw them, is predominantly governed by the amount of vertical space at the chosen scale, while also attempting to get a good representation of the data.** Outliers can be chosen to be included or excluded. Lines also might be drawn to pass directly through key places. Labels represent key towns and cities across the US and are coloured red or blue to denote the results of the counties nearest the place itself. On a map like this, you cannot represent every place equally, and avoiding overlapping the lines is also important to avoid congestion.

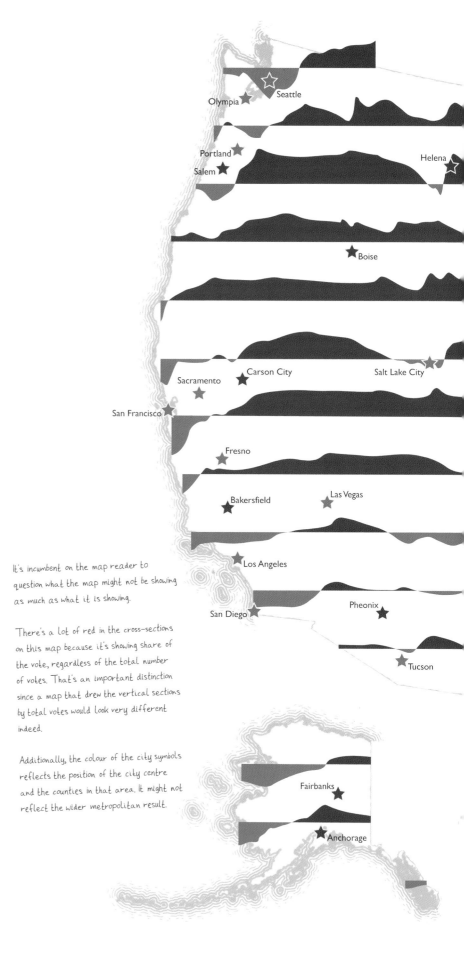

It's incumbent on the map reader to question what the map might not be showing as much as what it is showing.

There's a lot of red in the cross-sections on this map because it's showing share of the vote, regardless of the total number of votes. That's an important distinction since a map that drew the vertical sections by total votes would look very different indeed.

Additionally, the colour of the city symbols reflects the position of the city centre and the counties in that area. It might not reflect the wider metropolitan result.

Sometimes when you start out with an idea it leads you down an unintended path. The cross-sections were the intent of this map. Yet once plotted, they seemed to look like the flag of the United States with its famous stripes. The decision to place city 'stars' seemed a natural evolution and creates a striking image.

Bismark

Montpelier Augusta

Minneapolis St Paul Albany Boston
Pierre Hartford Providence

Madison Milwaukee Lansing Detroit
Chicago Cleveland New York
Des Moines Trenton
Omaha Harrisburg
Cheyenne Lincoln Columbus Pittsburgh Philadelphia
Denver Baltimore
DC Dover
Colorado Springs Indianapolis Cincinnati Charleston Richmond
Topeka Kansas City Springfield
St Louis
Jefferson City Frankfort Virginia Beach
Wichita
Raleigh
Santa Fe Tulsa Nashville Charlotte
Albuquerque Oklahoma City Memphis Columbia
Little Rock Atlanta
Fort Worth Dallas Montgomery
Jackson
Tallahassee Jacksonville
Austin Baton Rouge
Houston
San Antonio
Tampa
Honolulu
Miami

A different vignette treatment is used on this map compared to most others. The red and blue cross-sections work much better against a white background. A rippled effect is used for the surrounding water, with a single line for land boundaries.

Share of vote (%)

Republican 100
0
Democratic
100

City result is based on counties closest to city centre

★ Republican city
★ Democratic city

5. Cartograms

The basic requirement for a thematic map is to provide the reader with a visual mechanism which makes comparisons clear and discernible. It might sound paradoxical but geography can get in the way of this process because it introduces all sorts of complexity that is difficult to manage consistently across a map.

Rather than try to wrestle geography, an alternative is to reimagine or warp it, to nullify its impact on the core message of the map. This type of map is referred to as a *cartogram*.

Cartograms tackle the inherent visual biases in geographical maps by modifying the geometry of the geography, effectively normalising it to create a map in which each area takes on a new shape or size. Cartograms therefore depict geographical space diagrammatically as they lose their relationship with actual geographic coordinates.

Although cartograms deal well with unequal geographies in a technical sense, one disadvantage is that, by definition, they change the visual representation of geography. This has consequences as the map attempts to balance empirical accuracy, geographical accuracy, and topological accuracy. Some people simply have a visceral reaction to the manipulation of geography into a warped shape that is potentially harder to relate to.

Of course, strong reactions can work to your advantage because unfamiliarity can grab attention. Just go easy when you turn the next few pages.

Exploding the map

Non-contiguous cartogram

Non-contiguous cartograms are the simplest type of cartogram. **The shape of areas is retained but they are resized according to the data value being represented.** People's familiarity with well-known shapes helps them read the map in a relatively unhindered manner. This is particularly true when the shapes are familiar, such as the states of the United States.

The map is effectively a proportional symbol map, but rather than geometric symbols, it's the area of the geographical shape that is modified and scaled, relative to other areas on the map according to the data value being represented. The effect increases the size of areas with large values and reduces the size of areas with small values. If the areas remained in their geographical position (i.e. centred on their geometric centre), the map would have multiple overlaps. **To avoid overlaps, the reproportioned area shapes are moved into a new configuration with the intent of preserving general adjacencies as much as possible.** Keeping the map design as at least something approximating its real geography helps orient the map reader.

In a non-contiguous cartogram, some topology (adjacency and connectivity) is sacrificed to preserve shape and enable recognition of geographical areas. Although the recognition of individual shapes can be maintained, position warps location and distorts the overall appearance of the map.

In comparison to a geographical version of the same data, this version has benefits that are not at first easily recognised. **It immediately corrects for the visual skewness of looking at geographical areas that vary massively in size and, by virtue, means some areas are more visible than others.** On most maps in this book, Washington, DC, is hardly visible at the scale at which the maps are displayed. Yet on this sort of cartogram, it becomes enlarged and visible.

An important benefit of having geographical areas scaled relatively to one another is that the amount of ink (or pixels) that represents each data value (or range in a classified map) is in the correct relative proportion to everything else on the map. **So you're seeing the correct proportions of colour on the map relative to one another.** This resolves a shortcoming of the geographical map in which large areas in a dark colour naturally will dominate the visual and perceptual system, particularly when displayed in large areas. This often leads to a misinterpretation of what's going on because of the natural tendency to see larger areas and darker colours, as 'more', regardless of underlying areal extent or population densities.

The large states in the northern Midwest and Alaska have much lower populations and a lower population density. They appear much smaller than their true geographical size as a consequence.

Because Clinton won many of the urban battlegrounds, and Trump won much of the rural heartland, a geographical map will almost always look like it has a disproportionate amount of red on it. This map doesn't. It shrinks the sparsely populated states, and consequently it has a much more balanced amount of red and blue, representing the fact that the result was actually very close.

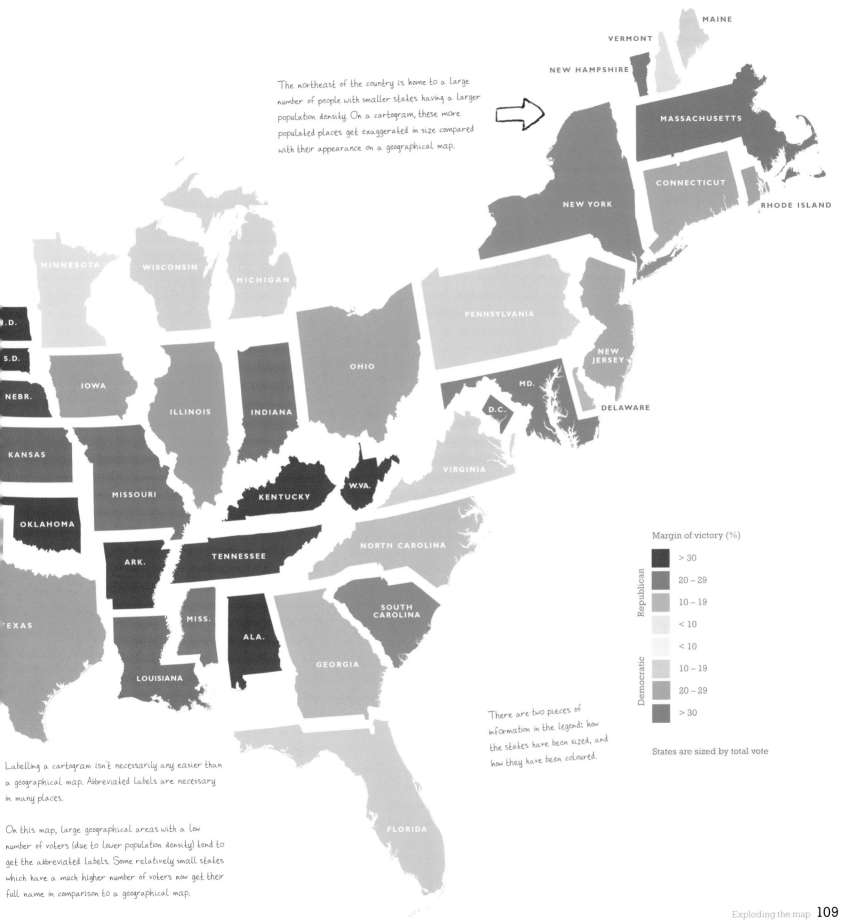

The northeast of the country is home to a large number of people with smaller states having a larger population density. On a cartogram, these more populated places get exaggerated in size compared with their appearance on a geographical map.

There are two pieces of information in the legend: how the states have been sized, and how they have been coloured.

States are sized by total vote

Margin of victory (%)

Republican
> 30
20 – 29
10 – 19
< 10

Democratic
< 10
10 – 19
20 – 29
> 30

Labelling a cartogram isn't necessarily any easier than a geographical map. Abbreviated labels are necessary in many places.

On this map, large geographical areas with a low number of voters (due to lower population density) tend to get the abbreviated labels. Some relatively small states which have a much higher number of voters now get their full name in comparison to a geographical map.

Out of shape

Population density–equalising cartogram

Contiguous cartograms maintain connectivity between adjacent geographical areas but tend to dramatically distort shape. Perhaps the most widely used is the Gastner-Newman cartogram (otherwise known as a *population density–equalising* cartogram), which does an excellent job of retaining the general shape of individual areas. **Alternative versions of this approach exist, and all seek to warp geography so that the shape of each individual area is scaled according to the input data value.** The result is effectively a unique map projection based on data values which skew, shift, and warp the vertices of the lines that make up each area. Its benefit is to account for the problems posed by unequally sized areas and population densities, which so often overawe choropleth maps of the same data and can lead to false interpretation.

In contrast to the non-contiguous cartogram, adjacencies are maintained, which at least allows people to understand what is happening to areas that surround somewhere they might be familiar with. Adjacency is a key requirement for navigation across a traditional set of geographical enumeration units, and the same principle exists here.

However, one of the biggest drawbacks of the contiguous cartogram is the degree of distortion of shapes rendered in the resulting map. **Distortions can be quite severe, especially if the data values have a large range or, worse, if the data has a few significantly different outliers.** It's therefore not recommended as a mapping technique unless your data is normally distributed and not going to cause unpalatable distortion. It's all a question of balance though, and where this sort of cartogram is most effective is where the original geographical shapes are still recognisable to an extent.

Regardless of the degree of distortion between the original geographical shapes and the cartogram, the result often renders the map difficult to interpret because of the abandonment of familiarity. **It's much harder for a map reader to overcome the lack of recognition of familiar shapes, and it takes much more effort to read and interpret what they're seeing.** This is perhaps one of the reasons this form of cartogram gives readers a visceral reaction. People tend either to love these map types or to hate them. There's little indifference.

Contiguous cartograms tend to grab attention by virtue of looking different from what people might expect of a map. Even the drawbacks of interpreting where places are, and the data values they represent, can be mitigated if the map is published as a web map with hover-over labels or pop-ups that can ameliorate the visual jarring.

Quite often a choropleth map doesn't need distinct boundaries because the symbol fill will often change from one adjacent area to another. But on this map, they really help give the map a structure. Heavier state boundaries are thicker and opaque. Thinner and slightly transparent county boundaries show the intra-state geography.

There's a huge compression of sparsely populated counties that almost splits the US in two.

On many maps it's possible to limit the amount of labels you add because of the familiarity of the shapes. On this map, despite it adding an element of visual clutter they really help orient the map reader, as well as see how the warped shapes relate to real-world geography.

One reviewer thought this map looked like the United Kingdom if you rotate it 90°. This goes to show that unless you're familiar with the geography of a cartogram they can be difficult to relate to known places.

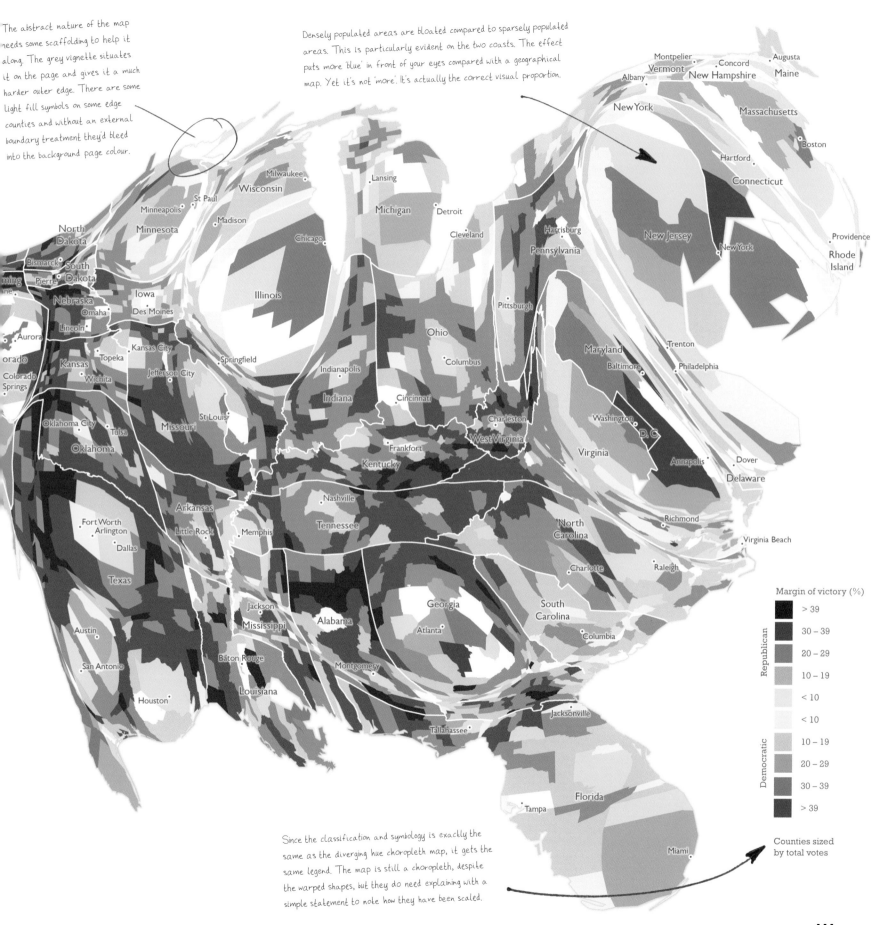

The abstract nature of the map needs some scaffolding to help it along. The grey vignette situates it on the page and gives it a much harder outer edge. There are some light fill symbols on some edge counties and without an external boundary treatment they'd bleed into the background page colour.

Densely populated areas are bloated compared to sparsely populated areas. This is particularly evident on the two coasts. The effect puts more 'blue' in front of your eyes compared with a geographical map. Yet it's not 'more'. It's actually the correct visual proportion.

Since the classification and symbology is exactly the same as the diverging hue choropleth map, it gets the same legend. The map is still a choropleth, despite the warped shapes, but they do need explaining with a simple statement to note how they have been scaled.

Margin of victory (%)

Republican
> 39
30 – 39
20 – 29
10 – 19
< 10

Democratic
< 10
10 – 19
20 – 29
30 – 39
> 39

Counties sized by total votes

Tessellated geographies

Gridded cartogram by hexagon

The gridded (or tessellated mosaic) cartogram is a compromise cartogram that ignores geographical shape but attempts to maintain a more uniform topology that enables you to find familiar places without the need for perfect maintenance of adjacencies.

Hexagons are a common shape for gridded cartograms because they are not too dissimilar from a circle, yet they fit together in a puzzle without the gaps that adjacent circles leave. **In a gridded cartogram, each unit area is symbolised by the same-size shape regardless of geographical reality, and unit areas sum to a total defined by a higher level of data aggregation.** For the map to the right, the unit area is the Electoral College vote, and states comprise the total number of Electoral College votes. Different symbol fills can show a second variable, using the same processes for classification and symbolisation as a choropleth map.

Using the same-size shape for each unit area gives them equivalent visual weight to every other symbol on the map. This goes a long way to dealing with the problem of real geographies being inherently unequal in size and population density. **Making unit areas equal results in a map in which the symbolisation is, by default, in the correct visual proportion across the map.** The amount of red and blue seen is not warped by the reality of geography and the mapping of land mass.

There is probably an almost infinite number of ways you could design a gridded cartogram. The way in which each unit is positioned relative to one another is a balance between maintaining roughly a recognisable shape and some adjacency with neighbouring areas.

The popular vote and Electoral College are different. Maps that show a percentage share of the vote, or margin of victory, or simply a red/blue result show who won, and by how much. But the process of electing a president turns the 'who won' into another metric – the Electoral College. This map shows that conversion.

The Electoral College consists of 538 electors and a majority of 270 is required to win the presidency. Each state has an allotment of Electoral College votes, or electors. A vote for a presidential candidate is therefore actually a vote for the state's electors. In most states, a "winner-takes-all" system exists, so once all votes are counted, whoever wins the state wins all the electors in that state.

The number of electors is equivalent to the number of members in the state's congressional delegation, which is one for each member in the House of Representatives, plus two for the state's senators. The effect is that states with much larger populations have more Electoral College votes and vice versa.

On this map, each state is represented by the number of Electoral College votes it holds, with one hexagon representing a single Electoral College vote. It's therefore possible to retrieve the number of Electoral College votes for each state through counting the hexagons.

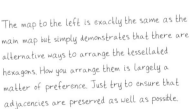

The map to the left is exactly the same as the main map but simply demonstrates that there are alternative ways to arrange the tessellated hexagons. How you arrange them is largely a matter of preference. Just try to ensure that adjacencies are preserved as well as possible.

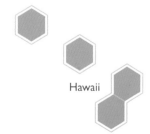

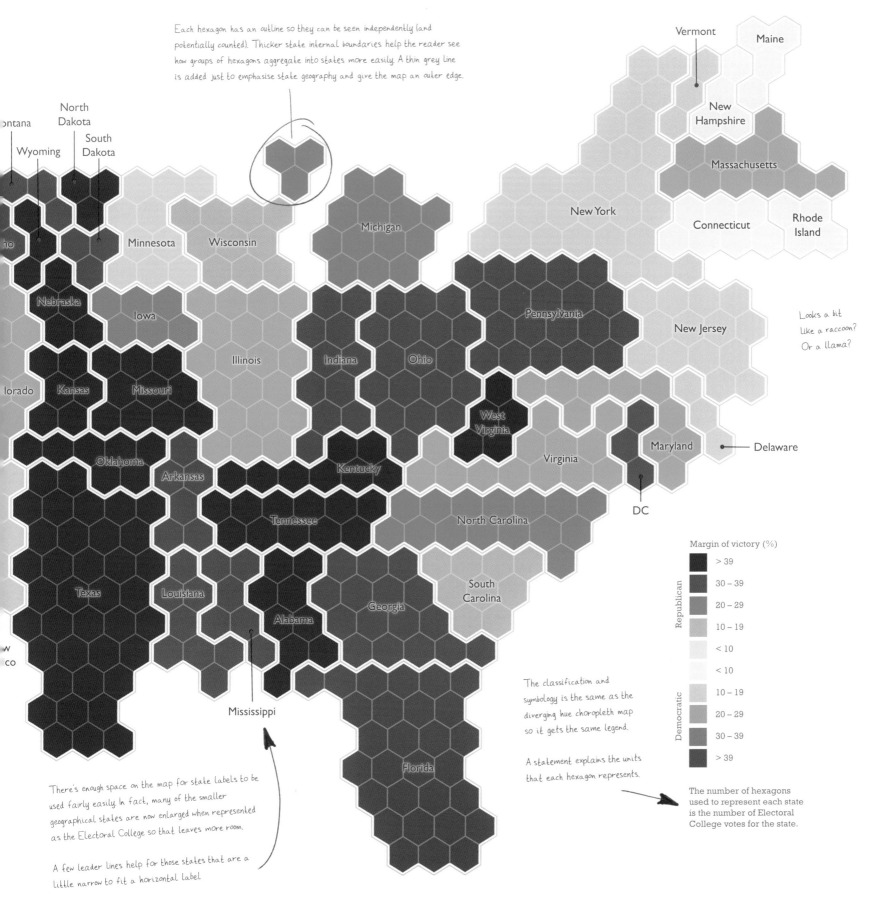

Each hexagon has an outline so they can be seen independently (and potentially counted). Thicker state internal boundaries help the reader see how groups of hexagons aggregate into states more easily. A thin grey line is added just to emphasise state geography and give the map an outer edge.

Vermont

Maine

North Dakota

ontana

Wyoming

South Dakota

New Hampshire

Massachusetts

Minnesota

Wisconsin

Michigan

New York

Connecticut

Rhode Island

ho

Looks a bit like a raccoon? Or a llama?

Nebraska

Iowa

Pennsylvania

New Jersey

lorado

Kansas

Missouri

Illinois

Indiana

Ohio

West Virginia

Virginia

Maryland

Delaware

Oklahoma

Arkansas

Kentucky

DC

North Carolina

W co

Texas

Louisiana

Tennessee

Georgia

South Carolina

Alabama

Mississippi

Florida

There's enough space on the map for state labels to be used fairly easily. In fact, many of the smaller geographical states are now enlarged when represented as the Electoral College so that leaves more room.

A few leader lines help for those states that are a little narrow to fit a horizontal label

Margin of victory (%)

Republican

> 39

30 – 39

20 – 29

10 – 19

< 10

< 10

10 – 19

Democratic

20 – 29

30 – 39

> 39

The classification and symbology is the same as the diverging hue choropleth map so it gets the same legend.

A statement explains the units that each hexagon represents.

The number of hexagons used to represent each state is the number of Electoral College votes for the state.

The honeycomb

Non-contiguous gridded cartogram

ALASKA

A non-contiguous cartogram traditionally uses recognisable shapes, positioned with gaps between them and resized according to the encoded variable. A gridded cartogram organises uniform shapes (hexagons, squares, triangles, and so on) without gaps so that each individual shape represents the same quantity of the data variable—in this case, one Electoral College vote. **But as with many mapping techniques, it's possible to take the best bits of other techniques and make an altogether new map, which is the example shown to the right.** Gridded hexagonal shapes tessellate into an aesthetically pleasing design, but they're exploded along state lines to make a non-contiguous version.

The non-contiguous approach supports the freedom to do several things in terms of design. Firstly, clusters of individual hexagons can be tessellated in a way that attempts to fit the general shape of the state itself. Secondly, clusters of individual states can be arranged to reflect their geographical position. The map itself is unprojected because it's only an arrangement of graphics, but that doesn't mean its design has to ignore a familiar shape. It therefore reflects the broad shape of an Albers equal-area projection.

On a traditional non-contiguous cartogram, the size of the individual shapes of states is resized, and the map exhibits an exaggerated northeast. The same is true on this version because of the number of Electoral College votes in those states but the clusters can be arranged more systematically. **For much of the Northeast and Eastern Seaboard, there's a gap of a single strip of hexagons between states that acts as a consistent state border.** The same approach to spacing could have been applied throughout the map but that might have truncated the West too much and led to a map that appeared too squashed from left to right. Instead, additional space in the West alludes to the sparseness that characterises the real geography. Retaining a good overall shape for the map was part of the challenge when considering spacing.

The context in which a map is to be displayed also has an impact on design decisions. For most maps in this book, Alaska is positioned as an inset in the bottom left of the map display to occupy the white space that exists as a function of the shape of the contiguous US. For cartograms (and other abstract maps), it makes more sense to position Alaska toward the northwest where it exists relative to the rest of the US. This placement reduces the cognitive load on the part of the map reader to figure out what real-world feature the abstract collection of shapes represents.

WASHINGTON
MONTANA
OREGON
IDAHO
WYOMING
SOUTH DAKOTA
NEBRASKA
NEVADA
UTAH
COLORADO
CALIFORNIA
ARIZONA
NEW MEXICO
HAWAII

Labels are often second-class citizens on many maps that have complex symbology, and often large disparities in the density of information. Their sizing and placement is often compromised.

On cartograms such as this, there's no excuse not to use the full name for each state. Positioning them makes clear the relationship with the cluster of hexagons they relate to. Going further, clear and consistent labels are critical on cartograms to bridge the gap between abstract shapes and the reader's ability to interpret meaning.

White space (or negative space) in graphic design is that space that surrounds other elements. It helps them breathe while not disconnecting them from the bigger picture.

Micro white space exists between small components like individual letters but here, macro white space exists to convey the sparseness of population density. White space can also be passive or active. The former approach, used here, improves legibility. It's subtle and will likely go unnoticed. Active white space, by contrast, is used to specifically draw the eye to a component. Whatever your approach, use white space with intent!

Maine and Nebraska have a different approach to allocating Electoral College votes. They allocate two electoral votes to the state popular vote winner, and then one electoral vote to the popular vote winner in each congressional district (2 in Maine, 3 in Nebraska). This is what led to a split in Maine in 2016.

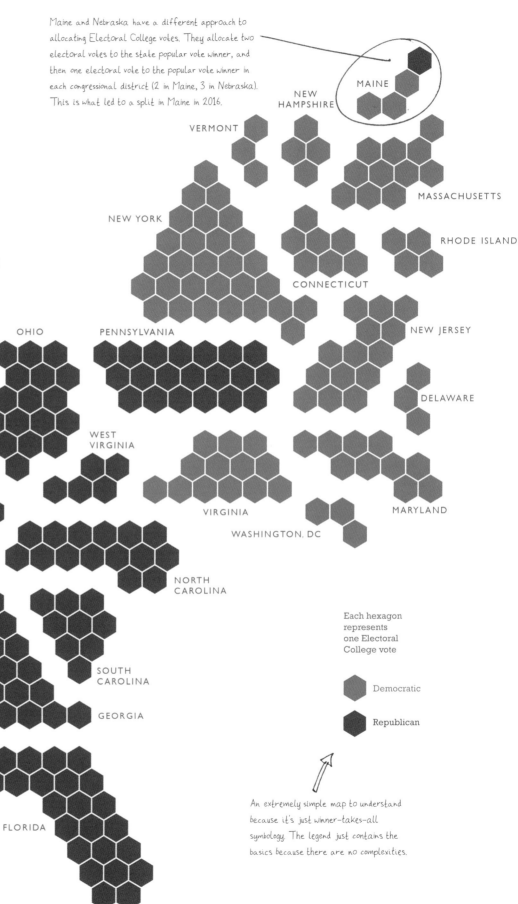

Each hexagon represents one Electoral College vote

Democratic

Republican

An extremely simple map to understand because it's just winner-takes-all symbology. The legend just contains the basics because there are no complexities.

The precise arrangement of tessellated shapes can vary. There's no perfect organisation. A different cartographer could have positioned these hexagons in differently shaped clusters and overall arrangement. Is one arrangement necessarily better than another? Likely not.

Round like a circle

Dorling cartogram

Proportional symbols are excellent devices for representing total (or absolute) data values across maps but often inconvenient artifacts of geography get in the way of making a good map. Overlaps cause problems particularly when large geographical areas have small values and small geographical areas have relatively large values. This problem often is compounded because of the arrangement of small, highly populated urban areas located in proximity. **If only geography were neat and tidy.**

The Dorling cartogram takes the circle, recognised as one of the best symbols for a proportional map, and dissociates it from its geographical position. Once a scaling factor is set—so the symbol sizes are known and there's a good variation between minimum and maximum sizes in relation to the data values— the position of the symbols is detached from geography and moved. **The algorithm to achieve this was created by British geographer Danny Dorling in the mid-1990s. It iteratively moves symbols to a position in which adjacency is retained as far as possible but individual symbols do not overlap. It therefore gets over the fundamental problem of a traditional proportional symbol map.** The algorithm also attempts to limit the distance of the final symbol in relation to its original location at least to approximate the original geographical shape, albeit with bulges and constrictions in certain places.

As with any cartogram, the abstract nature of the final map can be challenging for interpretation, yet it does a great job of mapping people (and their voting patterns, in this case) rather than land mass.

The larger coastal symbols are prominent on this map, and are in their correct proportion to the rest of the symbology. Geography is no longer mediating the perception of relative proportions of colour across the map due to the use of a consistently shaped symbol.

Sometimes the peculiar nature of geography splits clusters of symbols—e.g. New York becomes a thread of symbols throughout the upper right of the cartogram.

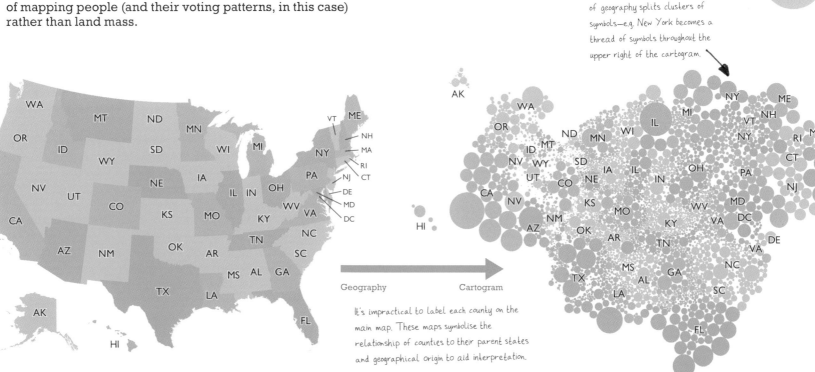

Geography → Cartogram

It's impractical to label each county on the main map. These maps symbolise the relationship of counties to their parent states and geographical origin to aid interpretation.

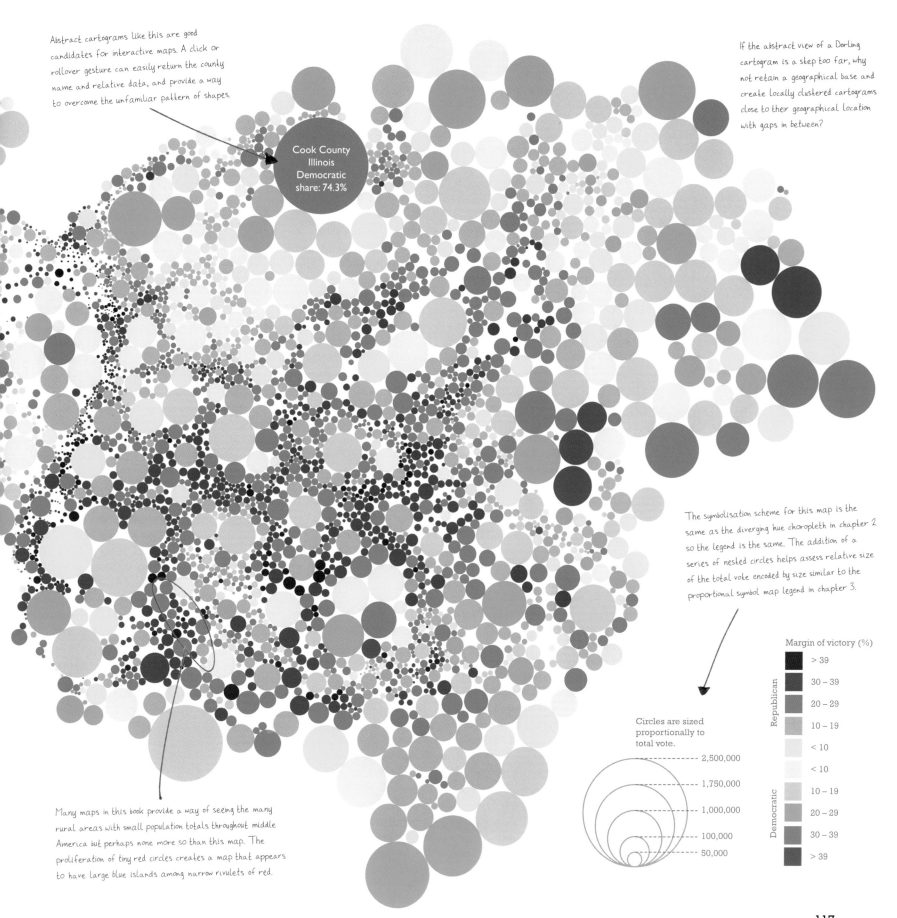

Abstract cartograms like this are good candidates for interactive maps. A click or rollover gesture can easily return the county name and relative data, and provide a way to overcome the unfamiliar pattern of shapes.

Cook County
Illinois
Democratic
share: 74.3%

If the abstract view of a Dorling cartogram is a step too far, why not retain a geographical base and create locally clustered cartograms close to their geographical location with gaps in between?

The symbolisation scheme for this map is the same as the diverging hue choropleth in chapter 2 so the legend is the same. The addition of a series of nested circles helps assess relative size of the total vote encoded by size similar to the proportional symbol map legend in chapter 3.

Many maps in this book provide a way of seeing the many rural areas with small population totals throughout middle America but perhaps none more so than this map. The proliferation of tiny red circles creates a map that appears to have large blue islands among narrow rivulets of red.

Circles are sized
proportionally to
total vote.

2,500,000
1,750,000
1,000,000
100,000
50,000

Margin of victory (%)

Republican
> 39
30 – 39
20 – 29
10 – 19
< 10

Democratic
< 10
10 – 19
20 – 29
30 – 39
> 39

It's hip to be square

Demers cartogram

An alternative to the Dorling cartogram is the map to the right, which uses the Demers approach. On the face of it, the map might simply appear to be composed of squares rather than the circles found on the Dorling version. To some extent, this is true. The area of the squares is scaled proportionally to the variable being mapped. But that's where the similarity ends.

A Demers cartogram attempts to maintain contiguity, and therefore adjacent relationships, as much as possible. It does so by reducing the gaps between shapes and trying to force the regular shapes into a regular tessellation.

However, maintaining contiguity doesn't occur without sacrifice, and the result is often that the distance between the resulting shape and its original position can be substantial. This leads to some peculiar-shaped maps, and without labels, it makes for a difficult map to interpret. **The extent to which the map can maintain contiguity also is compromised if the original map has a large data range, which will translate into a large variation between the symbol sizes.** The problems of sparse population data in large areas and dense population data in small areas becomes difficult to accommodate as large symbols jostle to fit together in some places and small symbols try to maintain proximity to their neighbour elsewhere.

Furthermore, keep in mind the sheer number of symbols that can be accommodated before contiguity breaks down. Where possible, most maps in this book use the county-equivalent data. **There are 3,141 counties and county equivalents in the United States, and there's no way to create a Demers cartogram of that many items and maintain contiguity.** Even at the state level as in the map here, there are already some gaps and issues emerging. Sometimes these gaps can be made to work well. For instance, visual cues help people orient themselves. There are large gaps among the states of Illinois, Michigan, Ohio, and Pennsylvania. They are not the Great Lakes, but they are certainly indicative of them. Similarly, the large states of California, Texas, and Florida jut out a little and balance the northeast expansion of the map because of the large predominance of Electoral College votes.

As with most maps, the key is to work with the technique to develop a satisfactory balance between what it can deliver and what drawbacks and issues you must accept. A Demers cartogram works well for the state data in this use case, but not the county data. If it's larger scale, more detailed, or a greater quantity of symbols to be mapped, a Demers cartogram is unlikely the mapping technique to use.

Because the symbol for each state is in the correct proportion relative to the number of Electoral College votes, the overall visual impact is to ensure exactly the correct proportions of red and blue are seen. With 538 Electoral College votes overall, and the Republicans gaining 306 to the Democrats' 232 there's 56.8% red to 43.2% blue.

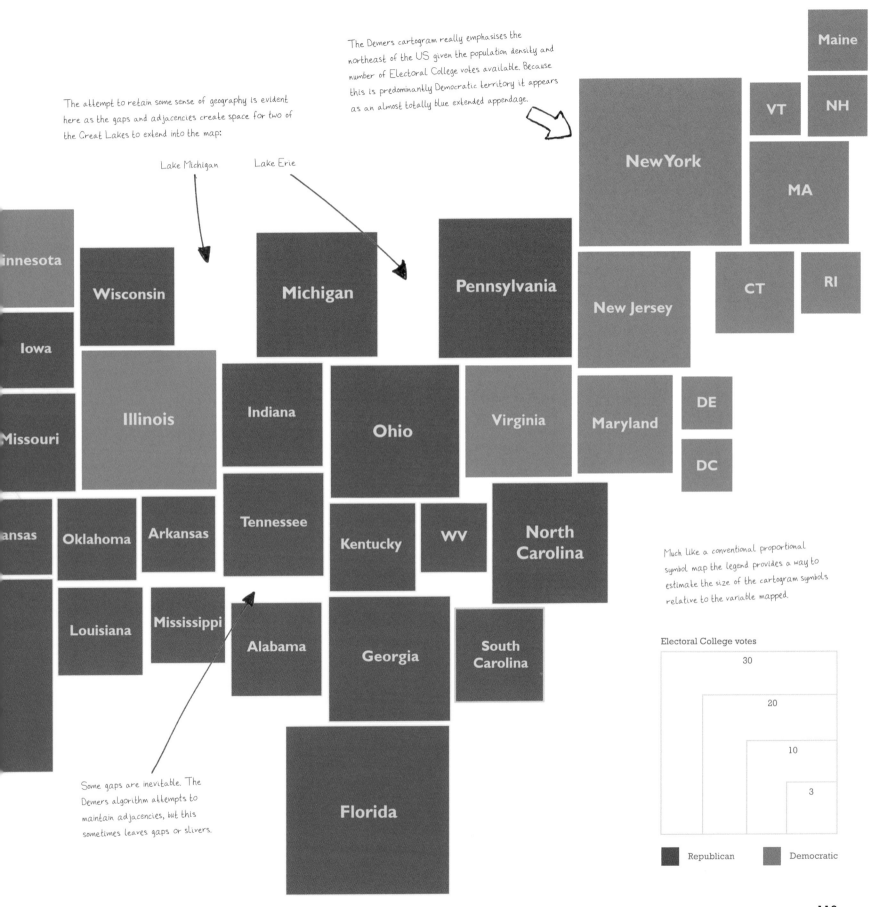

The attempt to retain some sense of geography is evident here as the gaps and adjacencies create space for two of the Great Lakes to extend into the map:

Lake Michigan

Lake Erie

The Demers cartogram really emphasises the northeast of the US given the population density and number of Electoral College votes available. Because this is predominantly Democratic territory it appears as an almost totally blue extended appendage.

Some gaps are inevitable. The Demers algorithm attempts to maintain adjacencies, but this sometimes leaves gaps or slivers.

Much like a conventional proportional symbol map the legend provides a way to estimate the size of the cartogram symbols relative to the variable mapped.

Minnesota

Wisconsin

Iowa

Michigan

Pennsylvania

New York

Maine

VT

NH

MA

Illinois

Missouri

Indiana

Ohio

Virginia

New Jersey

Maryland

CT

RI

DE

DC

ansas

Oklahoma

Arkansas

Tennessee

Kentucky

WV

North Carolina

Louisiana

Mississippi

Alabama

Georgia

South Carolina

Florida

Electoral College votes

30

20

10

3

Republican   Democratic

The ballot box

Pseudo-3D hexagonal mosaic cartogram

Maps perform plenty of different functions. Many are primarily conduits for sharing information with the reader. To optimise that information transfer, maps should be well designed. **But some maps are made more to delight than provide facts in a sensible format.** The main map to the right might lean more into that category, certainly compared with its smaller cousin to the bottom left.

Both maps on this page are mosaic cartograms. They use a single, uniform shape and size to represent each of the states, in comparison with the previous gridded hexagon example, which uses several smaller hexagons to form a larger whole unit which varies in size accordingly. **Here, the uniformly sized symbol makes each state equivalent in the visual plane across the map.** Beyond that, some artistic enhancements are applied so the eye sees a pseudo-3D cube more than a simple hexagonal shape. Because of that visual trick, data can be encoded to appear related to the volume of the symbol.

Volumetric symbology is generally not recommended in cartography. **Humans have greater difficulty assessing relative sizes among symbols that differ in volume.** Sometimes though, it's perfectly okay to bend the conventional wisdom, and this map demonstrates one example of that approach. It shows the relative quantity of Electoral College votes per state. The different-sized colour cubes do this, and because there are relatively few, the map isn't too detailed to cause an overwhelming problem of interpretation.

The map below shows the same information but with Electoral College votes as labels. It works well and is arguably easier to read than the main map. But it's perhaps not as visually immediate, or as striking, and the numbers must be read to recover the information. **Both maps have their place.**

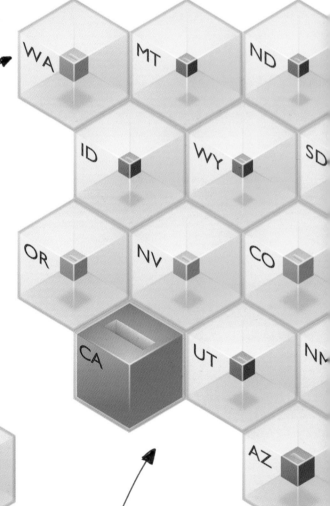

The state abbreviations are applied to reinforce the pseudo-3D graphical effects.

It helps make each cube seem to have similar properties other than the internal coloured cube.

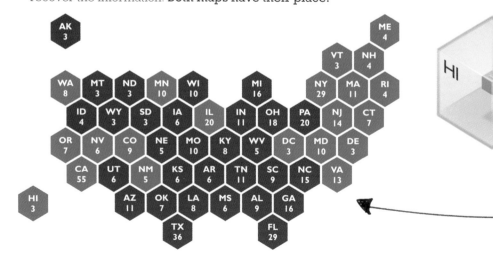

The main map takes a creative approach. The map to the left illustrates the simple flat version. It's the exact same gridded hexagons, but with a colour fill that simply shows each state's winner by colour, and a numerical value for the Electoral College votes.

They're the same maps but would they work as well if they were switched? The main map might seem bereft of visual interest and the inset map would have way too much detail, but on a mobile device, the smaller, simpler design would work better.

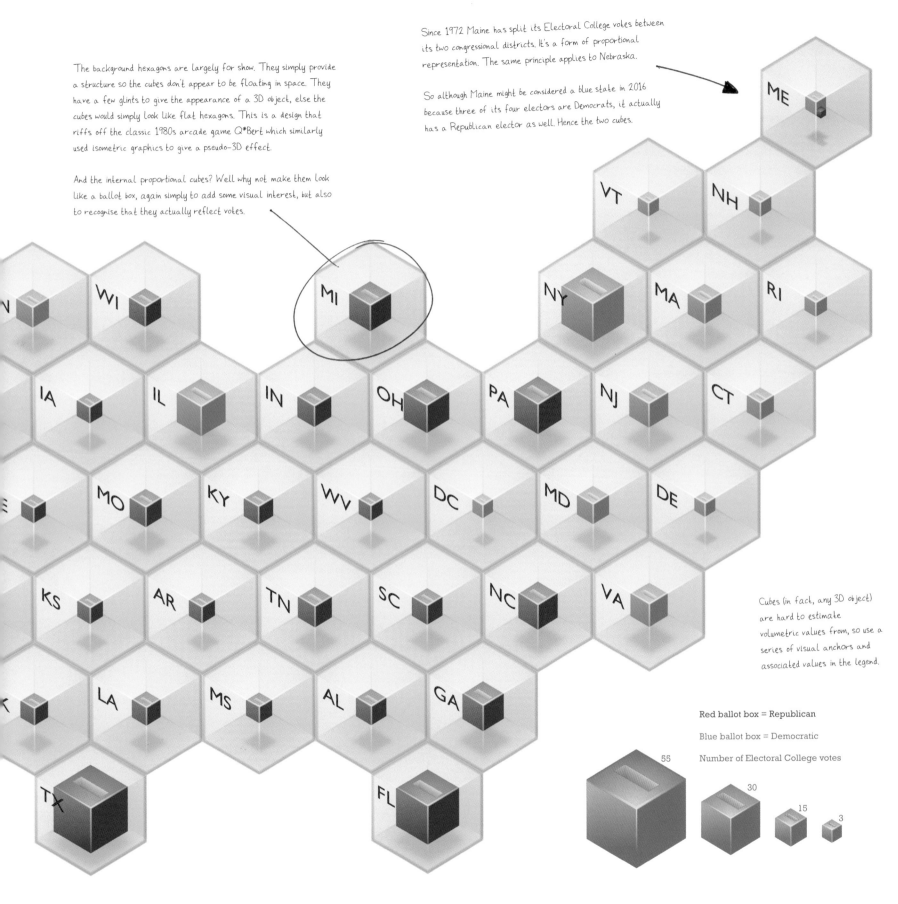

The background hexagons are largely for show. They simply provide a structure so the cubes don't appear to be floating in space. They have a few glints to give the appearance of a 3D object, else the cubes would simply look like flat hexagons. This is a design that riffs off the classic 1980s arcade game Q*Bert which similarly used isometric graphics to give a pseudo-3D effect.

And the internal proportional cubes? Well why not make them look like a ballot box, again simply to add some visual interest, but also to recognise that they actually reflect votes.

Since 1972 Maine has split its Electoral College votes between its two congressional districts. It's a form of proportional representation. The same principle applies to Nebraska.

So although Maine might be considered a blue state in 2016 because three of its four electors are Democrats, it actually has a Republican elector as well. Hence the two cubes.

Cubes (in fact, any 3D object) are hard to estimate volumetric values from, so use a series of visual anchors and associated values in the legend.

Red ballot box = Republican

Blue ballot box = Democratic

Number of Electoral College votes

55

30

15

3

Two colour counters

Compromise cartogram

One of the biggest criticisms of cartograms is that they lack the familiar structure of the geographical map. That's tautological because the entire point of a cartogram is to warp geography to resolve the perceptual and cognitive problems caused by the geography itself.

But it's also a fair criticism. **If maps are to be read easily, despite complex symbology and encoded meaning, the image of a cartogram is too challenging for many.** As with a lot of cartographic practice, compromise is at the heart of many decisions. So applying this maxim to the map itself is a compromise between a familiar geographical map and a cartogram.

As noted elsewhere, geography tends to get in the way of properly interpreting a more traditional election map, particularly in the US in which the population distribution is extremely heterogeneous. Large areas with sparse populations (e.g. Montana) contrast with small areas with relatively dense populations (e.g. Washington, DC). Add to that, that even when population densities might be similar, different-sized areas take on a false prominence. Montana and Vermont each have the same number of Electoral College votes, yet they are vastly different in size, population structure, and appearance on the map. In 2016, the red fill of Montana visually dominates the blue fill of Vermont despite the result leading to an equivalent number of Electoral College votes. **Maps work when a reader can make sensible comparisons and identify differences where they exist.** A simple colour fill doesn't allow such a comparison of Electoral College votes.

The compromise uses a geographical base. It's a familiar structure, and the size, location, and shapes are retained to enable rapid identification. **Instead of using a solid fill, counters are used to represent the number of Electoral College votes gained by each state.** The clusters reflect the correct amount of red and blue according to the outcome of the election. **The map is visually balanced according to the weight of the result, not the geographical surface area.**

A benefit of this approach also allows application of discrete colours to the dots in Maine and Nebraska to reflect the split electoral vote and potential of a faithless elector. This would require overprinting or some other treatment on a solid-fill unique values map or choropleth. Of course, there are always drawbacks, even with a compromise, and here it's in the crowded Northeast, where the density of dots leaves little room for labels. It also makes the entire map balance off-centre with most of the detail in the East. So even with a compromise, geography still has an impact on the map.

In order to make the dots take centre stage the map needs a background. White might be the obvious choice but that's reserved for the paper itself. Instead, a lighter tone of the beige colour used for the land edge vignette is used. This creates a clear visual heirarchy:

Counters

Background map

Page

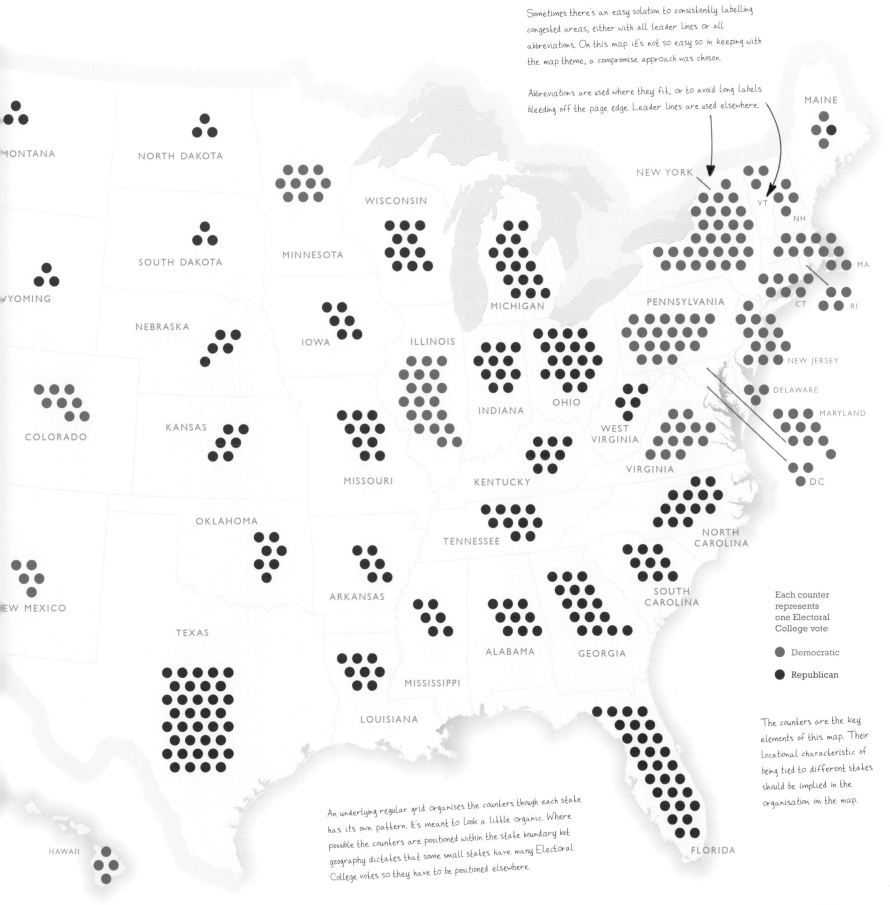

Sometimes there's an easy solution to consistently labelling congested areas, either with all leader lines or all abbreviations. On this map it's not so easy so in keeping with the map theme, a compromise approach was chosen.

Abbreviations are used where they fit, or to avoid long labels bleeding off the page edge. Leader lines are used elsewhere.

MAINE

MONTANA

NORTH DAKOTA

NEW YORK

VT

NH

WYOMING

SOUTH DAKOTA

MINNESOTA

WISCONSIN

MICHIGAN

PENNSYLVANIA

MA

CT

RI

NEBRASKA

IOWA

ILLINOIS

NEW JERSEY

DELAWARE

COLORADO

KANSAS

MISSOURI

INDIANA

OHIO

WEST VIRGINIA

VIRGINIA

MARYLAND

DC

NEW MEXICO

KENTUCKY

TENNESSEE

NORTH CAROLINA

OKLAHOMA

SOUTH CAROLINA

Each counter represents one Electoral College vote

● Democratic

● Republican

TEXAS

ARKANSAS

ALABAMA

GEORGIA

MISSISSIPPI

LOUISIANA

The counters are the key elements of this map. Their locational characteristic of being tied to different states should be implied in the organisation on the map.

HAWAII

An underlying regular grid organises the counters though each state has its own pattern. It's meant to look a little organic. Where possible the counters are positioned within the state boundary but geography dictates that some small states have many Electoral College votes so they have to be positioned elsewhere.

FLORIDA

Every vote's on the map

Dorling cartogram with embedded charts

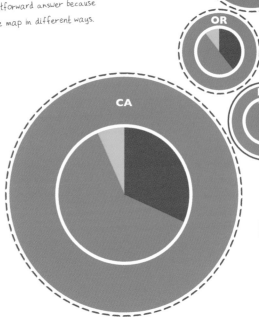

Circles are used frequently in cartography from point symbols to proportional symbols and as a way of representing base geography as a cartogram. A Dorling cartogram of the county-level election data is shown earlier in this chapter where the size of the circle is scaled proportionally to the turnout (number of voters) per county. **To the right, a Dorling cartogram, at state level, is used, with the individual circles proportional to the number of Electoral College votes per state.** The map to the right shows that California has the largest proportion of Electoral College votes, 55, and several of the smallest states on the map have 3 Electoral College votes each. But the Dorling cartogram is only part of the purpose of this map. If the symbols were left as basic circles, it'd show only a single variable. **Instead, a Dorling cartogram is used as a framework to add other data, encoded in different ways to reveal different aspects of the election results at a state level.**

An additional circle is shown for each state that identifies whether it has above (solid line) or below (dashed line) the average of 143,000 Electoral College votes per capita. Generally, the smaller states (in terms of Electoral College votes) have a higher per capita status than larger states, though it's not completely consistent.

The circles are shaded according to who gained all the Electoral College votes in the state because that's how the US electoral system works. So even if the winning candidate polled only 51 percent of the votes, they take all the Electoral College votes, as is the case of Michigan. But that popular vote is still an interesting metric and helps a reader understand how tight margins can be for certain states. **The internal pie chart (the hole in a doughnut) shows the popular-vote breakdown.**

The result is a map that doesn't show only the winners in a binary sense, but it shows the context, too: all the votes for all the candidates, even including those who didn't win. This reinforces the fact that often there are nearly as many Democratic votes in a red state, and vice versa. **The binary map identifies the winner, and in a presidential race, winning is what counts, but showing nuance and putting everyone's vote on the map gives everyone a place on the map.** That can be vitally important in showing people that every vote counts in a democratic election, even if your side didn't win this time.

If every vote counts and every vote matters, then using a map—either geographical or a cartogram, as in this example—can be designed to show every vote rather than just winners or winning margins. Perhaps it's a Dorling 'chartogram'?

When map symbols get complicated, and particularly when charts are used there's always the potential to mess up the map. Symbol complexity can lead to confusion, and difficulty in interpretation.

On this map, is the outer (lower) coloured circle symbol seen as a solid? Or does the pie chart make it seem like a doughnut? Would larger pie charts (leaving just a ribbon of colour) deal with this perception, or not? There's likely not a straightforward answer because different people will view the map in different ways.

Explainer for the pie chart symbol encoding

Explainer for symbol size

Electoral College votes

Dashed outer line = below average Electoral College votes per capita

Solid outer line = above average Electoral College votes per capita

Outer ring colour = state winner

Inner pie = proportion of popular vote
Republican
Democratic
Other

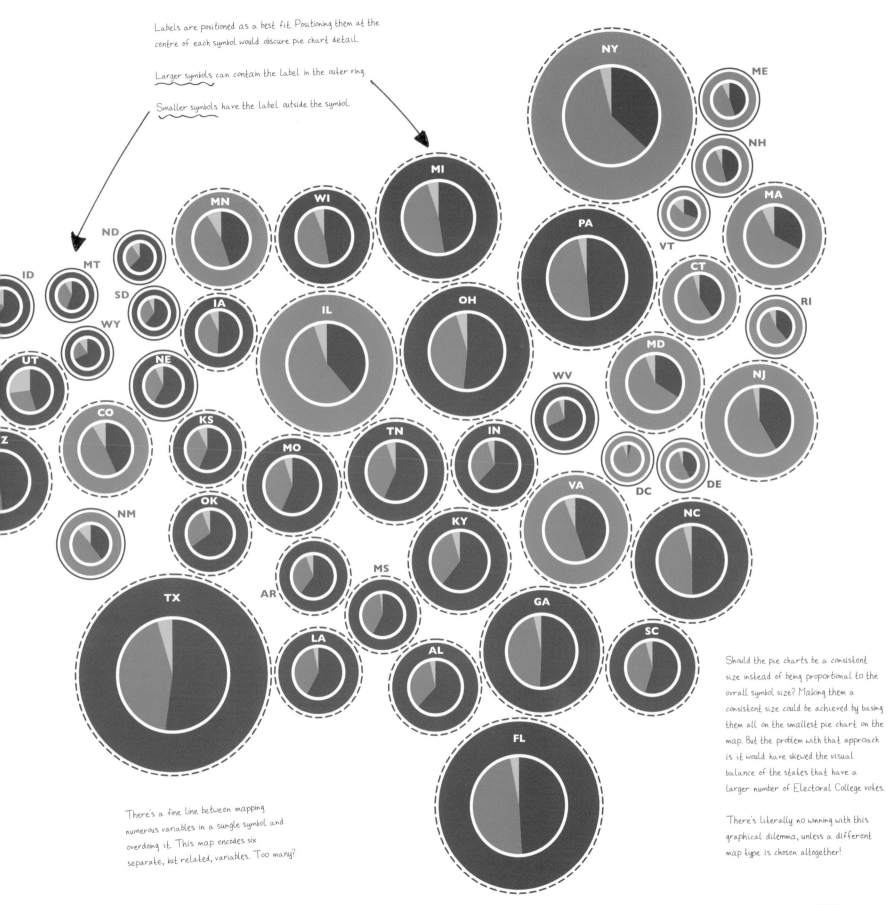

Labels are positioned as a best fit. Positioning them at the centre of each symbol would obscure pie chart detail.

Larger symbols can contain the label in the outer ring.

Smaller symbols have the label outside the symbol.

There's a fine line between mapping numerous variables in a sungle symbol and overdoing it. This map encodes six separate, but related, variables. Too many?

Should the pie charts be a consistent size instead of being proportional to the ovrall symbol size? Making them a consistent size could be achieved by basing them all on the smallest pie chart on the map. But the problem with that approach is it would have skewed the visual balance of the states that have a larger number of Electoral College votes.

There's literally no winning with this graphical dilemma, unless a different map type is chosen altogether!

6. Graphs, charts, and plots

Most of this book is focussed on exploring the rich variety of ways to represent thematic data cartographically. This chapter is a pause in that overarching framework but is important nonetheless and feeds into subsequent chapters. When exploring, analysing, and presenting data, it's important to remember that sometimes the spatial component isn't that special. In ditching the 'carto-' and focussing on the '-graphic' for graphical, a set of useful visual tools allows the presentation of data that teases out important components—namely, graphs, charts, and plots.

The terms often are used interchangeably, as synonyms. Strictly speaking, a graph includes some information about a mathematical relationship between two quantitative components of the data, perhaps when a line is drawn to indicate a trend (e.g. line graph). A chart is a graphic representation of data which uses shapes and symbols and tends to be for a quantitative component (e.g. bar chart, pie chart). A plot usually contains points marked on a coordinate system (e.g. scatter plot).

Graphs, charts, and plots are excellent ways to summarise thematic data. For an election, some clear summary pieces of information can be conveyed, such as who won? By how much did they win? And what are the headlines? It's unlikely that a map is needed for many of these summaries. Perhaps text or a summary chart becomes a more useful way to visually communicate in these situations.

Let the numbers tell the story

Tabulation

The focus of this entire book is on ways to visualise data to avoid the map reader having to wade through lists of numbers, to provide a visual summary, and to encode symbols with meaning to reveal patterns and information. Yet the raw data is the only way someone can get to the details. **Presenting the data in a table is a chart.** Information is arranged in rows and columns in a meaningful way to convey the exact data values. Tables are still compact visual forms and, in many ways, are the ultimate objective form of visual representation because there's little scope to modify the meaning through classification and symbolisation.

The reader is free to explore the data in their own way and make their own judgements about what is high or low and where patterns might exist. The cognitive load is greater because some numerate ability is required but the potential for misinformation through having to interpret a poorly designed map or graphic is lower. Tables also allow the reader to generate their own maps and graphics because of the access to raw data rather than relying on another person's interpretation.

Tables also require good design. They are not simply spreadsheets pasted into a presentation, email, or publication. They share common components such as an x- and y-axis, column and row headings and subheadings, horizontal and vertical rules to signify a section change, and baselines. **Arrangement and alignment are also critical in presenting the data in a way that makes sense but also creates a harmonious chart.** Usually, rows are ordered alphabetically or in some sensible numerical sequence. Designing a table requires as much care and thought as any other visual form of data representation to give the reader unencumbered access to the information.

The quantity of data often determines whether a table is a useful chart or not. In the example to the right, a table of the election results of 3,141 counties and county equivalents would be unwieldy. The typeface would need to be too small to fit the data on one page, and the alternative, running a table over several pages, would be profligate with space (though as an appendix, a table of the data used to make the maps might be useful). If the objective is to prepare a report for the states, or just a single state with a table of the counties of that state alone, a table works well. **As with a map, scale and amount of space combine to constrain what can be achieved graphically.** The table here presents the data for each state alphabetically with various summary values. A sense of the overall result can be seen, as well as allowing the reader to dig into a specific state and candidate's performance, or even make a map of the data themselves. This is the state data used in this book.

Horizontal dividing lines 'span' a collection of related columns. Here, for each candidate or set of summary columns. Along with decked headings and subheadings they help group columns to indicate some sort of relationship.

Omission of extraneous graphical marks is a good maxim. For instance, rather than position a % sign after each percentage value, the single sign in the subheading is sufficient.

The table has 23 columns. By the time the eye gets to the right-hand columns it's easy to lose track of rows. Alternating background shading helps guide the eye.

Colour helps identify which states went Republican, and which went Democratic. It also highlights the main candidates and data points.

Avoiding the use of gridlines outlining every cell in the table helps keep the overall look clean.

A slightly larger type size, and a horizontal dividing rule is used to demarcate the last row which contains the headline numbers.

State or District	Abbrev.	Hillary Clinton Democrat Votes	%	Electoral college
Alabama	AL	729,547	34.4	
Alaska	AK	116,454	36.6	
Arizona	AZ	1,161,167	45.1	
Arkansas	AR	380,494	33.7	
California	CA	8,753,788	61.7	55
Colorado	CO	1,338,870	48.2	9
Connecticut	CT	897,572	54.6	7
Delaware	DE	235,603	53.1	3
District of Columbia	DC	282,830	90.5	3
Florida	FL	4,504,975	47.8	
Georgia	GA	1,877,963	45.6	
Hawaii	HI	266,891	62.2	3
Idaho	ID	189,765	27.5	
Illinois	IL	3,090,729	55.8	20
Indiana	IN	1,033,126	37.9	
Iowa	IA	653,669	41.7	
Kansas	KS	427,005	36.1	
Kentucky	KY	628,854	32.7	
Louisiana	LA	780,154	38.5	
Maine	ME	357,735	47.8	3
Maryland	MD	1,677,928	60.3	10
Massachusetts	MA	1,995,196	60.0	11
Michigan	MI	2,268,839	47.3	
Minnesota	MN	1,367,716	46.4	10
Mississippi	MS	485,131	40.1	
Missouri	MO	1,071,068	38.1	
Montana	MT	177,709	35.8	
Nebraska	NE	284,494	33.7	
Nevada	NV	539,260	47.5	6
New Hampshire	NH	348,526	47.0	4
New Jersey	NJ	2,148,278	55.0	14
New Mexico	NM	385,234	48.3	5
New York	NY	4,556,124	59.0	29
North Carolina	NC	2,189,316	46.2	
North Dakota	ND	93,758	27.2	
Ohio	OH	2,394,164	43.6	
Oklahoma	OK	420,375	28.9	
Oregon	OR	1,002,106	50.1	7
Pennsylvania	PA	2,926,441	47.5	
Rhode Island	RI	252,525	54.4	4
South Carolina	SC	855,373	40.7	
South Dakota	SD	117,458	31.7	
Tennessee	TN	870,695	34.7	
Texas	TX	3,877,868	43.2	
Utah	UT	310,676	27.5	
Vermont	VT	178,573	56.7	3
Virginia	VA	1,981,473	49.7	13
Washington	WA	1,742,718	52.5	8
West Virginia	WV	188,794	26.4	
Wisconsin	WI	1,382,536	46.5	
Wyoming	WY	55,973	21.6	
US total	US	65,853,514	48.2	227

d Trump olican			Gary Johnson Libertarian			Jill Stein Green			Evan McMullin Independent			Others			Margin		Total
%		Electoral college	Votes	%	Electoral college	Votes	%	Electoral college	Votes	%	Electoral college	Votes	%	Electoral college	Votes	%	Votes
3,255	62.1	9	44,467	2.1		9,391	0.4					21,712	1.0		588,708	27.7	2,123,372
887	51.3	3	18,725	5.9		5,735	1.8					14,307	4.5		46,933	14.7	318,608
2,401	48.7	11	106,327	4.1		34,345	1.3		17,449	0.7		1,476	0.1		91,234	3.5	2,573,165
672	60.6	6	29,829	2.6		9,473	0.8		13,255	1.2		12,712	1.1		304,378	26.9	1,130,635
8,810	31.6		478,500	3.4		278,657	2.0		39,596	0.3		147,244	1.0		4,269,978	30.1	14,181,595
2,484	43.3		144,121	5.2		38,437	1.4		28,917	1.0		27,418	1.0		136,386	4.9	2,780,247
215	40.9		48,676	3.0		22,841	1.4		2,108	0.1		508	0.0		224,357	13.6	1,644,920
27	41.7		14,757	3.3		6,103	1.4		706	0.2		1,518	0.3		50,476	11.4	443,814
23	4.1		4,906	1.6		4,258	1.4					6,551	2.5		270,107	86.4	311,268
7,886	49.0	29	207,043	2.2		64,399	0.7					25,736	0.3		112,911	1.2	9,420,039
9,104	50.8	16	125,306	3.1		7,674	0.2		13,017	0.3		1,668	0.0		211,141	5.1	4,114,732
847	30.0		15,954	3.7		12,737	3.0					4,508	1.1	1	138,044	32.2	428,937
055	59.3	4	28,331	4.1		8,496	1.2		46,476	6.7		8,132	1.2		219,290	31.8	690,255
5,015	38.8		209,596	3.8		76,802	1.4		11,655	0.2		1,627	0.0		944,714	17.1	5,536,424
7,286	56.8	11	133,993	4.9		7,841	0.3					2,712	0.1		524,160	18.9	2,734,958
983	51.2	6	59,186	3.8		11,479	0.7		12,366	0.8		28,348	1.8		147,314	9.4	1,566,031
018	56.7	6	55,406	4.7		23,506	2.0		6,520	0.6		947	0.1		244,013	20.6	1,184,402
2,971	62.5	8	53,752	2.8		13,913	0.7		22,780	1.2		1,879	0.1		574,117	29.8	1,924,149
3,638	58.1	8	37,978	1.9		14,031	0.7		8,547	0.4		9,684	0.5		398,484	19.6	2,029,032
593	44.9	1	38,105	5.1		14,251	1.9		1,887	0.3		356	0.1		22,142	3.0	747,927
469	33.9		79,605	2.9		35,945	1.3		9,630	0.4		35,169	1.3		734,759	26.4	2,781,446
0,893	32.8		138,018	4.2		47,661	1.4		2,719	0.1		50,559	1.5		904,303	27.2	3,325,046
9,543	47.5	16	172,136	3.6		51,463	1.1		8,177	0.2		19,126	0.4		10,704	0.2	4,799,284
2,951	44.9		112,972	3.8		36,985	1.3		53,076	1.8		51,113	1.7		44,765	1.5	2,944,813
714	57.9	6	14,435	1.2		3,731	0.3					5,346	0.4		215,583	17.8	1,209,357
4,511	56.8	10	97,359	3.5		25,419	0.9		7,071	0.3		13,177	0.5		523,443	18.6	2,808,605
240	56.2	3	28,037	5.6		7,970	1.6		2,297	0.5		1,894	0.4		101,531	20.4	497,147
961	58.8	5	38,946	4.6		8,775	1.0					16,051	1.9		211,467	25.1	844,227
058	46.0		37,384	3.3								36,683	3.2		27,202	1.5	1,125,385
790	46.6		30,777	4.2		6,496	0.9		1,064	0.1		11,643	1.2		2,736	0.4	744,296
,933	41.0		72,477	1.9		37,772	1.0					13,586	1.2		546,345	14.0	3,874,046
567	40.0		74,541	9.3		9,879	1.2		5,825	0.7		3,173	0.4		65,567	8.2	798,319
9,534	36.5		176,598	2.3		107,934	1.4		10,373	0.1		50,890	0.7		1,736,590	22.5	7,721,453
2,631	49.8	15	130,126	2.7		12,105	0.3					47,386	1.0		173,315	3.7	4,741,564
794	63.0	3	21,434	6.2		3,780	1.1					8,594	2.5		123,036	35.7	344,360
,005	51.7	18	174,498	3.2		46,271	0.8		12,574	0.2		27,975	0.5		446,841	8.1	5,496,487
36	65.3	7	83,481	5.8											528,761	36.4	1,452,992
403	39.1		94,231	4.7		50,002	2.5					72,594	3.6		219,703	11.0	2,001,336
0,733	48.2	20	146,715	2.4		49,941	0.8		6,472	0.1		65,176	1.1		44,292	0.7	6,165,478
543	38.9		14,746	3.2		6,220	1.3		516	0.1		9,594	2.1		71,982	15.5	464,144
5,389	54.9	9	49,204	2.3		13,034	0.6		21,016	1.0		9,011	0.4		300,016	14.3	2,103,027
721	61.5	3	20,850	5.6								4,064	1.1		110,263	29.8	370,093
2,925	60.7	11	70,397	2.8		15,993	0.6		11,991	0.5		16,026	0.6		652,230	26.0	2,508,027
5,047	52.2	36	283,492	3.2		71,558	0.8		42,366	0.5		8,895	0.1	2	807,179	9.0	8,969,226
231	45.5	6	39,608	3.5		9,438	0.8		243,690	21.5		12,787	1.1		204,555	18.1	1,131,430
69	30.3		10,078	3.2		6,758	2.1		639	0.2		23,650	7.5		83,204	26.4	315,067
9,443	44.4		118,274	3.0		27,638	0.7		54,054	1.4		33,749	0.9		212,030	5.3	3,984,631
,747	36.8		160,879	4.9		58,417	1.8					133,258	4.0	4	520,971	15.7	3,317,019
371	68.5	5	23,004	3.2		8,075	1.1		1,104	0.2		4,075	0.6		300,577	42.1	714,423
5,284	47.2	10	106,674	3.6		31,072	1.0		11,855	0.4		38,729	1.3		22,748	0.8	2,976,150
419	67.4	3	13,287	5.1		2,515	1.0					9,655	3.7		118,446	45.8	255,849
984,828	46.1	304	4,489,341	3.3		1,457,218	1.1		731,991	0.5		1,154,084	0.8	7	2,868,686	2.1	136,669,276

Off the charts

Sometimes, the best map is no map

Graphs, charts, and plots visually encode quantitative information of a variable against a reference (e.g. a scaled axis) or other variables. **Collectively, they are efficient vehicles for plotting large amounts of information, which can be more easily reviewed graphically.** Patterns in the data are often far more visible compared with the tabulation of the same data. Beyond simple patterns in the distribution of the data values, deviations, outliers, and trends can be made apparent. **Visually encoding data forms a picture which numbers alone cannot convey.**

The number of types of charts and their variations easily would take up a book, but in the context of this book, they are a useful alternative (or often a complement) to a map. **Sometimes, a map isn't what's needed to convey the essence of the question being asked or of the message you want to pass to your reader.** Headlines often can be best expressed using common graphs, charts, or plots with the specific type driven by the data and need.

Visual representations of data of this type tend to share characteristics. They have a plot area that contains the visual encoding, but they also often have axis lines, labels, and other marks to lead the reader to understanding. Here, then, are a few alternatives that process the data in slightly different ways and support different goals.

The bar chart is perhaps the most common type of chart, representing quantitative data through a series of rectangles with length scaled along a graduated axis. They may be horizontal (bars) or vertical (often called *column graphs*). In the examples to the right, the two pieces of data per state are stacked but alternatively they could have been placed side by side (grouped). By changing the central axis, a small modification helps present a slightly different reading of the data.

A scatter plot shows the relationship between two variables with each data point positioned relative to two axes. Scatter plots are excellent at supporting the exploration of relationships between two variables in a dataset.

Pie charts are often useful for headlines, but with increasing data items, they quickly can become overloaded. Because there's no axis, they are also sometimes difficult to understand and rely on data values and labels being plotted or a separate legend. For a quick visual overview though, they can be helpful.

Line graphs or timelines are useful for showing change over time, or between two points in time as shown here. The magnitude of change can be illustrated by the length of the line or alternatively through thickness or colour coding.

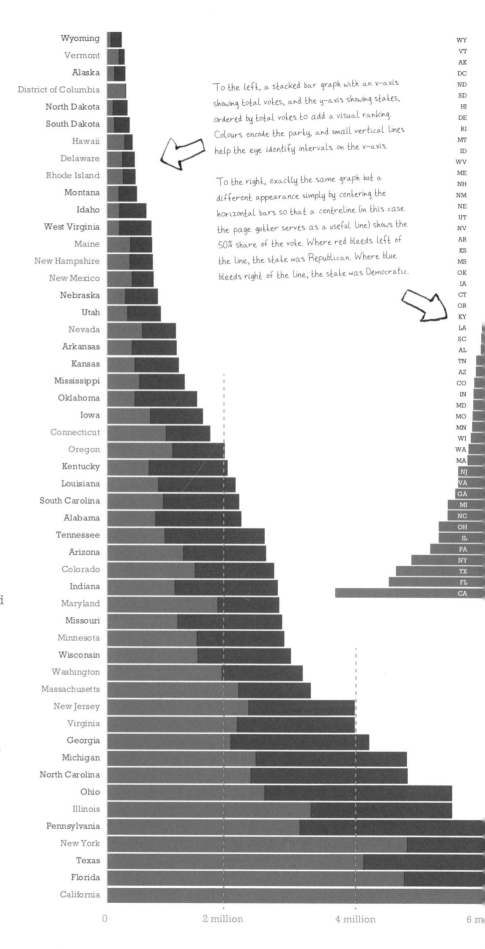

To the left, a stacked bar graph with an x-axis showing total votes, and the y-axis showing states, ordered by total votes to add a visual ranking. Colours encode the party, and small vertical lines help the eye identify intervals on the x-axis.

To the right, exactly the same graph but a different appearance simply by centering the horizontal bars so that a centreline (in this case the page gutter serves as a useful line) shows the 50% share of the vote. Where red bleeds left of the line, the state was Republican. Where blue bleeds right of the line, the state was Democratic.

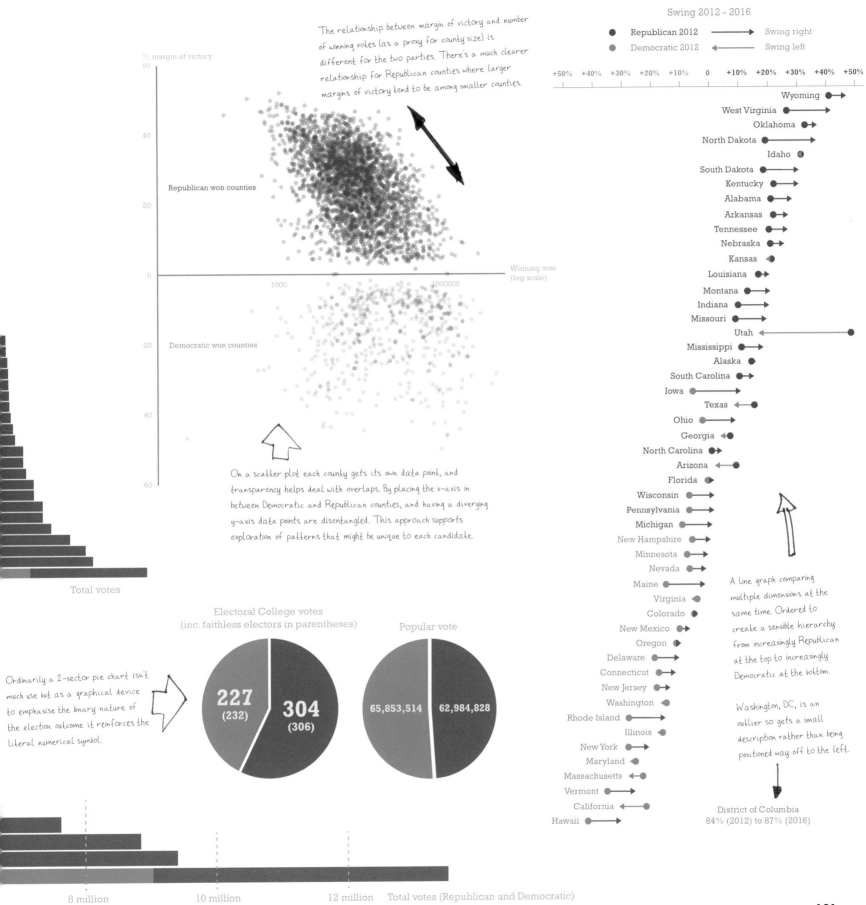

The relationship between margin of victory and number of winning votes (as a proxy for county size) is different for the two parties. There's a much clearer relationship for Republican counties where larger margins of victory tend to be among smaller counties.

% margin of victory

Republican won counties

Winning vote (log scale)

Democratic won counties

On a scatter plot each county gets its own data point, and transparency helps deal with overlaps. By placing the x-axis in between Democratic and Republican counties, and having a diverging y-axis data points are disentangled. This approach supports exploration of patterns that might be unique to each candidate.

Total votes

Electoral College votes
(inc. faithless electors in parentheses)

Popular vote

Ordinarily a 2-sector pie chart isn't much use but as a graphical device to emphasise the binary nature of the election outcome it reinforces the literal numerical symbol.

227 (232)   304 (306)

65,853,514   62,984,828

8 million   10 million   12 million   Total votes (Republican and Democratic)

Swing 2012 - 2016

● Republican 2012 → Swing right
● Democratic 2012 ← Swing left

+50% +40% +30% +20% +10% 0 +10% +20% +30% +40% +50%

Wyoming
West Virginia
Oklahoma
North Dakota
Idaho
South Dakota
Kentucky
Alabama
Arkansas
Tennessee
Nebraska
Kansas
Louisiana
Montana
Indiana
Missouri
Utah
Mississippi
Alaska
South Carolina
Iowa
Texas
Ohio
Georgia
North Carolina
Arizona
Florida
Wisconsin
Pennsylvania
Michigan
New Hampshire
Minnesota
Nevada
Maine
Virginia
Colorado
New Mexico
Oregon
Delaware
Connecticut
New Jersey
Washington
Rhode Island
Illinois
New York
Maryland
Massachusetts
Vermont
California
Hawaii

A line graph comparing multiple dimensions at the same time. Ordered to create a sensible hierarchy from increasingly Republican at the top to increasingly Democratic at the bottom.

Washington, DC, is an outlier so gets a small description rather than being positioned way off to the left.

District of Columbia
84% (2012) to 87% (2016)

Repeatable pictures

Isotype-styled unit chart

Communication of information through encoding data into a visual form usually demands an appreciation of how that form will be read and interpreted. **In its simplest form, if one data point is encoded with one symbol, and then repeated to represent the total, the whole story of that dataset is told.** This is the essence of isotype, short for the International System Of TYpographic Picture Education, developed by Otto Neurath in Austria in the 1920s.

Neurath's motivation was to find a way to bring to life what he considered 'dead statistics' and to make use of the human ability to understand pictures better than numbers. **Isotype therefore uses a pictorial symbol that in some way represents the phenomenon being illustrated, and then uses it in a repeatable way.** Symbols are normally in silhouette or profile and heavily stylized. There's no space for graphical embellishment such as drop shadows, feathered edges, or bevelling to give it a raised appearance. Isotype is flat and static to support the clarity of communication.

Quantities are encoded through repetition, and colour is applied only where necessary. Symbols are maintained at a consistent size and spacing, so 'more' is understood by seeing more symbols of the same type. A benefit of the approach is it's simple to count the symbols to retrieve the actual data values.

The example to the right represents the 2016 Electoral College votes. Democratic votes are symbolised as a blue-suited female to represent Hillary Clinton. Republican votes are represented by a red-suited male to represent Donald J. Trump. Of the 538 Electoral College votes available, including faithless electors, Clinton won 232 and Trump won 306. To win the election and become president, a candidate needs 270 Electoral College votes. **There's little additional detail or information required to assist the reader in interpreting the overall graphic as it's designed to be self-explanatory.**

Ordinarily, symbols would be arranged to read left to right and then top to bottom, but the natural tendency to view Democrats on the left and Republicans on the right leads to this layout design where it's read top to bottom and then left to right. The 270th data point is noted in the centre of the graphic. It helps to have a reference point which makes it easier to see how big the Republican win was. All filled red symbols above and to the left of the 270th value indicate the winning majority.

Isotype itself is an art form in which keeping the graphics as simple as possible is the objective. It creates a striking visual and works well for headline summaries of population data.

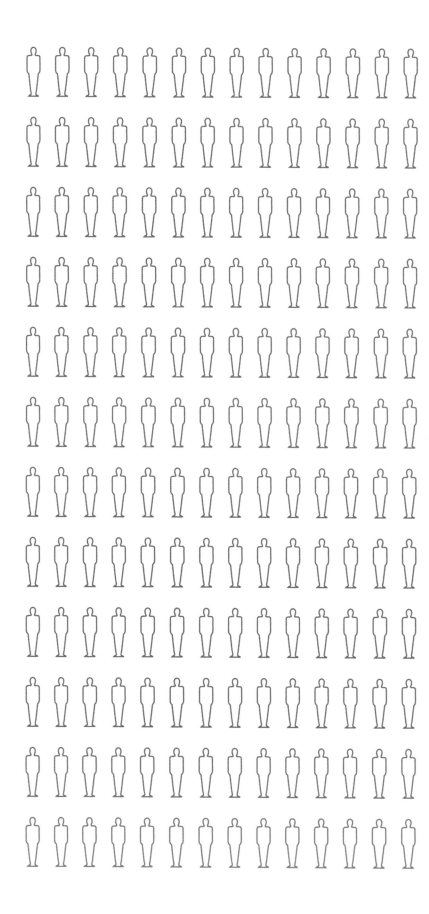

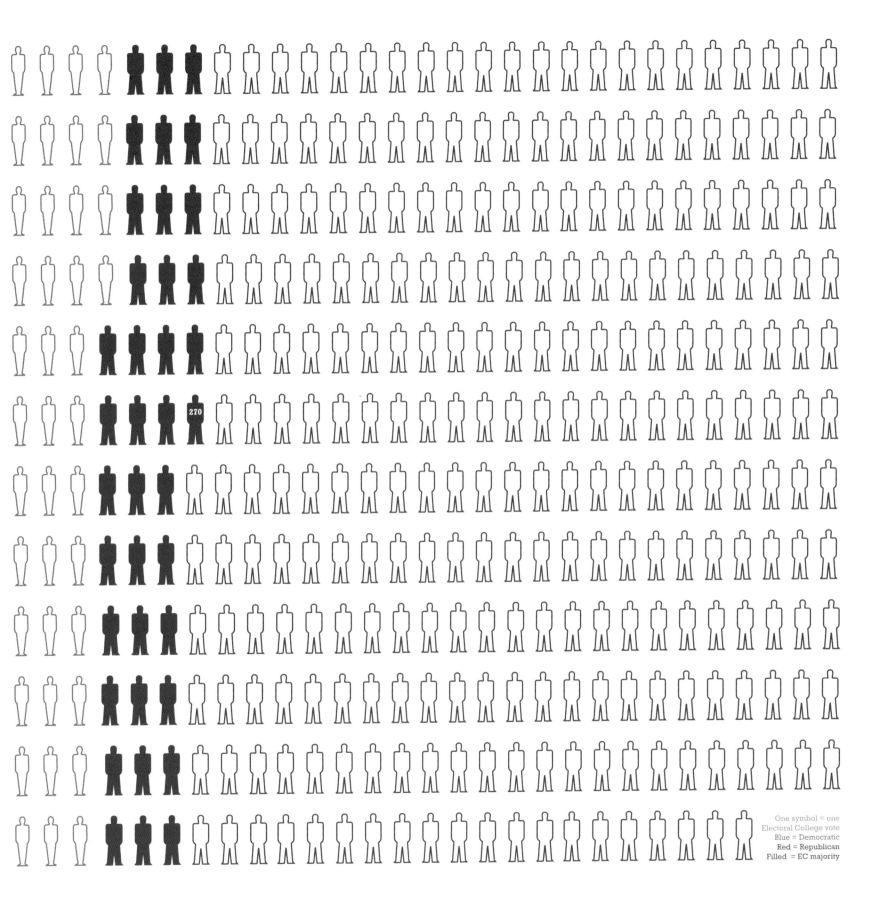

One symbol = one
Electoral College vote
Blue = Democratic
Red = Republican
Filled = EC majority

The river

Line graph as a path to victory

An election is a race. It has a start, an often protracted middle, and an end. In many electoral systems, the winner is the first past the post by virtue of taking a majority of votes. This may be by popular vote, a number of electoral constituencies, or in the case of the United States, 270 of the 538 Electoral College votes spread among the winners of each state.

Winning a state adds the requisite number of Electoral College votes to a candidate's total. **The race to the finish line therefore can be imagined as a linear path.** Each time a state is won, a certain number of Electoral College votes is added, and the race continues until one of the candidates reaches the magic threshold and crosses the line.

There are two versions illustrated here. The main one, to the right, is a meandering pair of rivers flowing from two watersheds at either end, representing the Republican base or the Democratic base. This acts as a nice visual metaphor for the circuitous route taken by the candidates desperate to tick off states and amass 270 Electoral College votes. At the origins of each river are the states that give the candidate the largest share of the vote. These would be likely the most partisan states, and almost certain wins. As the rivers both move toward the centre, the share of the vote reduces, and marginal states exist where a small percentage swing can gift the candidate many of the all-important Electoral College votes.

For campaigning, this sort of graphic also might be a useful mechanism to understand the key battlegrounds. With a finite amount of time, energy, and funds, it's important to identify where resources might be best spent. It's in these bipartisan battlegrounds where much of the most keenly contested campaigning takes place. It's where small margins can result in an overall victory.

The graphic at the foot of the page is virtually the same but has a different stylistic treatment. It's a straight line with solid red and blue. Numbers and pictures add to the detail, and the halfway point of 270 Electoral College votes lies in the page gutter to make it easy to see the extension of the red line on the left page. Eagle-eyed readers will have noticed the same red and blue proportions are used on this book's cover. It makes the cover itself a data visualisation. If you didn't see it the first time, you will now.

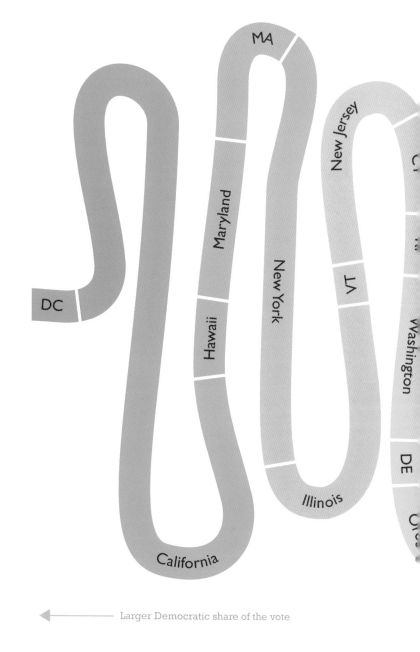

Larger Democratic share of the vote

A red and blue line is pretty boring on its own. Small graphics, photographs, cartoons or similar just give the reader a little more to capture the attention.

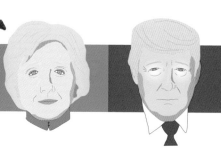

Clinton 232

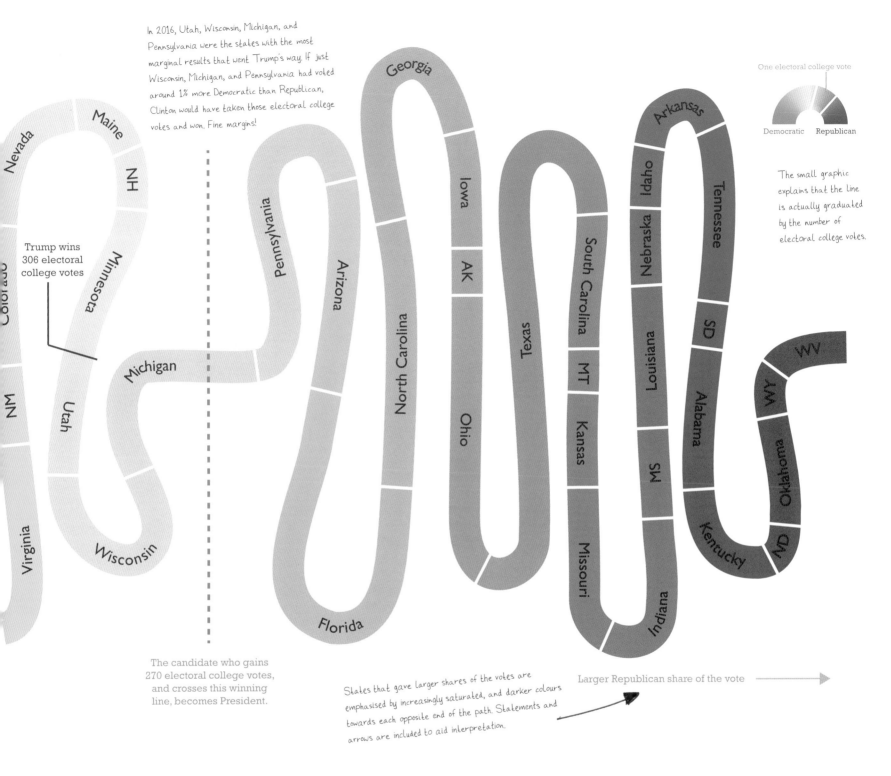

In 2016, Utah, Wisconsin, Michigan, and Pennsylvania were the states with the most marginal results that went Trump's way. If just Wisconsin, Michigan, and Pennsylvania had voted around 1% more Democratic than Republican, Clinton would have taken those electoral college votes and won. Fine margins!

Trump wins 306 electoral college votes

One electoral college vote

Democratic   Republican

The small graphic explains that the line is actually graduated by the number of electoral college votes.

The candidate who gains 270 electoral college votes, and crosses this winning line, becomes President.

States that gave larger shares of the votes are emphasised by increasingly saturated, and darker colours towards each opposite end of the path. Statements and arrows are included to aid interpretation.

Larger Republican share of the vote →

270 electoral college votes

Trump 306

Creating order from chaos

Treemap

The word *map* is used in so many contexts. In its purest form, and as a noun, it can be defined broadly as a representation of features of the earth or activities upon it. So what does that actually mean? **Usually, maps connote a geographical representation but maps can take on many forms.** There are also grey areas. For instance, a schematic map of a subway network—is it a map or a diagram? Countless debates toss this back and forth. Indeed, London calls its subway diagram a map, and New York calls its map a diagram. So what's a treemap, and is it a map?

A treemap is a technique for visually representing data in a hierarchical way, most commonly through nested rectangles. The use of the tree metaphor expresses the tree-structured approach. If each data value is considered a branch, represented by a rectangle proportional to the value itself, it is tiled in the overall graph with smaller adjacent rectangles representing subbranches. For geographical data, a treemap can be useful to organise data values to reveal patterns among the data that might not be apparent from their geographical arrangement.

A treemap can be tiled in several ways but the key characteristic is preserving a sense of the order in which the data is arranged. Ideally, an aspect ratio of 1:1 (a square) is optimum for ease of perception but that makes ordering harder. If the aspect ratio parameter is relaxed, with the treemap becoming more rectangular, ordering becomes more stable. To the right, the aspect ratio is formatted to best fit the dimensions of the page, which gives a good order to the arrangement. **Each rectangle represents a state, with the winning number of votes used to scale the rectangles.** Each rectangle uses a classified colour scheme similar to a geographical choropleth, which encodes further detail and leads to ordered clusters emerging.

California is disaggregated into individual counties to demonstrate how it's possible to create further tiered nesting within a higher order treemap. This design works well for rectangles that have large data values and which are subdivided by relatively few larger scale geographies. But unless the treemap is presented in an interactive environment that supports zooming, the amount of subdividing that's possible, although technically infinite, will be limited by what can be shown meaningfully. For some of the states with relatively low population numbers, shown toward the bottom right of the treemap, it's impossible to see anything more detailed than the state-level geography at the size shown here.

Map, graph, diagram? You decide.

California illustrates how county-level results can be nested inside the parent treemap but it becomes difficult to label as states and counties get smaller.

The classification scheme for the parent treemap also encompasses the data ranges for nested detail in California, which has a wider range of data values.

Size of rectangle proportional to total winning votes

Margin of victory (%)

Republican

■	50 +
■	30 – 49.9
■	20 – 29.9
■	10 – 19.9
■	5 – 9.9
	< 5

Democratic

	< 5
■	5 – 9.9
■	10 – 19.9
■	20 – 29.9
■	30 – 49.9
■	50 +

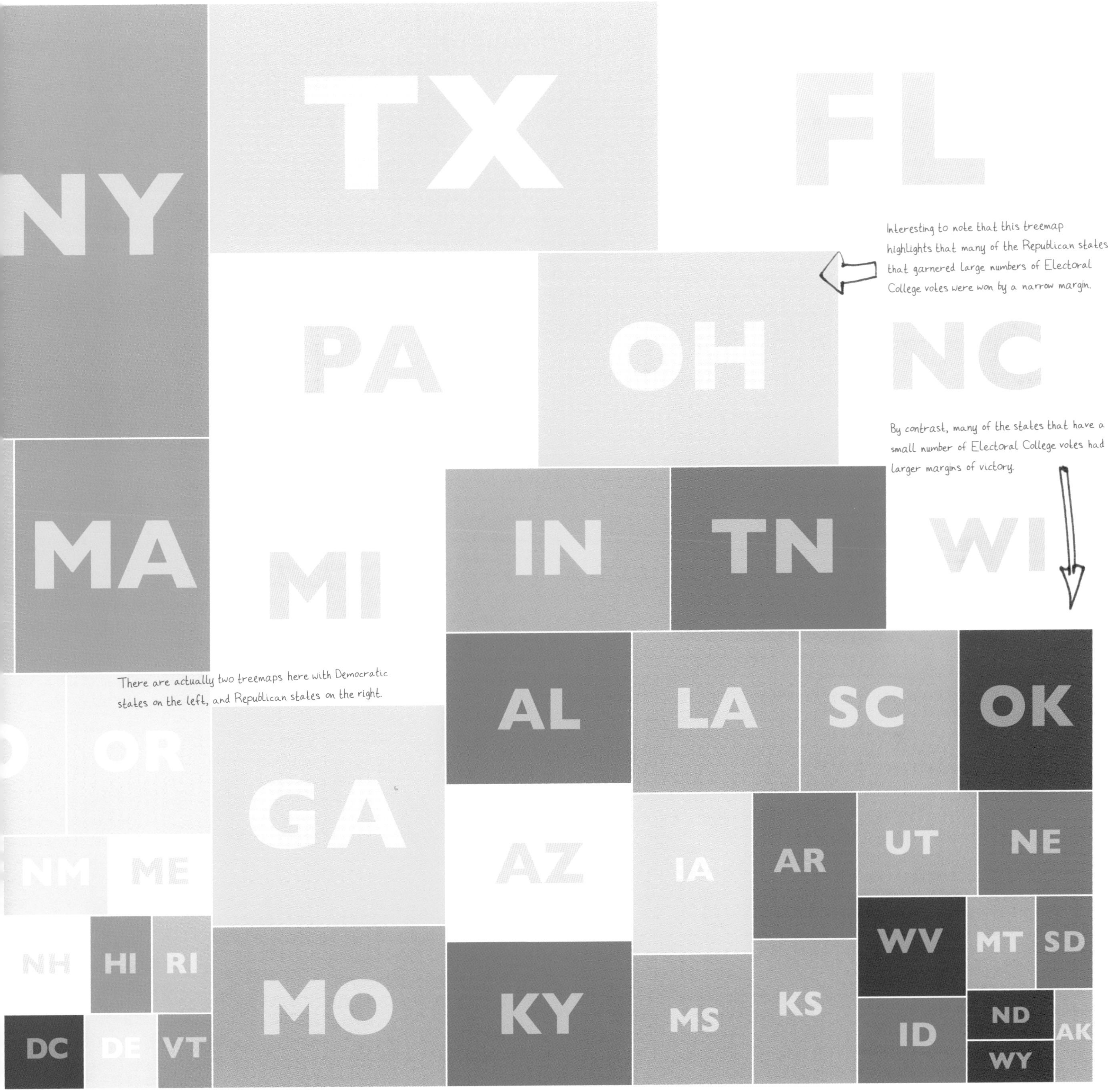

NY

TX

FL

Interesting to note that this treemap highlights that many of the Republican states that garnered large numbers of Electoral College votes were won by a narrow margin.

PA

OH

NC

By contrast, many of the states that have a small number of Electoral College votes had larger margins of victory.

MA

MI

IN

TN

WI

There are actually two treemaps here with Democratic states on the left, and Republican states on the right.

AL

LA

SC

OK

OR

GA

AZ

IA

AR

UT

NE

NM

ME

WV

MT

SD

NH

HI

RI

MO

KY

MS

KS

ID

ND

AK

DC

DE

VT

WY

Creating a new order

Spatially ordered treemap

The point of a treemap is to bring visual order to a set of data, except there isn't a perfect technique to ensure that data order and visual order are consistently applied. Compromise between the aspect ratio of the treemap and the order of rectangles must be made so that rectangles tessellate with no gaps. If this compromise is difficult to resolve, the treemap won't necessarily display the data in a sensible visual order with larger data values to the top left and smaller to the bottom right. That in itself reduces its cognitive plausibility and makes the treemap redundant. Can the treemap itself be improved upon to make the ordering more visually intuitive?

The additional cognitive issue for data with a geographical component is the innate desire of readers to want to see a geographical map. **Reorganising data, whether in a cartogram or some other form, tends to loosen the relationship between data and geography.** It potentially creates a larger hurdle for map readers to negotiate when trying to interpret what they are seeing. One way around this might seem counterintuitive because it's based on reimagining the largely graphical treemap into a more geographical form.

For population data such as the election data, making a spatially ordered treemap might improve people's ability to relate the ordering of data values but in an arrangement that more closely relates to where they expect the data to appear on a traditional map. **In comparison with the previous 1D treemap, ordered by the ranked position of data values, a spatially ordered treemap is modified to become 2D.** Data still is ordered through the proportionally scaled rectangles but the spatial dimension is used to tessellate and position rectangles more geographically.

Put simply, readers of a map of the US would expect to see California to the west of Colorado and New York to the north of Florida. These are basic and well-known geographical realities that the spatially ordered treemap can preserve compared with the regular non-spatial treemap.

Of course, what is gained by reintroducing some geography is lost in terms of simple ordering from large rectangles in one corner of a treemap to small rectangles in the opposite corner. And considering the difficulties in thinking of a treemap as a map or diagram, what exactly is this version to be called? It perhaps could be described as a type of cartogram. It's certainly more maplike than a conventional treemap but constraining it to a square or rectangle in overall appearance leaves it even more abstract than most cartograms. Perhaps it is a *chartogram*, which is why it appears in the chapter on graphs, charts, and plots.

In organising a spatial treemap there's a constant tension between a regular shape and the need to tessellate proportional shapes.

The larger states are generally where they should be and act as visual anchors. Most of the coastal and border states are in a good relative position.

There's no way to fit everything where they go in a spatial treemap, else it would be a contiguous cartogram, or a geographical map!

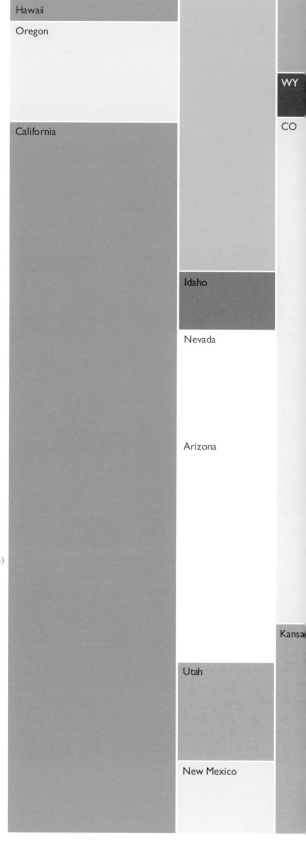

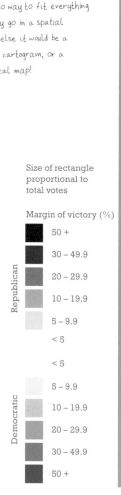

Size of rectangle proportional to total votes

Margin of victory (%)

Republican
50 +
30 – 49.9
20 – 29.9
10 – 19.9
5 – 9.9
< 5

Democratic
< 5
5 – 9.9
10 – 19.9
20 – 29.9
30 – 49.9
50 +

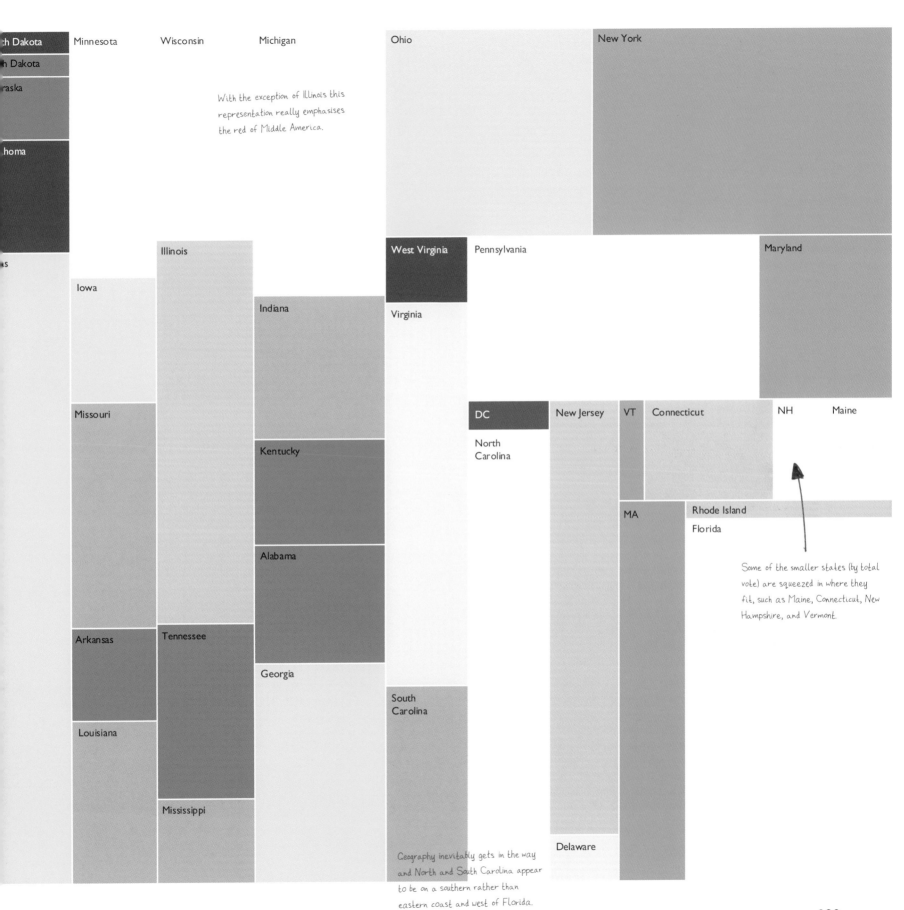

th Dakota

h Dakota

raska

homa

s

Minnesota

Wisconsin

Michigan

With the exception of Illinois this
representation really emphasises
the red of Middle America.

Ohio

New York

Illinois

West Virginia

Pennsylvania

Maryland

Iowa

Indiana

Virginia

Missouri

DC

New Jersey

VT

Connecticut

NH

Maine

North
Carolina

Kentucky

MA

Rhode Island

Alabama

Florida

Some of the smaller states (by total
vote) are squeezed in where they
fit, such as Maine, Connecticut, New
Hampshire, and Vermont.

Arkansas

Tennessee

Georgia

South
Carolina

Louisiana

Mississippi

Delaware

Geography inevitably gets in the way
and North and South Carolina appear
to be on a southern rather than
eastern coast and west of Florida.

Make some noise

Violin and beeswarm plots

In the same way that a choropleth map is the go-to for many people making a thematic map, conventional graphs, charts, and plots are more commonly seen when graphically showing data. Bar charts, line graphs, and scatter plots are the graphical equivalent of the choropleth, proportional symbol, and dot density map types. They are widely used, widely understood, and easy to construct. But as with maps, there is a plethora of alternate chart types, each with many parameters that can be altered to convey a version of truth for the data under consideration.

A violin plot is a modification of the more conventional box plot. A box plot graphically depicts the range of numerical data along a graphical continuum while overprinting regular width boxes that depict quartiles, sometimes with whiskers that extend to interquartile ranges (which also may be the minimum and maximum values). **A violin plot takes the same basic premise as a box plot and extends the graphical summary to include a kernel density plot on either side of the plotted data values.** In this sense, a violin plot illustrates the density of data values for a similar value across the range of values. It can be more informative than a box plot, which illustrates only mean, median, and interquartile ranges, because it shows the full distribution of the data. The shape of the plot can illustrate different modal distributions—for instance, if there is more than one peak in the data frequency or skewness across the range.

A beeswarm plot (also known as a *dot plot*) is a 1D scatter plot, and it plots data values on two axes. **Whereas a conventional scatter plot uses two numeric variables for each axis, a beeswarm uses one numeric and one categorical variable, like the violin plot.** It functions like a strip chart with closely packed but non-overlapping points representing data values. This wouldn't be possible on a conventional scatter plot because data values that are similarly scaled on both axes inevitably would be overprinted. The position of points away from the central line is arbitrary and merely an artifact of the graph type, ensuring no overlaps. This can create unwanted impressions of the data because there's no relationship between the distance from the central line and its horizontal position (for a vertical beeswarm plot).

Both violin plots and beeswarm plots provide a way to view the frequency of data values across the range of the dataset and the population density at a given data value. They also allow the easy identification of outliers at either end of the distribution of data. The Democratic and Republican plots to the right evidence that the relationships between share of the vote and turnout against counties won by each candidate are quite different.

Share of vote (%)

100
95
90
85
80
75
70
65
60
55
50
45
40

Trump counties won

For the violin plot of Trump's counties the distribution is reasonably close to a normal distribution with a mean of around 74%. Of the counties that Trump won, most were with a strong share of the vote.

100
95
90
85
80
75
70
65
60
55
50
45
40

Clinton counties won

Clinton's violin plot is somewhat different in shape to Trump's. Clinton's distribution is heavily skewed around a mean of 52%, and the majority of the counties she won were by a fairly low margin. She did take some counties with a high share of the vote, but not to the same extent as Trump.

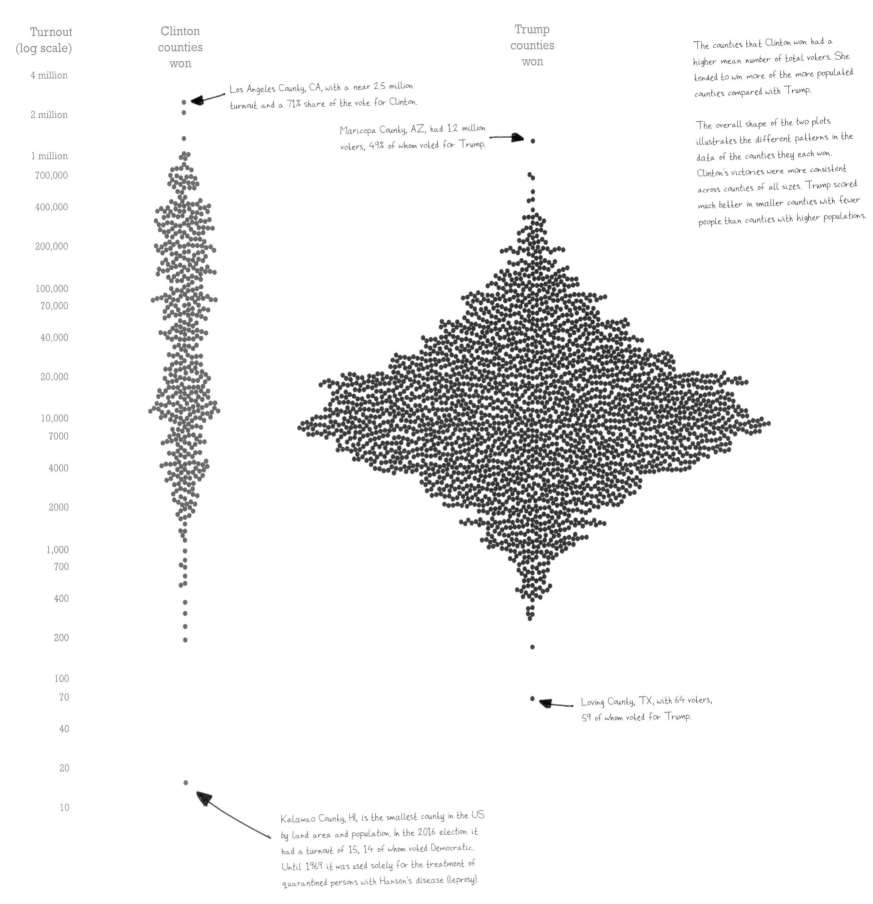

Turnout
(log scale)

Clinton
counties
won

Trump
counties
won

4 million

2 million

Los Angeles County, CA, with a near 2.5 million turnout and a 71% share of the vote for Clinton.

1 million
700,000

Maricopa County, AZ, had 1.2 million voters, 49% of whom voted for Trump.

400,000

200,000

100,000
70,000

40,000

20,000

10,000
7000

4000

2000

1,000
700

400

200

The counties that Clinton won had a higher mean number of total voters. She tended to win more of the more populated counties compared with Trump.

The overall shape of the two plots illustrates the different patterns in the data of the counties they each won. Clinton's victories were more consistent across counties of all sizes. Trump scored much better in smaller counties with fewer people than counties with higher populations.

100
70

Loving County, TX, with 64 voters, 59 of whom voted for Trump.

40

20

10

Kalawao County, HI, is the smallest county in the US by land area and population. In the 2016 election it had a turnout of 15, 14 of whom voted Democratic. Until 1969 it was used solely for the treatment of quarantined persons with Hansen's disease (leprosy).

Start the party

Infographics

There's always a raft of potential constraints when visually representing a dataset through a map, graph, chart, or plot. In a perfect world with no such constraints, there also would be endless time to craft a visualisation without deadlines or, perhaps, the need to meet specific customer or organisational requirements. Except that's not the way the world works, and all too often, a last-minute request or a tight deadline is the cue to create a standard map or graph. This also can lead to hurried reliance on software defaults for colours, typography, map projections, and so on. An alternative is to take an infographic approach.

An infographic is designed to show information quickly and clearly. It attempts to make better use of human cognition by using familiar objects and graphical language to express the data in visual form. Designed for mass consumption, it relies less on a reader's prior knowledge or understanding not only of the subject matter but also of the mode of visual representation that mediates the message. Infographics also are designed to support short attention spans.

Having a limited amount of time or making simpler graphics doesn't lead necessarily to inferior work. Think of journalism and newsrooms, whose work usually is dictated by the daily news cycle. There's also a huge demand on visual content for newspapers and websites, often with extremely short time frames for design and production. **Meeting tight deadlines with high-quality visual content is not the time to attempt a complex and picture-perfect cartographic or graphical masterpiece.** Those may take several days or longer to piece together and often will be based on longer term needs. But the election results are in, and a quick graphic is needed to explain who won—hence, party balloons.

Here's an example of how thinking tangentially can provide inspiration for a highly visual and impactful graphic that conveys the headline message. **Donald J. Trump won, and the party has started.** He won certain states, while Hillary Clinton took others. Both celebrate their state victories, and any victory party will have bunting, banners, and balloons so the graphic riffs off those visuals to show the result as a collection of party balloons. They are sized proportionally according to the number of Electoral College votes won by each candidate, per state. Trump has a larger handful of balloons, so this is the essence of the headline result.

It's the sort of graphic you might put out front on a website. Interactively, the balloons could wave in a breeze and move when the map reader hovers over them to reveal some empirical detail for each state. It's a fun visual, and so captures attention on that basis alone.

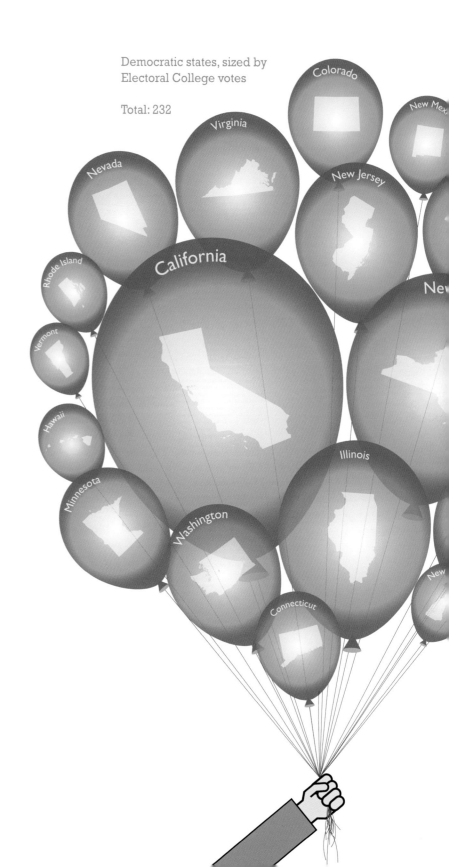

Democratic states, sized by Electoral College votes

Total: 232

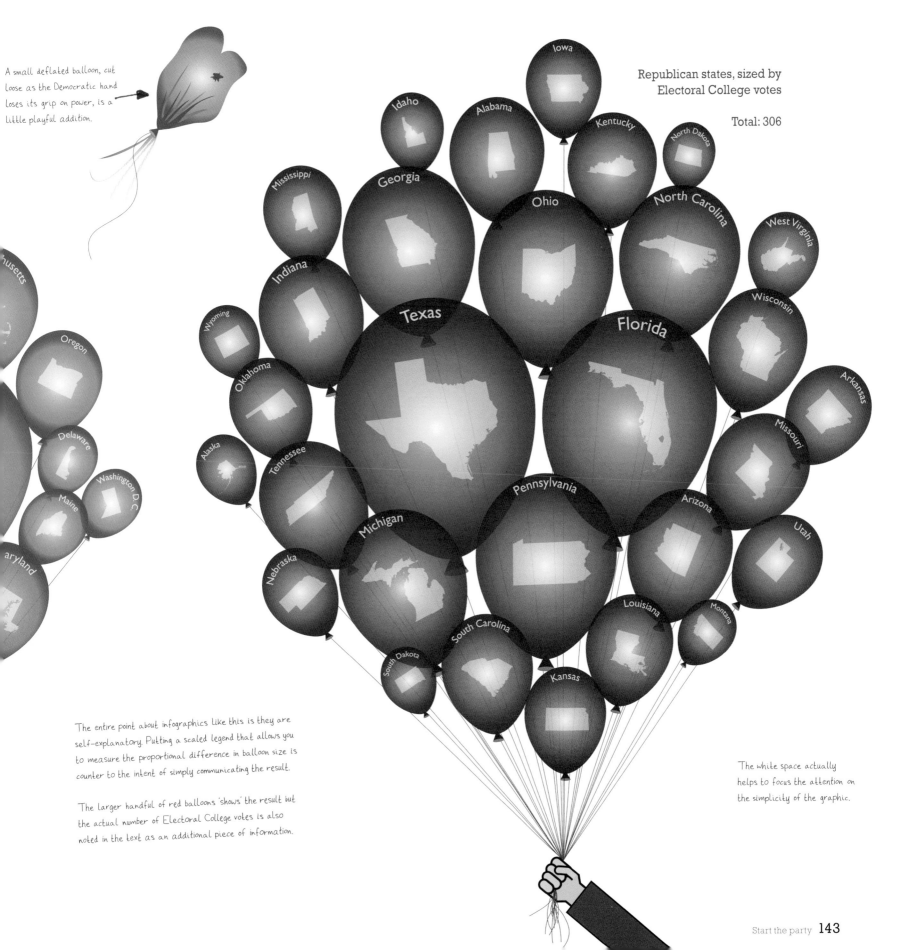

A small deflated balloon, cut loose as the Democratic hand loses its grip on power, is a little playful addition.

Republican states, sized by Electoral College votes

Total: 306

Iowa
Idaho
Alabama
Kentucky
North Dakota
Mississippi
Georgia
Ohio
North Carolina
West Virginia
Indiana
Wisconsin
Wyoming
Texas
Florida
Oklahoma
Arkansas
Alaska
Tennessee
Missouri
Pennsylvania
Arizona
Michigan
Utah
Nebraska
Louisiana
Montana
South Dakota
South Carolina
Kansas

husetts
Oregon
Delaware
Washington D.C.
Maine
aryland

The entire point about infographics like this is they are self-explanatory. Putting a scaled legend that allows you to measure the proportional difference in balloon size is counter to the intent of simply communicating the result.

The larger handful of red balloons 'shows' the result but the actual number of Electoral College votes is also noted in the text as an additional piece of information.

The white space actually helps to focus the attention on the simplicity of the graphic.

Trump's tower

Building an Electoral College histogram

There's so much potential for marrying metaphors and visuals in an election infographic. And when one of the candidates has a background in real estate, there's opportunity for using that as part of the visual narrative as well. **The basic requirement for winning an election is building a victory.** The candidate builds a solid foundation, and that comes about from core support in areas in which they have a good majority. States are built on top of the foundation, summing the Electoral College votes along the way to create a tower.

To take the building metaphor into a visual structure, it's akin to laying down a block at the base of a graph and adding blocks, state by state, until all states are counted. When there are two candidates, and they represent left- and right-leaning parties, the blocks similarly can be on the left and the right and coloured accordingly. The width and the height of the blocks have meaning, and they each become rectangular, covering an area that, in relative terms, gives an impression of the extent of victory.

These election towers have been used occasionally in the US media to report the results of elections. The *New York Times* used the graphical device in 2000 to report the results of the Bush versus Gore election. **The graphic was used to highlight the fact that relatively small margins of victory can result in large numbers of Electoral College votes in a slim statewide contest.** It is often these marginal results that ultimately return a candidate to victory.

In the graphs to the right, one for Barack Obama's 2012 victory and one for Donald J. Trump's 2016 victory, the height of each state represents the number of Electoral College votes won. The width is scaled to the margin of victory, and the states are stacked with those that had a larger margin of victory at the base. In this way, the states are stacked so they decrease in width by the taller the stack so that, at the peak, it's the marginal states that sit atop.

The graph is effectively a vertically stacked histogram that incorporates a measure of the strength of victory in the horizontal direction. There's an argument that states with large margins of victory appear to be more important by virtue of their width. For instance, Hawaii was strongly Democratic in both 2012 and 2016 but yields only three Electoral College votes, the same number as Delaware, also Democratic but with a much lower margin of victory. In a winner-takes-all system, the margin of victory is irrelevant.

Putting graphs side by side such as this also gives an opportunity to reflect on the change from one election cycle to another by seeing where states move, both side to side and vertically, and who takes the Electoral College votes.

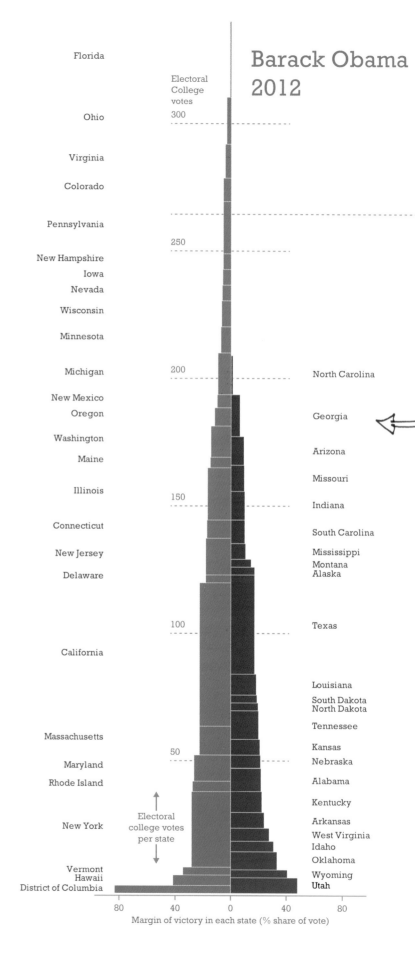

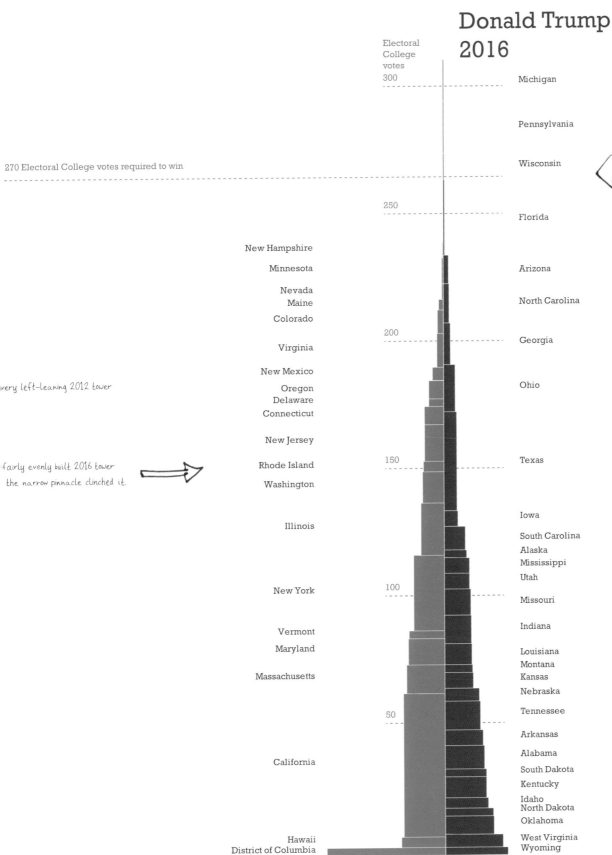

Donald Trump
2016

Electoral
College
votes

300 .... Michigan

Pennsylvania

In 2016 one in four of Donald Trump's
Electoral College votes were won in states
that he gained by a margin of 2% or less.

270 Electoral College votes required to win .... Wisconsin

250 .... Florida

By comparison, in 2012, Barack Obama
had similarly small margins for only one
in ten Electoral College votes.

Fine margins got Trump over the line.

New Hampshire

Minnesota   Arizona

Nevada
Maine   North Carolina
Colorado

200 .... Georgia
Virginia

very left-leaning 2012 tower   New Mexico
Oregon   Ohio
Delaware
Connecticut

New Jersey

fairly evenly built 2016 tower   150 .... Texas
the narrow pinnacle clinched it.   Rhode Island

Washington

Iowa

Illinois   South Carolina
Alaska
Mississippi
Utah

New York   100 .... Missouri

Indiana

Vermont
Maryland   Louisiana
Montana
Massachusetts   Kansas

Nebraska

50 .... Tennessee

Arkansas

Alabama

California   South Dakota

Kentucky

Idaho
North Dakota
Oklahoma

Hawaii   West Virginia
District of Columbia   Wyoming

80   40   0   40   80

Margin of victory in each state (% share of vote)

Plotting the x, y, and z

Space-time cubes

Representing geography often requires plotting more than only x- and y-dimensions. Thematic data often is thought of as occurring in a specific place, but many datasets are also of a specific time or period. **Elections occur at defined points in time, and the result provides a snapshot of the circumstances at polling.** Ignoring that geographies can and do change (through boundary changes), the major difference between election cycles is time. Exploring changes in the results at different times or over time is an important component of electoral analysis. **Space-time cubes can provide a useful mechanism for visualising change in data over time.**

Phenomena can be plotted planimetrically using x,y coordinates with the z-dimension encoding time. **Each data point occupies a specific position in a 3D cube in which the x,y extent fits the geographical study area, and the z-extent encompasses all periods in the data.** For example, if the location of a person's movement is plotted, it would be possible to see how their x,y location differs over time by connecting locations vertically. Connecting lines reveal additional information. If the z-axis is linear, a steeper line between adjacent data points would indicate slower movement, and a shallower line would indicate faster movement. Vertical lines, in which the x,y coordinates are the same but the z-coordinate changes, indicate a stationary period.

Space-time cubes don't have to be fixed to geography in the x,y plane though. **In the example to the right, time is encoded on the z-axis, but share of the vote is on the x-axis and population is on the y-axis.** The data within the space-time cube represents each state with data points showing the relationship between the three variables for each state, joined by a line. The line, therefore, plots the changing position of the x,y paired coordinates through time with 1920 toward the bottom of the cube and 2016 toward the top. The steepness of connecting lines indicates the magnitude of swing between election cycles.

Space-time cubes rapidly can become congested and are toward the more challenging end of the spectrum of graphicacy. They require the reader to interact with them to a greater degree to recover information. They do not have the same level of immediacy in understanding that much simpler forms of visual representations might. Interactivity also helps considerably with these more complex 3D forms, which cannot be similarly achieved through print. The ability to rotate the space-time cube to see different angles, overcome occlusion, and click on various components to reveal information is helpful.

For a static version, picking the viewing angle and avoiding overloading the space-time cube is the most prudent approach.

The more lines that are displayed, the more legibility is compromised. With 51 lines it's not easy to differentiate between states easily even though they are each coloured differently.

Adding labels isn't really an option either as it'd simply add to the visual mess. The complexity outweighs utility.

Extracting the signal from the noise can be achieved by focussing on only a subset of the data.

Comparing Nevada, Iowa, and California

Line thickness sized by winning margin (total votes)

These four views of the same space-time cube show how the viewing angle plays an important role in how the data is seen.

Nevada's swing between parties has the largest range but all have narrowed over time.

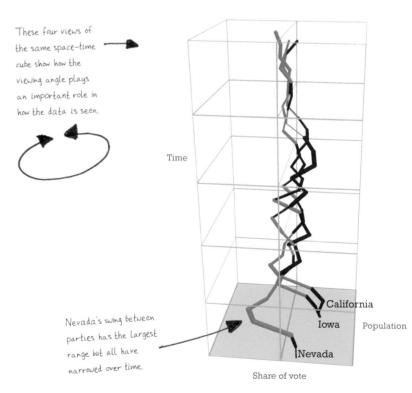

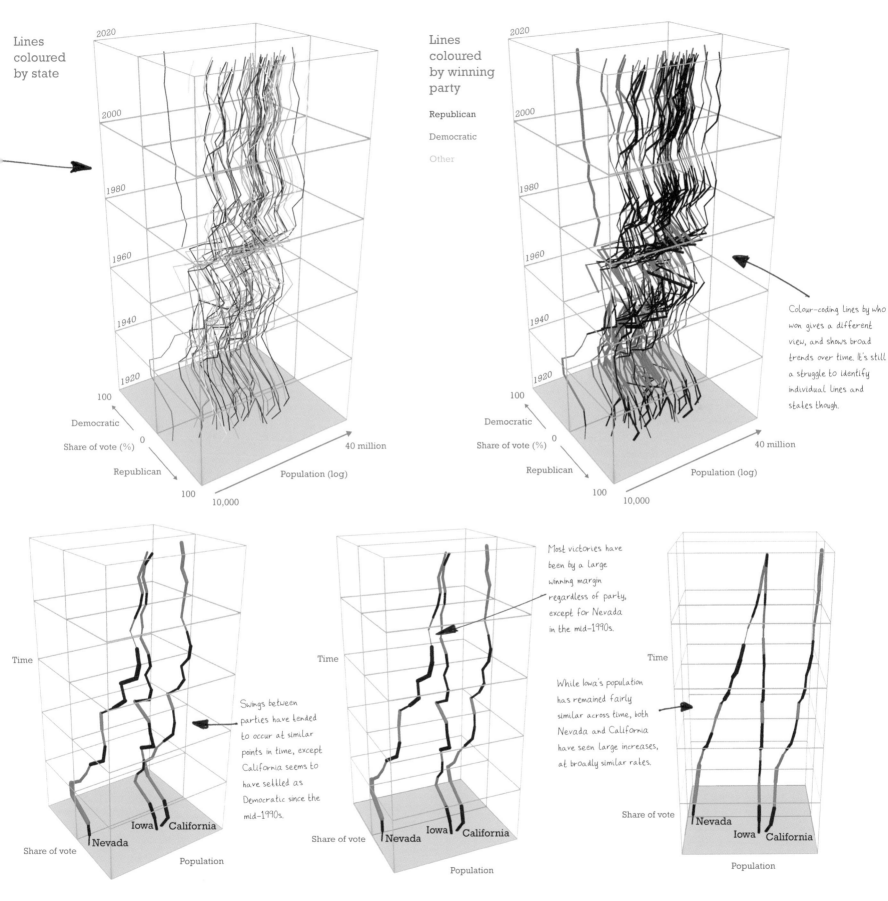

Lines coloured by state

2020
2000
1980
1960
1940
1920

100
Democratic

0
Share of vote (%)

Republican

100

40 million

Population (log)

10,000

Population (log)

Lines coloured by winning party

Republican

Democratic

Other

2020
2000
1980
1960
1940
1920

100
Democratic

0
Share of vote (%)

Republican

100

40 million

Population (log)

10,000

Colour-coding lines by who won gives a different view, and shows broad trends over time. It's still a struggle to identify individual lines and states though.

Time

Swings between parties have tended to occur at similar points in time, except California seems to have settled as Democratic since the mid-1990s.

Share of vote

Nevada   Iowa   California

Population

Time

Most victories have been by a large winning margin regardless of party, except for Nevada in the mid-1990s.

Share of vote

Nevada   Iowa   California

Population

Time

While Iowa's population has remained fairly similar across time, both Nevada and California have seen large increases, at broadly similar rates.

Share of vote

Nevada   Iowa   California

Population

7. Chartmaps

In a book predominantly about maps, it might seem odd to suggest that maps are not necessarily the most useful tool to visualise data in all circumstances. As the previous chapter explored, graphs, charts, and plots are valuable as ways to represent visually patterns and trends in data. But put graphs and maps together, and it creates a powerful mechanism to explore not only patterns in the data (via the graphs), but also how those patterns exist in different places. This combined approach might be referred to as *chartmaps*.

Because of the inherent problems of different-sized geographies making it difficult to position graphs on a map in a sensible order, avoiding overlaps, leader lines, and congestion, chartmaps often resort to abstract geographies. Abstraction was a key feature of the cartogram chapter, and most of the maps in this chapter are an extension of that idea. Some use a geographical background whereas others use a gridded cartogram to provide consistently sized graphical containers. They are effectively a combination of some of the techniques from chapters 5 and 6, marrying cartograms with graphs, charts, and plots.

For the maps that use a gridded cartogram, states are arranged to mimick the shape of the United States, but inevitably some topology is lost. This compromise is made in lieu of a framework for the positioning of small but highly detailed graphs that communicate many hundreds of pieces of information.

Counting votes

Waffle grid on a geographical base

Map design involves encoding data into a visual mechanism. More complex datasets often require more complex designs. But sometimes the simplest approach, although not necessarily the most obvious, offers a sound way forward. **Instead of relying on people's ability to interpret colour or size or tonal shading or any number of other visual variables, falling back on counting can be useful.**

Basic graphical units can be counted to support simple data recovery. This is the function of a waffle grid, which shows quantities of categories. **Waffle grids are similar in some respects to a pie chart, but whereas slices of a pie are often difficult to estimate empirically, a waffle grid has a square appearance populated by smaller squares.** Each square acts as a data cell representing a specific value, and repetition of cells in a grid provides a proportional representation among data categories. This results in differently sized waffle grids that have a different number of grid cells according to the sum of the categories of data being mapped, albeit usually rounded to the nearest whole unit. These cells can be counted, and if cells are categorically encoded using colour, such as shown to the right into parties, the overall balance of colour shows the broad pattern of the distribution.

Alternatively, a regular matrix can be constructed based on percentages, not absolute values, which yields a similarly sized grid for each state. This variant of the waffle grid approach removes the ability to count and recover data and falls back on the need for the reader to visually estimate relative proportions of colour within each waffle grid and between them. Whether this has a higher cognitive load than counting is debatable. Both serve their purpose in slightly different ways.

If the United States had proportional representation, this map might be very useful as it shows not only the mix of the popular vote, but how much each candidate polled per state.

Washington

Oregon

Idaho

Nevada

Utah

California

Arizona

Alaska

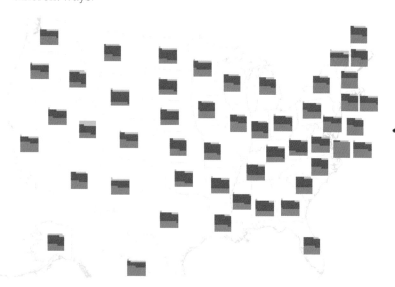

It would be impossible to see individual cells on the percentage version of the map so block colour is used and the background simplified to make the charts more figural.

Partisan states such as North and South Dakota and Wyoming, plus DC, have a much larger proportion of the winning coloured grid cells.

Waffles still have to be moved a little to accommodate the congested Northeast.

States where there was a marginal victory such as Wisconsin, Michigan, Minnesota, and Florida have a waffle grid of fairly even numbers of blue and red grid cells, regardless of the overall number.

Montana
North Dakota
Minnesota
Michigan
Maine
Vermont
New Hampshire
New York
South Dakota
Wisconsin
Massachusetts
Wyoming
Iowa
Pennsylvania
Connecticut
Rhode Island
Nebraska
Illinois
Ohio
Indiana
New Jersey
Colorado
West Virginia
Maryland
DC
Kansas
Missouri
Kentucky
Delaware
Virginia
New Mexico
Oklahoma
Tennessee
North Carolina
Arkansas
South Carolina
Texas
Mississippi
Alabama
Georgia
Louisiana

Each cell = 20,000 votes
■ Republican
■ Democratic
■ Other

Hawaii

Florida

Double dessert

Pie charts on a waffle grid

The regular structure of a gridded cartogram lends itself to being used as a framework for many forms of small multiples. **Overall trends can be seen by gridding small pie charts across the map.** Although they can be viewed individually, it's the overall patterns that are of most interest.

On the map to the right, each state is represented by a square tessellated to approximate the shape of the United States, although without the maintenance of perfect adjacencies. Each square then is subdivided into a 5 × 5 grid. In each of the cells is a pie chart representing the share of the vote between Democratic and Republican Parties across the last 25 elections.

Starting from 1920 in the bottom left to the 2016 election in the top right, it's possible to see how the share of the vote has changed across time at a state level and nationally. **There's a lot of information packed into this seemingly simple chartmap.** The regular geometry helps give order but there's a fine balance between a map that is useful and one that tries to do too much. Some maps suffer from information overload by attempting to encode too much information into each symbol or by having too many symbols.

Ordinarily, a two-segment pie chart wouldn't be particularly useful on its own. It'd be far more useful to state the two numbers. But when 1,244 pie charts are placed in a grid, the two segments almost work. There isn't room to include the remaining candidates, who always pick up a few percentage points from the main two parties.

In a web map environment, a hover or click action for each pie chart could lead to the precise breakdown of results, making this a simple interface for detailed information recovery.

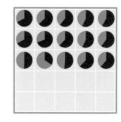

Alaska was granted statehood in 1959. Richard Nixon (Republican) polled 50.9%, beating John F. Kennedy (Democrat) at 49.1% though the latter won the race to the White House.

A dark outline for each state helps demarcate state groupings of pie charts.

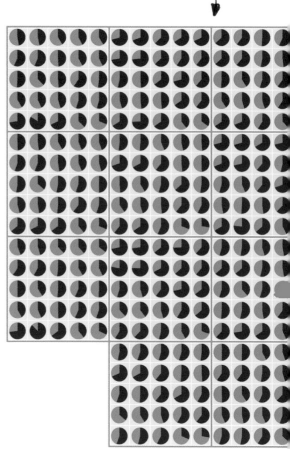

There are 1,244 separate pie charts on the main map. That's a lot of information. Double it because each pie has two slices. Now add in that each is spatially located, and in a time series, so there's a considerable amount of detail.

It's doubtful whether anyone would want to use the map to find something out about a particular election but it shows that you do not have to always omit detail. Bringing clarity to complexity is just as important a design tenet.

At a smaller scale the same principle of repeated shapes can be used to focus on a single dataset. Here, to the left, the 2016 result. The grey background grids have been removed to reduce clutter because with fewer pieces of information and a single symbol for each state, the structure holds up without additional detail.

2016 presidential election results

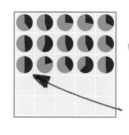

Hawaii became the 50th state in 1959 so its first presidential election was 1960. John F. Kennedy (Democrat) defeated Richard Nixon (Republican) by the slimmest of margins: 115 votes (0.06%) on his way to overall victory.

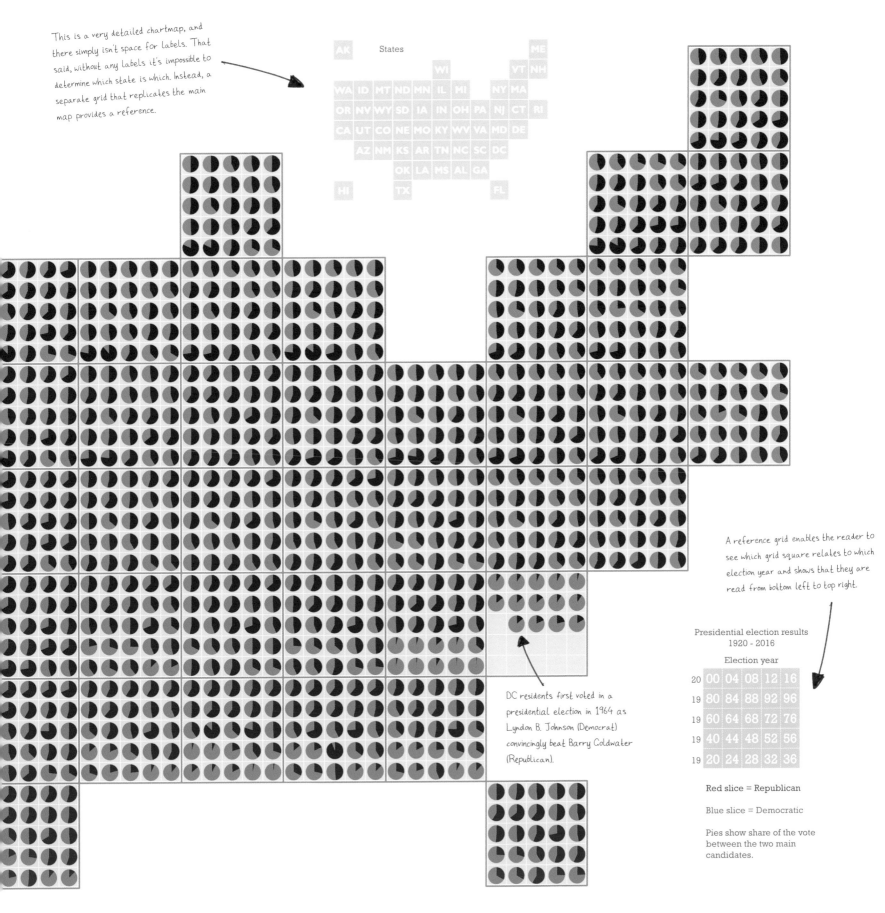

This is a very detailed chartmap, and there simply isn't space for labels. That said, without any labels it's impossible to determine which state is which. Instead, a separate grid that replicates the main map provides a reference.

States

AK										ME	
			WI							VT	NH
WA	ID	MT	ND	MN	IL	MI		NY	MA		
OR	NV	WY	SD	IA	IN	OH	PA	NJ	CT	RI	
CA	UT	CO	NE	MO	KY	WV	VA	MD	DE		
	AZ	NM	KS	AR	TN	NC	SC	DC			
			OK	LA	MS	AL	GA				
HI		TX					FL				

A reference grid enables the reader to see which grid square relates to which election year and shows that they are read from bottom left to top right.

DC residents first voted in a presidential election in 1964 as Lyndon B. Johnson (Democrat) convincingly beat Barry Goldwater (Republican).

Presidential election results
1920 – 2016

Election year

20	00	04	08	12	16
19	80	84	88	92	96
19	60	64	68	72	76
19	40	44	48	52	56
19	20	24	28	32	36

Red slice = Republican

Blue slice = Democratic

Pies show share of the vote between the two main candidates.

Square pegs in square holes

Unique values waffle grid

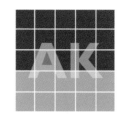

A thematic map commonly is used to show a single piece of information or possibly a few pieces of information. **When the data has multiple pieces of information or the point of the map is to show data in a time series, things can get complicated.** The challenge is to find a graphically prudent way of encoding the data without overcomplicating the visual.

Most of the maps in this book have explored a single election, but elections are part of an ongoing narrative, and historical election data gives context for deciphering the most immediate results. Questions such as 'Did the incumbent party retain control?' and 'How have parties fared over time?' are handled only with a technique that supports the display of multiple pieces of information. A waffle grid can help achieve that goal.

Waffle grids show categories or quantities for small pieces of a whole. In this case, the state is the whole, and the map shows change in the winning party over the last 25 election cycles in a 5 × 5 grid. **The equally divided cells create a waffle pattern with different hues used to encode the two winning parties.** On some waffle grid maps, such as the one on page 150, a cell is a part of the whole. On the map to the right, each cell is a moment in time, not a category.

The grid form supports data recovery through counting how many times a Democrat or Republican has won in a state. It shows trends over time and predominance in which a party appears to have a high win rate across many elections. And the opposite is true in that it's possible to see states in which the winning party tends to flip often. These are referred to as *swing states* and often become the focus of intense campaigning in the run-up to an election because winning swing states is often the key to winning the race to the White House.

Waffle grids dissolve geography into an abstract framework. The basis of repetition is, in large part, the reason it is a successful visualisation. Humans are good at seeing repeated patterns and assessing how they vary. **Grids are easy to read, and so the information recovery is discoverable.** It'd be difficult to show this level of change over time using a geographical map and still maintain clarity and legibility.

This map is arguably more readable than the previous one that used gridded pie charts. You could have encoded the winning share of the vote by changing the shade (value) of red or blue per cell. But all of that increases graphical complexity in a chartmap that already contains hundreds of pieces of information. It's the cartogram equivalent of an equal values map, and this version shows only one piece of information per cell.

The map employs the same designation as elsewhere in the book with red for Republican and blue for Democratic. It's a conventional use of colour to represent the two parties.

However, before 2000 there was no uniformity in the colours used to represent the parties across television, newspapers, and magazines. Most organisations used red and blue, but not necessarily for the same parties, and not always between election cycles.

The reasons are unclear but a former chief White House correspondent for NBC News said, "During the Cold War, who wanted to be red?" So, often the opposition was shown in red, regardless of who they were. From the mid-1970s to the late 1980s Republicans were more often than not shown in blue and Democrats in red. And that colour coding reflects the system used in Great Britain, a potential reason for the choice as conservatives are blue and the more liberal alternative (Labour) is red.

In 2000, the New York Times and USA Today began printing maps in colour. Both used red for Republican and blue for Democratic. Archie Tse, chief graphics editor for the Times, says it was simply because red began with 'r' and Republican began with 'R' so there was a natural association. And that scheme has stuck.

It's also used in this book and on this map, though consistently red for Republican and blue for Democratic for every election from 1920 to 2016 because it would make no sense to switch colours half way across the map!

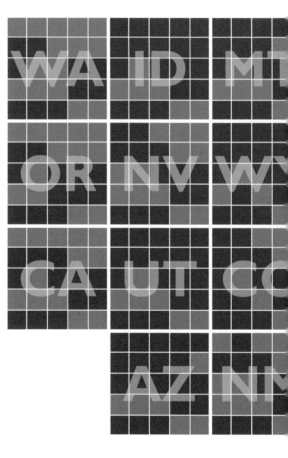

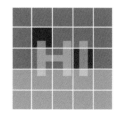

The addition of a thin white line emphasises the distinction between each cell in a state. A further, thicker white line demarcates states from one another. These small decisions help to organise the data effectively and makes the graphic easier to navigate for the map reader.

The arrangement of the states is designed to appear like the geographical map, albeit hugely generalised. But to do so is at the expense of some adjacencies which are impossible to deal with. So some states are not neighbouring their neighbours on a long edge, but might be touching at a corner instead.

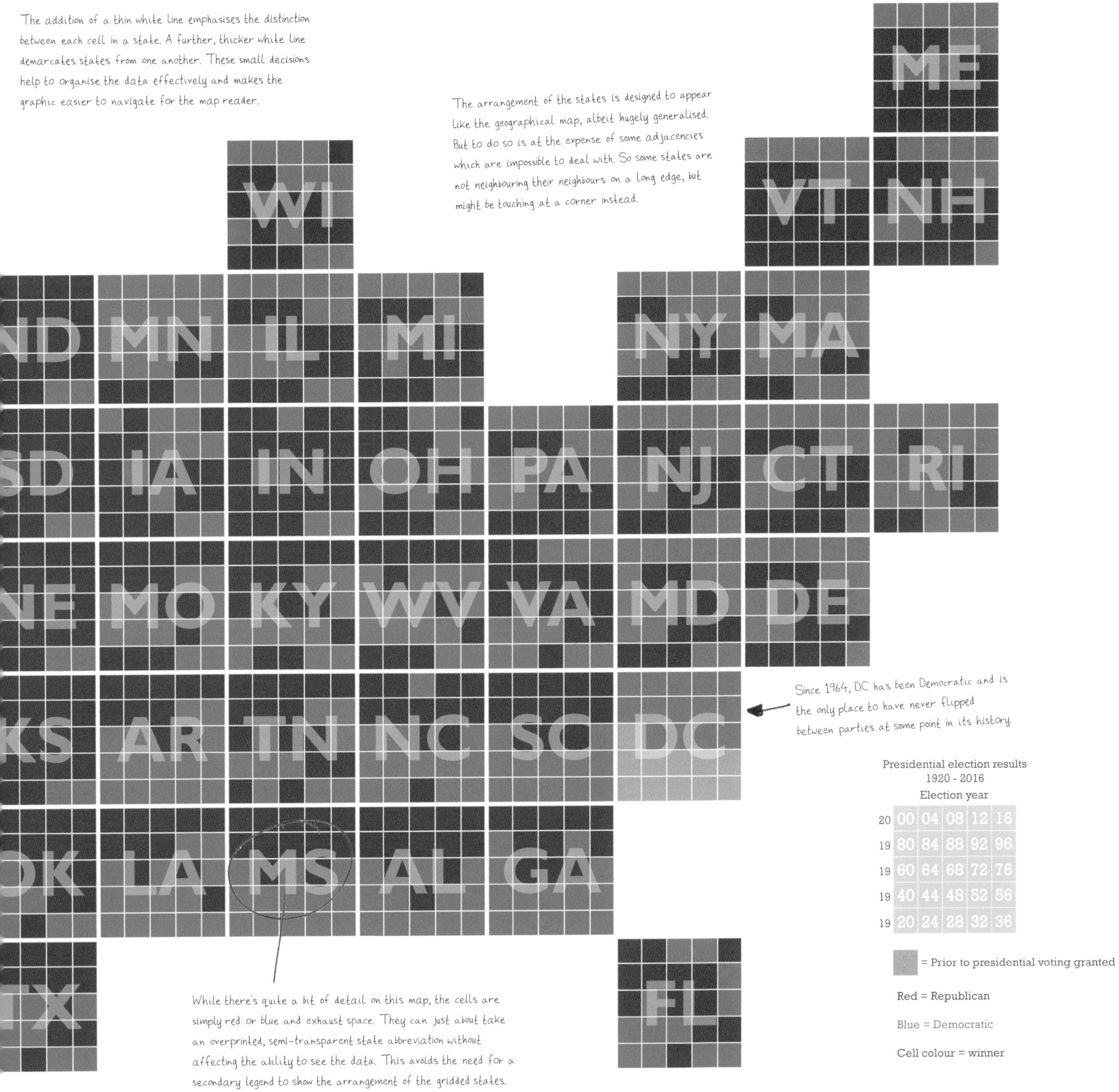

ME

WI

VT NH

ND MN IL MI NY MA

SD IA IN OH PA NJ CT RI

NE MO KY WV VA MD DE

KS AR TN NC SC DC

Since 1964, DC has been Democratic and is the only place to have never flipped between parties at some point in its history.

OK LA MS AL GA

TX

FL

While there's quite a bit of detail on this map, the cells are simply red or blue and exhaust space. They can just about take an overprinted, semi-transparent state abbreviation without affecting the ability to see the data. This avoids the need for a secondary legend to show the arrangement of the gridded states.

Presidential election results
1920 - 2016
Election year

20	00	04	08	12	16
19	80	84	88	92	96
19	60	64	68	72	76
19	40	44	48	52	56
19	20	24	28	32	36

= Prior to presidential voting granted

Red = Republican

Blue = Democratic

Cell colour = winner

Pulse

Small multiples line graphs

Line graphs are wonderful graphical devices for exploring change over time. **Coupling line graphs with the gridded layout of the US allows the state cells to be used as spaces for mini line graphs.** These build into an array of small multiples graphs, which allows the map reader to see change over time in a single state and compare patterns between states across the period mapped.

The last 25 election cycles are shown to the right with a red line for the Republican trend and a blue line for the Democratic trend. The x-axis of the line graph is time, with the most recent year, and data point, toward the right of the graph. The y-axis is scaled to show the share of the winning vote by the two main parties, from 0 percent at the foot of the y-axis to 100 percent at the top.

The lines, and the patterns they make together, can reveal several aspects of the data. The line that is toward the top at any point in the graph shows the winner and the incumbent party. Where the lines cross, it's indicative of a flip in the voting at a state level. The gap between the two lines at any given point indicates the difference in the share of the vote, with a larger gap showing a much wider difference. The steepness of the lines from one vertex to another shows the strength of the swing between election cycles. A steeper line indicates a swing of much larger proportion.

Inevitably, the two lines are reflections of one another because of the nature of the dataset. When one candidate polls 70 percent, the other will poll 30 percent. Although this figuring does not account for some of the independent candidates who often take a few percent of the total share of the vote, the story is really about the way in which the two main parties fare over time.

The graphs almost take on the sense of a pulse, beating across time. **None of the states flatline, and the changes demonstrate that there's no such thing as a completely safe seat**. As the discussion to the right of the Solid South states notes, change in voting patterns is much more than who might be the best candidate each year. Greater issues often not only will have a profound effect on an election in a specific year but also might shift the political system fundamentally for a longer period. These sorts of patterns become evident when using line graphs to show change over time.

This sort of map is useful not only for highlighting contextual issues that help explain political shifts, but also the map invites the inquisitive reader to look into why shifts may have taken place at a specific point in time and in a specific place.

The passage of time reveals much more than simply who won the election. The southern states, historically referred to as the Solid South, were an electoral voting bloc which tended to vote along lines that were beneficial to Democrats from the mid-1800s to the mid-1900s.

By 1960, Democratic President Kennedy was increasingly in support of civil rights. By the time of Martin Luther King's arrest in October 1960 at a peaceful sit-in, Kennedy's intervention sealed his release. Along with many other issues of the day, this act had the effect of causing Democrats to lose ground with White voters in the South. The Democratic Party was no longer a Whites-only party. Republicans won a number of states in 1960.

Following Kennedy's assassination in November 1963, and the passage of the Civil Rights Act in 1964, African Americans were able to register and vote. They tended to vote Democratic while many White conservatives shifted their allegiance to the Republican Party. Despite Lyndon B. Johnson winning a landslide victory in 1964, much of the Solid South switched to Republican due to their anger over Johnson and the Civil Rights Act. In fact, it was a complete inversion of many previous decades of Democratic control.

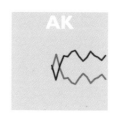

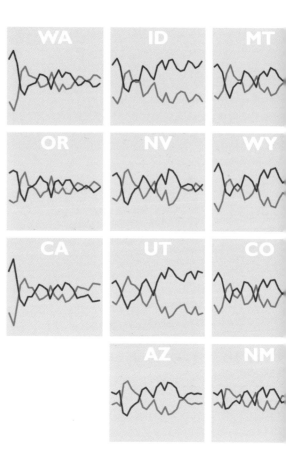

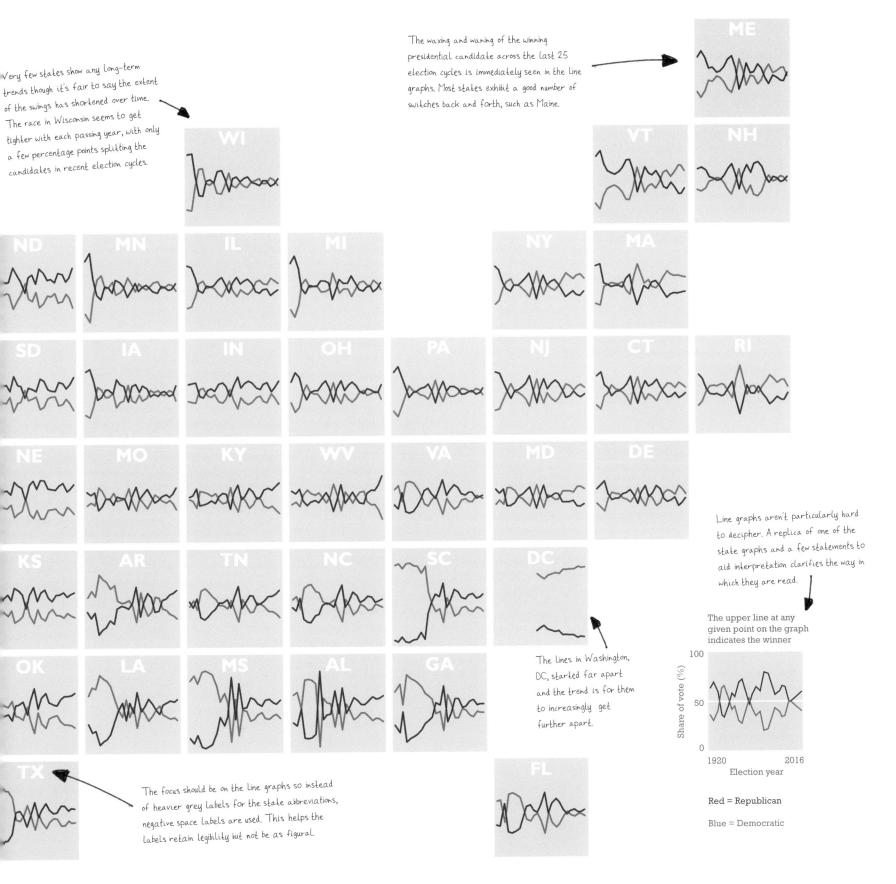

Very few states show any long-term trends though it's fair to say the extent of the swings has shortened over time. The race in Wisconsin seems to get tighter with each passing year, with only a few percentage points splitting the candidates in recent election cycles.

The waxing and waning of the winning presidential candidate across the last 25 election cycles is immediately seen in the line graphs. Most states exhibit a good number of switches back and forth, such as Maine.

Line graphs aren't particularly hard to decipher. A replica of one of the state graphs and a few statements to aid interpretation clarifies the way in which they are read.

The upper line at any given point on the graph indicates the winner

Share of vote (%)

100

50

0

1920   2016
Election year

Red = Republican

Blue = Democratic

The lines in Washington, DC, started far apart and the trend is for them to increasingly get further apart.

The focus should be on the line graphs so instead of heavier grey labels for the state abbreviations, negative space labels are used. This helps the labels retain legibility but not be as figural.

WI · ME · VT · NH · ND · MN · IL · MI · NY · MA · SD · IA · IN · OH · PA · NJ · CT · RI · NE · MO · KY · WV · VA · MD · DE · KS · AR · TN · NC · SC · DC · OK · LA · MS · AL · GA · TX · FL

No.1 at the end of the bar

Small multiples stacked bar charts

Bar charts present categorical data as an extended, thickened line, with the length proportional to some empirical value. Of all the types of graphs available to visualise data, they are one of the simplest to understand. In comparison to, say, a pie chart, they also tend to lead to a better estimate of the values represented in the graph. This is because the human perceptual and cognitive system can better estimate lengths of objects compared against a scale, and with each other, than estimating quantifiable or relative differences in area between objects.

Bars can be plotted vertically or horizontally. The more common orientation is for bars to be oriented vertically with the categories along the x-axis and the empirical values on the y-axis. **In the example to the right, bars are positioned horizontally to reflect the leanings of the two main parties, with Democrats to the left and Republicans to the right.** In this sense, the properties of the bar graph are modified to support the theme being mapped.

These bar graphs are in a stacked form. Because the percentage of the share of the vote always sums to 100, the Democratic, Republican, and other parties' share of the vote can be shown in a single, stacked bar, one for each election cycle. Stacking variables in the same sequence ensures continuity between individual bars and makes the task of reading and interpretation easier. Stacked bar charts provide an excellent way to see how the share of the vote was distributed each year. The consistency of the bars always being the same total length allows people to make a good estimate of the vote value for any given year.

Categories in this example are election years, with 25 bars positioned in a stacked sequence, from 1920 at the bottom to 2016 at the top. **By positioning each stacked bar in sequence, they collectively display time-series information at the same time.** Additional patterns formed by the blocks of colour across time are revealed by this technique.

The same idea could be used to show a single election in which either a stacked or grouped bar graph could be used, per state, as illustrated below.

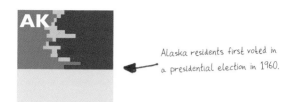

Alaska residents first voted in a presidential election in 1960.

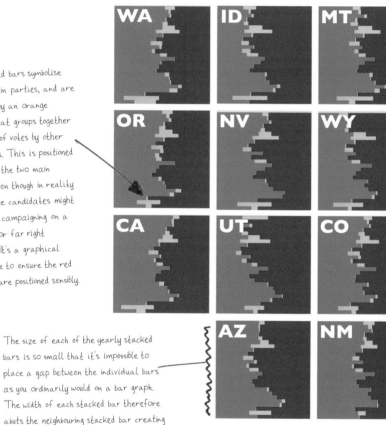

Blue and red bars symbolise the two main parties, and are supported by an orange category that groups together the share of votes by other candidates. This is positioned in between the two main parties even though in reality some of the candidates might have been campaigning on a far left or far right platform. It's a graphical compromise to ensure the red and blue are positioned sensibly.

The size of each of the yearly stacked bars is so small that it's impossible to place a gap between the individual bars as you ordinarily would on a bar graph. The width of each stacked bar therefore abuts the neighbouring stacked bar creating uniform blocks of colour across the years. Another graphical compromise.

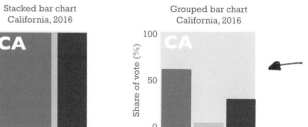

Stacked bar chart
California, 2016

Grouped bar chart
California, 2016

Which of these alternatives makes it easier for the reader to form an accurate visual impression of the result?

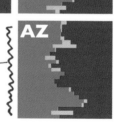

Hawaii residents first voted in a presidential election in 1960.

Graphs provide an excellent way to explore trends over time. When placed in some sort of geographical context they also allow the identification of spatio-temporal dimensions in the data. For instance, each of the northeastern states shows a general reduction in Republican support over time indicated by the general slope of the ratio between blue and red. This pattern broadly repeats across the wider region.

ME

VT

NH

WI

D

MN

IL

MI

NY

MA

O

IA

IN

OH

PA

NJ

CT

RI

E

MO

KY

WV

VA

MD

DE

Enough information to decipher the individual stacked bars, as well as their overall arrangement.

S

AR

TN

NC

SC

DC

Voting in presidential elections in DC began in 1964.

K

LA

MS

AL

GA

Each line in the graph represents the share of vote for an election from 1920 (bottom) to 2016 (top).

2016
2000
1980
1960
1940
1920

Year

Share of vote (%)
10% intervals

X

FL

Labels are positioned to cause the least amount of interference with the underlying graphs. Positioning them in the middle would compromise the ability to see crucial zones of change between parties.

Red = Republican

Blue = Democratic

Orange = Other

Ebb and flow

Small multiples Sankey diagrams

Lines are excellent for showing change over time. A conventional line graph shows a single variable per line over time, and it is up to the reader to interpret any relationship among the lines. **But rather than display variables and their change in value over time, Sankey diagrams show how data might flow among categories between points in time.**

Flow maps have a long and storied history in cartography, with Charles Minard's classic map below of Napoleon's Russian Campaign of 1812, published in 1869, perhaps one of the most famous, and often cited.

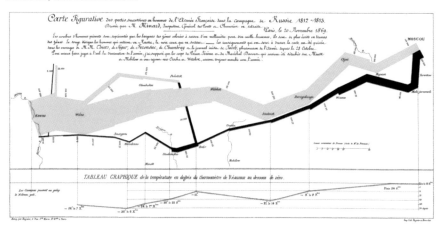

The map shows how a total at one end of the line gradually becomes narrower to reflect the decimation of troops over time. **As data values become less, so the line is proportionally less, and a distributive element shows how different categories of data split off from the main group and some later rejoin.** At these points, the lines either split or merge. Minard's map is effectively a Sankey chart.

Sankey charts first were proposed by Matthew Sankey in 1898 in a diagram showing the efficiency of the steam engine. The flow of steam was plotted through the engine, with different thicknesses of the line and the same distributive characteristic.

For the US election data, the flow of votes between election cycles among the two main parties and a third category that combines other candidates can be shown using the technique. The 2012 result is plotted to the left of each chart and the 2016 result to the right. The charts distribute votes lost by one party proportionally among those that gain votes. This is likely not the case, and it's also impossible to know how many more voters changed their votes from one cycle to the next. **The charts show the difference between two periods and how the voting pattern changed by state in aggregate form.** Assumptions about voting at a finer resolution cannot be made.

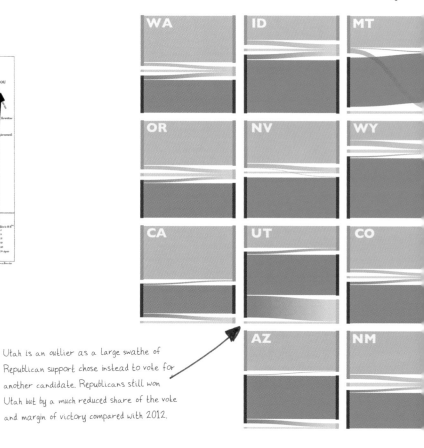

Montana, North Dakota, and South Dakota illustrate how the flow of vote was lost from Democrats, and redistributed partly among Republicans but also to other candidates. This pattern of flow is repeated across the West and Midwest.

Utah is an outlier as a large swathe of Republican support chose instead to vote for another candidate. Republicans still won Utah but by a much reduced share of the vote and margin of victory compared with 2012.

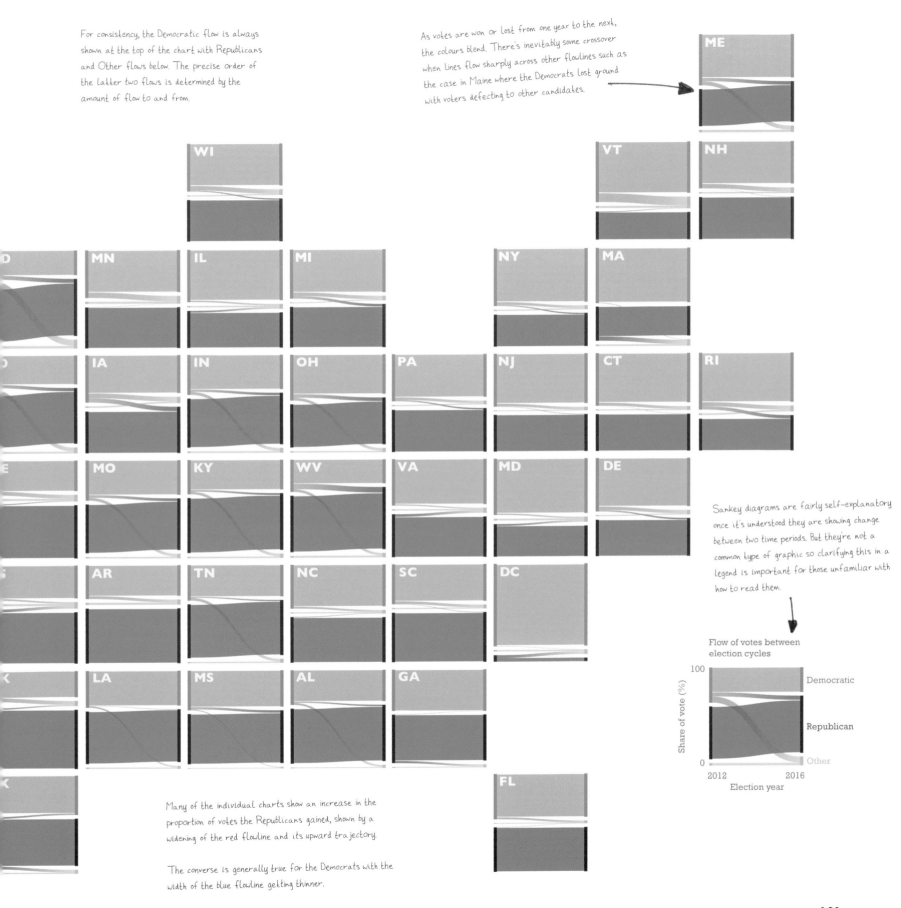

For consistency, the Democratic flow is always shown at the top of the chart with Republicans and Other flows below. The precise order of the latter two flows is determined by the amount of flow to and from.

As votes are won or lost from one year to the next, the colours blend. There's inevitably some crossover when lines flow sharply across other flowlines such as the case in Maine where the Democrats lost ground with voters defecting to other candidates.

ME

WI

VT **NH**

D **MN** **IL** **MI** **NY** **MA**

O **IA** **IN** **OH** **PA** **NJ** **CT** **RI**

E **MO** **KY** **WV** **VA** **MD** **DE**

Sankey diagrams are fairly self-explanatory once it's understood they are showing change between two time periods. But they're not a common type of graphic so clarifying this in a legend is important for those unfamiliar with how to read them.

S **AR** **TN** **NC** **SC** **DC**

Flow of votes between election cycles

K **LA** **MS** **AL** **GA**

100

Democratic

Republican

0

Other

K **FL**

Many of the individual charts show an increase in the proportion of votes the Republicans gained, shown by a widening of the red flowline and its upward trajectory.

The converse is generally true for the Democrats with the width of the blue flowline getting thinner.

2012   2016
Election year

Share of vote (%)

Ticking away the moments

Small multiples radar charts

AK

Radar charts are a form of line graph but their structure is circular rather than rectangular. Both line graphs and radar charts have two axes. A typical line graph has the two axes perpendicular to one another. On a radar chart, one axis extends around the chart, normally read in a clockwise direction. The other axis extends outward. **Radar charts are used to display multivariate data in which the relationship between variables can be seen along a straight line from the centre of the chart to the outer edge.**

In the example to the right, time is plotted around the chart. As such, this chart type is referred to as a *data clock* because the chart mimics a clock face and uses that metaphor to help the reader interpret the information using a familiar mechanism—reading the time around a clock face. In addition to data clocks, radar charts are also referred to as *spider charts* or *polar charts*. When not showing temporal data, they often illustrate categorical variables along each radial axis.

The position of data points on each radial axis (or spoke) is proportional to the magnitude along the axis radiating outward—in this case, the strength of the share of the vote for the two main parties and a sum of all the other candidates. The outer data point is the maximum. The line drawn connecting a data point on one spoke to another on the neighbouring spoke illustrates an interconnectedness that is sometimes unrelated—a popular criticism of the radar chart. **In this example, the lines show how the share of the vote for each party changed from one election cycle to the next.** They are not necessarily related other than the steepness of the line makes it possible to see the strength of gains or losses and swings between elections.

Radar charts are good at showing outliers in a dataset, in which a line takes a sharp deviation inward or outward from a data point on a neighbouring spoke. The concentric circles showing the measured scale help readers understand the magnitude of variation from one data point to the next.

Are radar charts preferable to a more traditional line graph? It depends. They certainly have visual appeal, but they have limitations, as noted above. The cyclic appearance inevitably positions the last spoke in the sequence next to the first. Proximity of these data points may not be appropriate in the context of the dataset, and seeing the 1920 data point adjacent to the 2016 result is useful only in a general sense. Additionally, the lengths of interconnecting lines are much greater toward the outer edge, often accentuating the appearance and implied importance of those lines compared with some at the centre.

On many of the maps in this chapter squares are used as the framework for each graph. But here, the circular grids of each radar chart provide a regular framework.

The key here is making a decision to omit as much as possible without compromising the map. The squares would simply have been unnecessary.

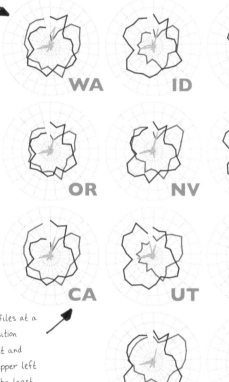

WA   **ID**

OR   **NV**

CA   **UT**

AZ

Because of the very different voting profiles at a state level there's no easy place to position labels. South Carolina makes upper right and upper centre very difficult. DC makes upper left difficult. On balance, lower right has the least potential conflicts with other graphical detail.

It's pretty easy to see how the Republican, and Democratic parties fared in most of the graphs since their share of the vote is normally pretty high. But what of the 'others' shown in the orange? Because the share of the vote is normally fairly low, it's harder to see how things changed from one election cycle to another simply by being plotted in the space-constrained centre of the graph.

HI

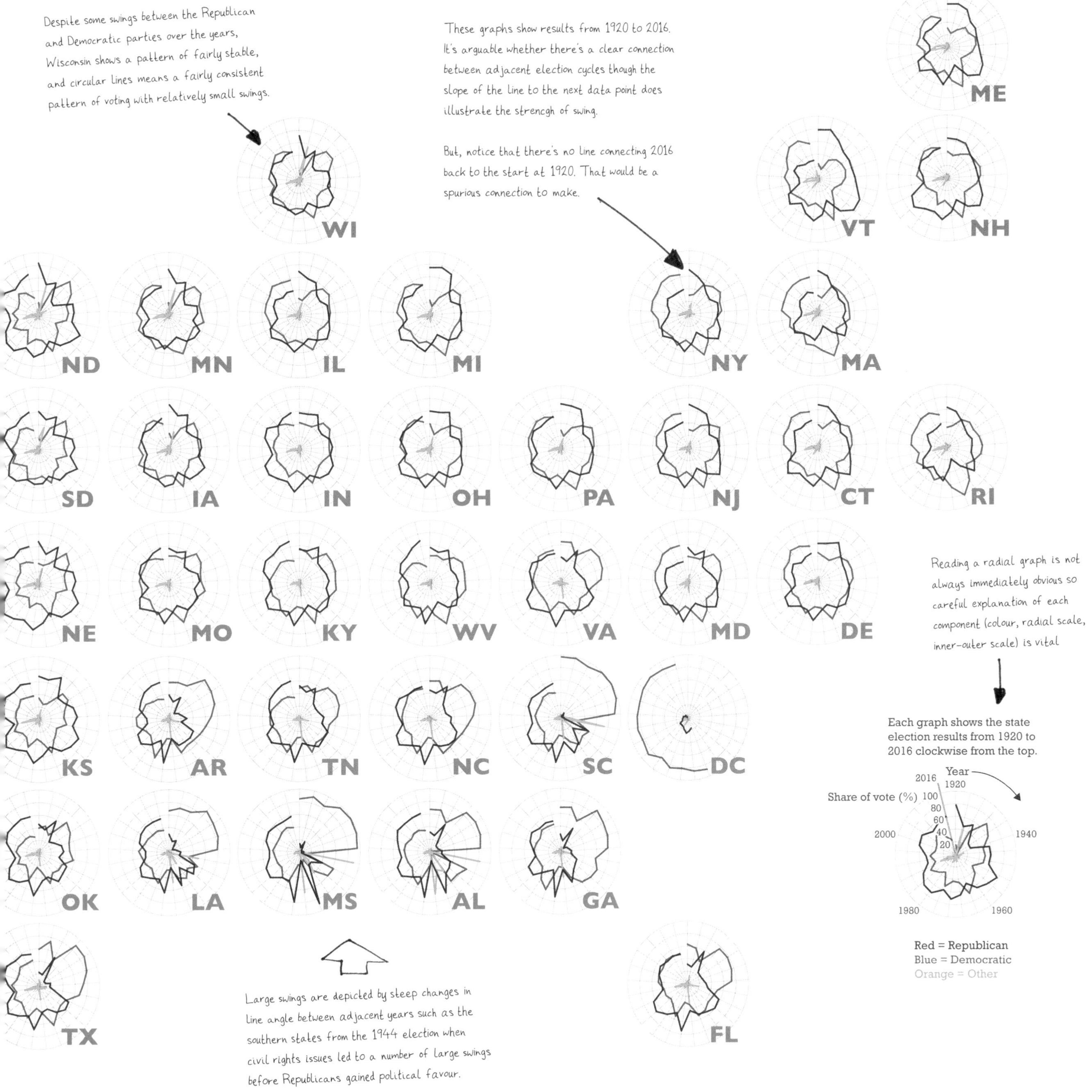

Despite some swings between the Republican and Democratic parties over the years, Wisconsin shows a pattern of fairly stable, and circular lines means a fairly consistent pattern of voting with relatively small swings.

These graphs show results from 1920 to 2016. It's arguable whether there's a clear connection between adjacent election cycles though the slope of the line to the next data point does illustrate the strength of swing.

But, notice that there's no line connecting 2016 back to the start at 1920. That would be a spurious connection to make.

ME

WI

VT   NH

ND   MN   IL   MI   NY   MA

SD   IA   IN   OH   PA   NJ   CT   RI

Reading a radial graph is not always immediately obvious so careful explanation of each component (colour, radial scale, inner-outer scale) is vital

NE   MO   KY   WV   VA   MD   DE

KS   AR   TN   NC   SC   DC

Each graph shows the state election results from 1920 to 2016 clockwise from the top.

OK   LA   MS   AL   GA

Large swings are depicted by steep changes in line angle between adjacent years such as the southern states from the 1944 election when civil rights issues led to a number of large swings before Republicans gained political favour.

TX

FL

2016   Year
Share of vote (%) 100   1920
                    80
2000   60   1940
                    40
                    20

1980   1960

Red = Republican
Blue = Democratic
Orange = Other

American pie

Small multiples polar area charts

Inspiration for data visualisation can come from many places. In the mid-1800s, a British nurse, Florence Nightingale, cared for soldiers in the Crimean War. She was also a pioneer in data visualisation to bring attention to the many soldiers who were dying of illness brought about by poor sanitary conditions as opposed to directly from wounds inflicted on the battlefield. She collected data on deaths from various causes and is credited as one of the first to use a polar area chart (colloquially called a *coxcomb chart*) to present the data, shown below.

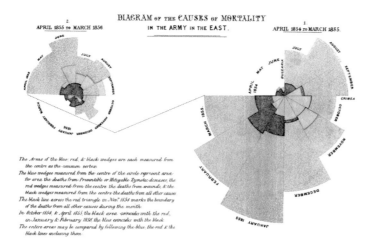

The polar area chart is a variation of a pie chart but has some important differences. A pie chart is a circular chart in which data values are sliced into segments. The length of the arc at the outer edge (and therefore internal angle) of each slice is proportional to the quantity represented by the slice. **Instead, a polar area chart keeps constant the arc length of slices and varies the area of each segment.** The radius of each segment differs depending on the data value, with larger segments representing larger data values. They commonly are used to show temporal data, in a similar way to a radar chart, with time plotted clockwise from the top of the chart. The arc length is purely a function of the number of segments shown.

Polar area charts also can show categorical information as well as proportional differences. **In the example to the right, the results of each election cycle are shown as three segments that overlay one another such that the smallest (third place) sits on top so as not to be occluded.** This works well, although in some instances, it can seem as if the segment representing the runner-up is more prominent than the winner because it sits atop the winning segment and occludes some of the colour associated with the winner. This might compromise a map reader's ability to make sensible assessments of proportions of the category within each segment considering that some occlusion is inevitable.

AK

There is an alternative approach to the construction of coxcombs. Instead of simply overprinting smaller segments on top of larger ones, and assuming people can make the perceptual adjustment required to estimate proportions of colour, each segment could be stacked. This would avoid all occlusion and show all data at once.

The reason this approach hasn't been used here is twofold. Firstly, given the map uses percentage share of the vote each coxcomb would be a circle of the same size and simply show different proportions of categorical colour internal to each yearly segment. That simply makes it a circular version of the previous stacked bar chartmap.

Secondly, if the scale was totals, different coxcombs would be very differently sized, and with such a large variation in data values across the map, and between states, there would be too many unwieldy small and very large coxcombs relative to one another.

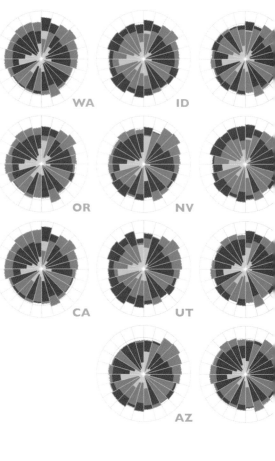

WA   ID

OR   NV

CA   UT

AZ

HI

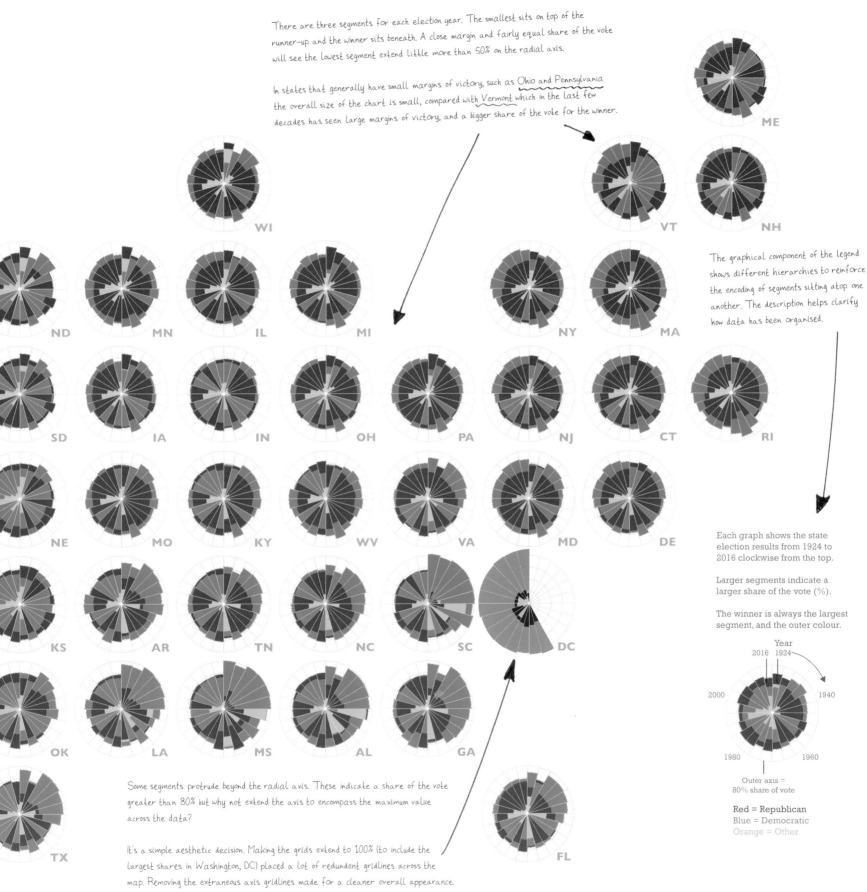

There are three segments for each election year. The smallest sits on top of the runner-up and the winner sits beneath. A close margin and fairly equal share of the vote will see the lowest segment extend little more than 50% on the radial axis.

In states that generally have small margins of victory, such as Ohio and Pennsylvania the overall size of the chart is small, compared with Vermont which in the last few decades has seen large margins of victory, and a bigger share of the vote for the winner.

The graphical component of the legend shows different hierarchies to reinforce the encoding of segments sitting atop one another. The description helps clarify how data has been organised.

Each graph shows the state election results from 1924 to 2016 clockwise from the top.

Larger segments indicate a larger share of the vote (%).

The winner is always the largest segment, and the outer colour.

Year
2016   1924
2000   1940
1980   1960

Outer axis = 80% share of vote

Red = Republican
Blue = Democratic
Orange = Other

Some segments protrude beyond the radial axis. These indicate a share of the vote greater than 80% but why not extend the axis to encompass the maximum value across the data?

It's a simple aesthetic decision. Making the grids extend to 100% (to include the largest shares in Washington, DC) placed a lot of redundant gridlines across the map. Removing the extraneous axis gridlines made for a cleaner overall appearance.

Just for the record

Small multiples tree-ring charts

Tree-ring charts are so named because of their similarity to a sawn tree trunk, with the rings indicating annual growth cycles. The first year of a tree's growth is shown by the core, central circle, and subsequent rings show each consecutive year of growth. They have many other names too, such as *concentric ring* or *circle char*ts. They also look like classic vinyl records.

Tree-ring charts share certain characteristics with the pie chart, but instead of segments radiating from the centre showing different categories of data, it's the entire circular ring that is used. **Each concentric ring displays one piece of data from inner, earlier periods to outer, later periods.** In this case, the tree-ring metaphor is literal by showing an early election cycle (1920) at the core and the 2016 election cycle at the outer edge. Each is symbolised by category to show the winning party.

To show how patterns change over time, tree-ring charts provide an easy visual mechanism to communicate change. They also allow the reader to see patterns across time for each individual chart. They are not without difficulties for interpretation, though. **Inevitably, outer rings are longer, even though they have the same line width. They therefore assume a more significant visual importance compared with inner rings.** It also can be difficult to compare a specific ring from one chart to another because of the number of rings and the way in which the eye struggles to settle on a specific ring. This difficulty is compounded by charts that have little variation from one ring to another.

Tree-ring charts easily can become overcomplicated, which tends to reduce their ability to communicate. Consider the version shown below of a tree-ring chart for California that splits each ring into three categories for Republicans, Democrats, and other candidates. Although there's now more information encoded for each election cycle, it's arguably an order of magnitude harder to disentangle and therefore interpret. In this case, it's possible that a square chart would work better, with each line indicating an election cycle with a uniform width. **Keeping these sorts of charts simple is key to their value, and assessing when one might be preferable to another for a specific dataset is inherent in the design process.**

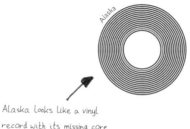

Alaska looks like a vinyl record with its missing core from pre-voting years.

Alaska, Hawaii and Washington, DC, all have holes in lieu of rings up until they were granted voting rights in the mid 1960s

Circle

Years as concentric rings

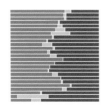

Square

Years as uniform bars

Wisconsin's results from 1920 to 1984 are characterised by flipping. From the 1988 George H. W. Bush (Republican) / Michael Dukakis (Democratic) election the state has been solid blue.

The exception is the 2016 outlier where Donald J. Trump (Republican) defeated Hillary Clinton (Democratic) by a narrow margin of 0.77% leaving it with an outer ring of red.

There's 25 rings in each of the graphs representing the electoral cycles from 1920 to 2016. Some states have been predominantly red over time, some blue and some tend to flip back and forth regularly or in blocks.

Because the rings are thin lines, at a distance it's possible that simultaneous contrast leads to seeing different hues of red and blue. This is by virtue of the contrasting colours they immediately abut.

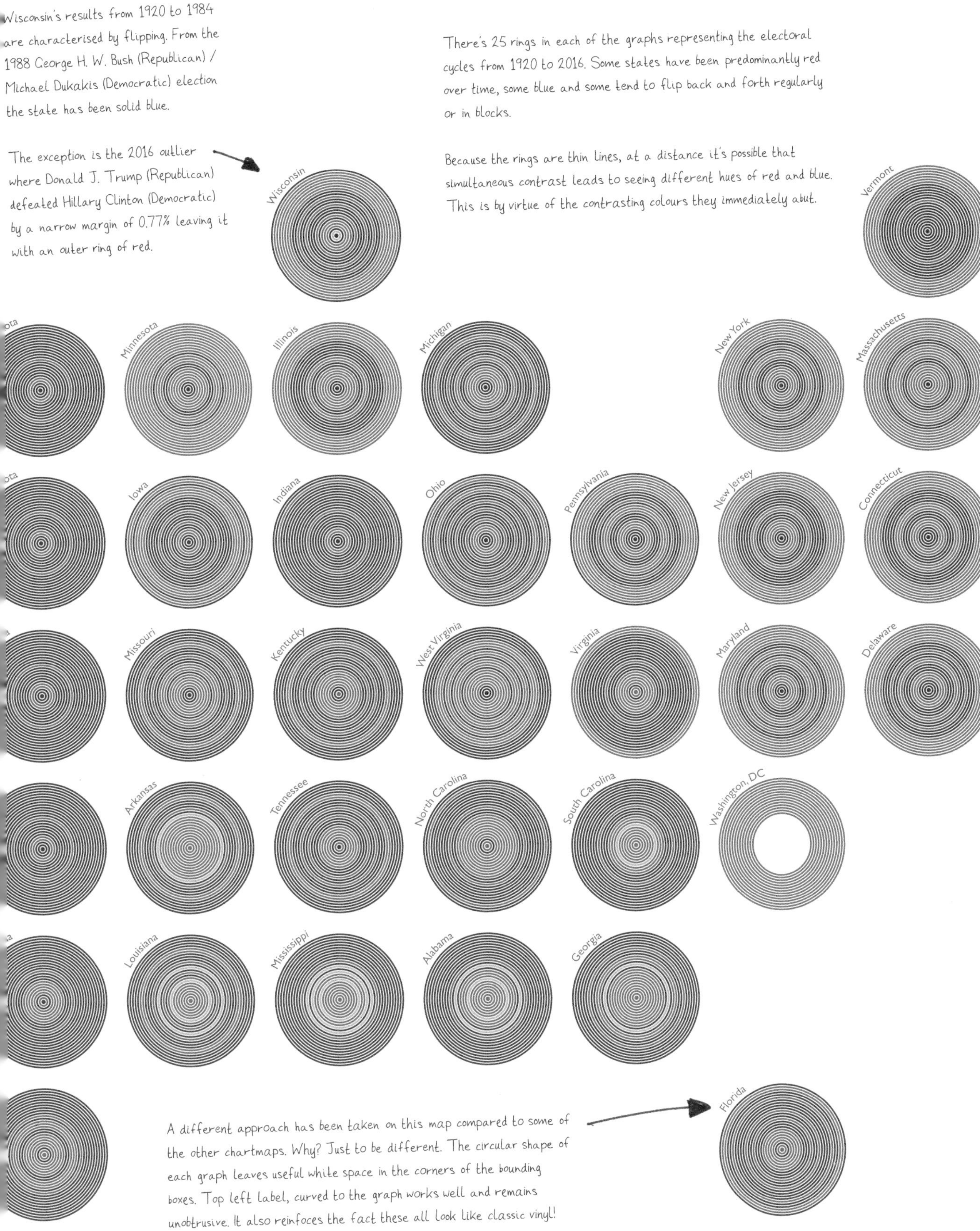

Maine

Wisconsin

Vermont

New Hampshire

ota

Minnesota

Illinois

Michigan

New York

Massachusetts

Iowa

Indiana

Ohio

Pennsylvania

New Jersey

Connecticut

Rhode Island

Missouri

Kentucky

West Virginia

Virginia

Maryland

Delaware

All that's needed is a way to understand how the inner rings differ from the outer rings and what the colours mean.

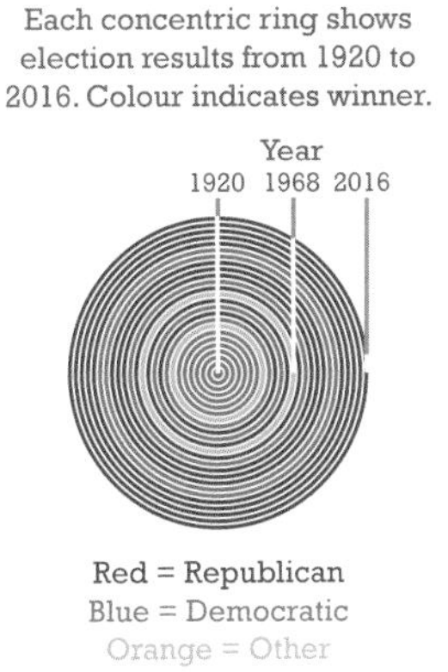

Arkansas

Tennessee

North Carolina

South Carolina

Washington, DC

Each concentric ring shows election results from 1920 to 2016. Colour indicates winner.

Year
1920   1968   2016

a

Louisiana

Mississippi

Alabama

Georgia

A different approach has been taken on this map compared to some of the other chartmaps. Why? Just to be different. The circular shape of each graph leaves useful white space in the corners of the bounding boxes. Top left label, curved to the graph works well and remains unobtrusive. It also reinfoces the fact these all look like classic vinyl!

Florida

Red = Republican
Blue = Democratic
Orange = Other

Common people

Trump and Clinton as Chernoff faces

A lot of population data has multiple facets. Maps often focus on a specific aspect, or perhaps show the relationships between two aspects. The more that is encoded into the map's symbols, the more the risk of symbol and information overload. **One technique literally stands head and shoulders above the rest in encoding multivariate map symbols with meaning—the Chernoff face technique.**

Chernoff faces are multivariate charts that encode multiple pieces of information into a symbol on the basis of modifying aspects of a human face. They were developed by Herman Chernoff in the early 1970s. Face shape, eyes, nose, mouth, ears, hair, and so on all can be varied so they show presence or some sort of change on an empirical scale, perhaps with a mouth slowly turning into a smile. **Modifications to individual facial features can give them a different position, size, shape, or orientation.** The overall effect is a face specific to an area, which combines the characteristics of that area into a unique symbol. Chernoff suggested up to 18 separate variables could be encoded into a single Chernoff face multivariate symbol. That clearly is outside the range of what can be encoded into a geometric shape.

Chernoff faces give a human representation as a counterpoint to the more common geometric approach to cartographic symbology. **Chernoff suggested that the symbology worked because humans can process the complexities of emotional response from a human face relatively easily.** Small changes in the shape of an eye or a raised eyebrow, for instance, might be more easily recognised than changes in the size of a circle or a segment of a pie chart.

If they are so effective at encoding information and people are so good at interpreting human faces, why is this map type not more prevalent? There are three main reasons. Firstly, they are difficult to construct, and to construct well. They often require considerable manual effort to achieve a good outcome. Secondly, there's a fine line between making a Chernoff face that seems objective and a stereotypical face that easily might cause offense. There are a host of potential missteps that a Chernoff face might succumb to, often unintentionally. Finally, they require complex legends that need constant reference by the map reader to understand the subtleties of the different facial features, much less their overall combination.

To the right are Chernoff face likenesses of Donald J. Trump and Hillary Clinton encoded with several election data variables. The combination of symbol variations brings different personalities to the faces across the US as a multivariate chartmap.

Chernoff faces are often abstract with little resemblance to actual people. Yet politics is deeply associated with the personalities of the main contenders so in this case, using the likeness of Trump and Clinton provides an immediate link to the subject matter.

The use of red and blue on the collar and earrings and neck tie helps to provide a subtle link to the party colours.

Without a detailed legend the subtleties in the way each facial element has been altered would be lost on the reader.

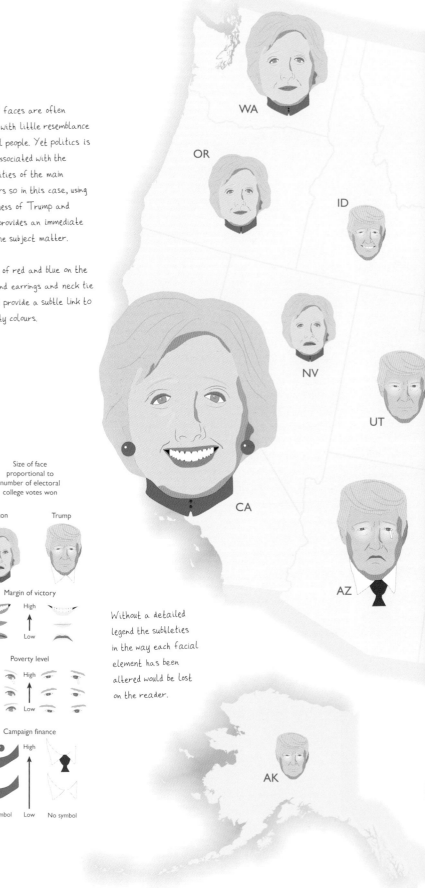

Size of face proportional to number of electoral college votes won

Clinton   Trump

Margin of victory

High

Low

Poverty level

High

Low

Campaign finance

High

No symbol   Low   No symbol

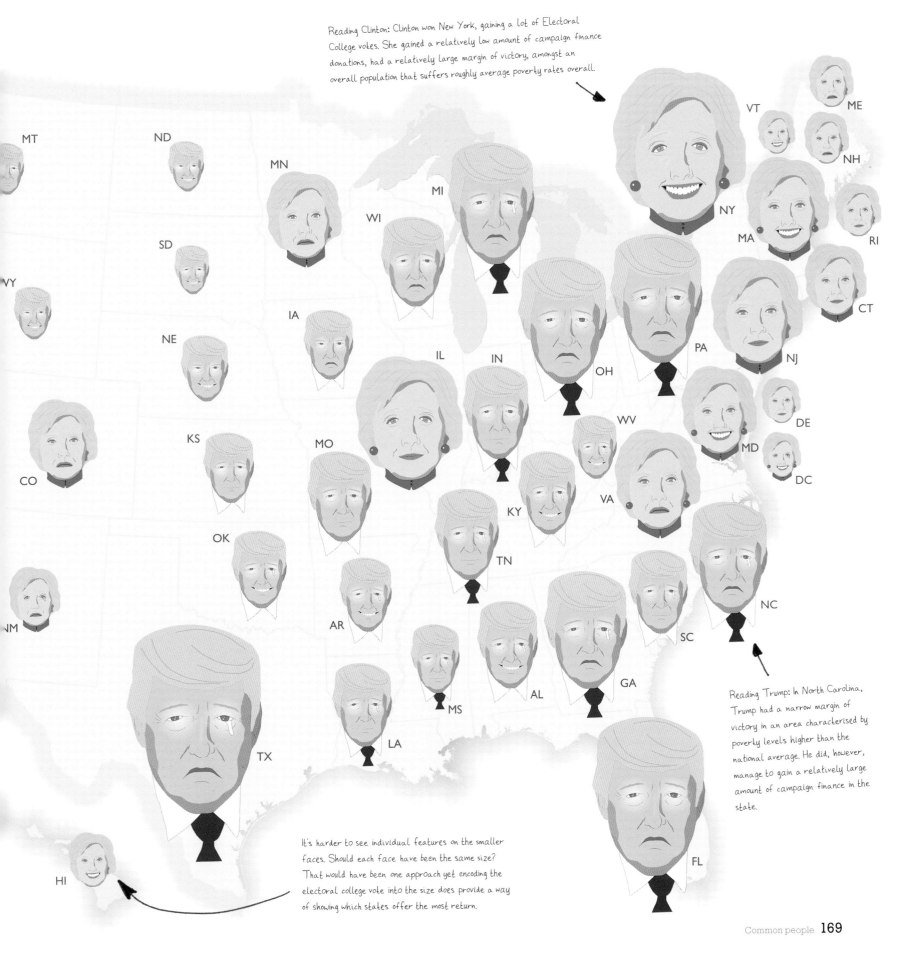

Reading Clinton: Clinton won New York, gaining a lot of Electoral College votes. She gained a relatively low amount of campaign finance donations, had a relatively large margin of victory, amongst an overall population that suffers roughly average poverty rates overall.

MT

ND

MN

MI

WI

VT

ME

NH

SD

NY

MA

RI

WY

IA

NE

IL

IN

OH

PA

NJ

CT

CO

KS

MO

WV

VA

DE

MD

DC

KY

OK

TN

NC

AR

SC

Reading Trump: In North Carolina, Trump had a narrow margin of victory in an area characterised by poverty levels higher than the national average. He did, however, manage to gain a relatively large amount of campaign finance in the state.

NM

TX

AL

GA

MS

LA

It's harder to see individual features on the smaller faces. Should each face have been the same size? That would have been one approach yet encoding the electoral college vote into the size does provide a way of showing which states offer the most return.

HI

FL

Common people **169**

Faces of America

Chernoff face Dorling cartogram

The map on the previous page showed that, at a state level, it was possible to use a representation of both candidates because the map needs to contain only 51 individual Chernoff faces. But with 3,141 areal units at a county-equivalent level, and considering the scale limitations of the page size of this book, to use detailed likenesses wouldn't work. **This chartmap reverts to Herman Chernoff's original idea of an abstract face in which individual features are varied but there's no attempt to make the faces look like a particular person.**

Additionally, this chartmap goes one stage further to escape the clutches of geography by using a cartogram. On the preceding page, the geographical basemap provided a foundation but it didn't contain any data. It was merely a scaffold for placing the Chernoff faces on top. **Here, to the right, a Dorling cartogram of the counties provides the scaffold.** It differs from the Dorling cartogram of counties shown in chapter 5 because each county is scaled equivalently to one another. They therefore have equal visual importance in terms of the amount of space they have on the page, though precise symbol combinations make each symbol slightly different.

Although Chernoff suggested that up to 18 separate variables can be mapped in a Chernoff face, with 3,141 separate faces on this map, the space for each is still relatively small. Instead, the variables are limited to five: head shape, head fill, eye size and shape, nose size and shape, and mouth curvature. In essence, it's the basic idea of a Chernoff face but the features are amalgamated (e.g. both eyes are part of the same encoding rather than separate), and certain features are discarded altogether (e.g. ears, hair, and eyebrows).

Each of the four physical features has three separate states representing high, medium, and low on the basis of the different indicators. **Combined with being either a red or a blue face, there are potentially 625 symbol combinations, and therefore that many different possible faces.** Yet instead of seeing randomly positioned different faces, there are clearly clusters. This is spatial autocorrelation, in which populations and voting characteristics of immediate neighbours tend to be more similar. This illustrates Waldo Tobler's First Law of Geography that 'everything is related to everything else, but near things are more related than distant things'. Different symbols occur occasionally among neighbours, which tends to show large urban and metropolitan area regions abutting rural communities.

The positives of this type of map: it looks interesting and invites scrutiny. The drawbacks: it's almost impossible to disentangle the data from the symbol to work out exactly what's going on.

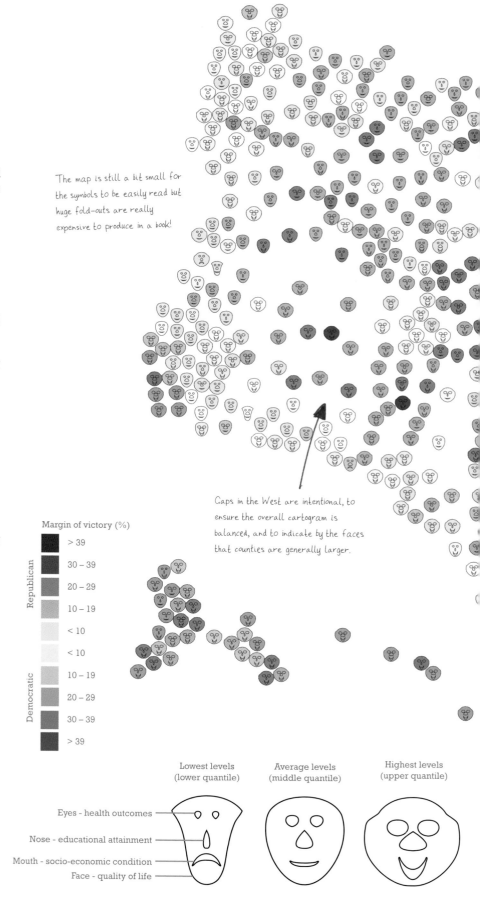

The map is still a bit small for the symbols to be easily read but huge fold-outs are really expensive to produce in a book!

Gaps in the West are intentional, to ensure the overall cartogram is balanced, and to indicate by the faces that counties are generally larger.

Margin of victory (%)

Republican
> 39
30 – 39
20 – 29
10 – 19
< 10

Democratic
< 10
10 – 19
20 – 29
30 – 39
> 39

Lowest levels (lower quantile)   Average levels (middle quantile)   Highest levels (upper quantile)

Eyes - health outcomes
Nose - educational attainment
Mouth - socio-economic condition
Face - quality of life

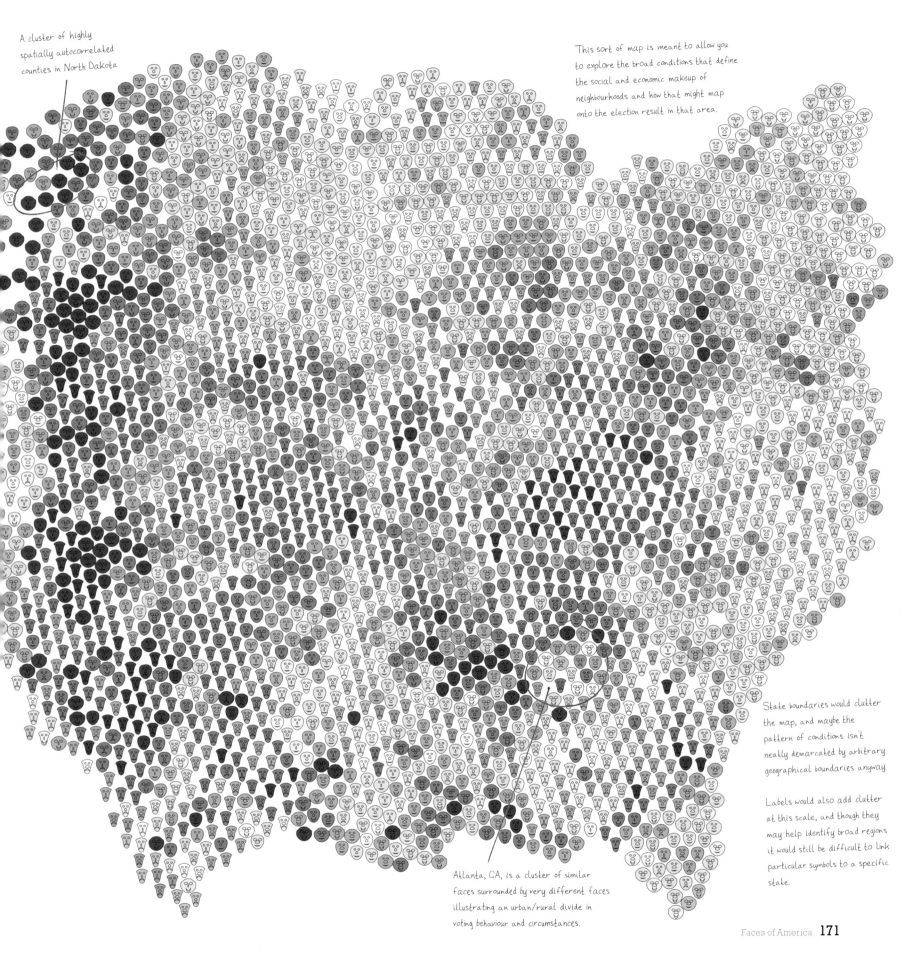

A cluster of highly spatially autocorrelated counties in North Dakota

This sort of map is meant to allow you to explore the broad conditions that define the social and economic makeup of neighbourhoods and how that might map onto the election result in that area.

State boundaries would clutter the map, and maybe the pattern of conditions isn't neatly demarcated by arbitrary geographical boundaries anyway.

Labels would also add clutter at this scale, and though they may help identify broad regions it would still be difficult to link particular symbols to a specific state.

Atlanta, GA, is a cluster of similar faces surrounded by very different faces illustrating an urban/rural divide in voting behaviour and circumstances.

Just the basic facts

Minimalist sparklines

It's been mentioned before, but one of the basic tenets of good cartography is omission. That is the removal of as much as possible from the map, not just for the sake of removal but to improve clarity on the map and the potential for readers to see the data more clearly. It's not simplification, but instead, it's showing complexity through good graphic design. The same is true for graphs, charts, and plots, which often can suffer from having too much detail, such as too many axis labels or additional lines across from each numeric point on the axis. **Sparklines are a type of graph or chart that employs the bare minimum of graphical marks while bringing clarity through design.**

A sparkline is a miniature graph or chart, although it omits most of what you ordinarily would see on a larger version. There are no axis labels or coordinates. In fact, the only information you see are data points that often are used to show change, usually over time. They commonly are used to report changes in stock market prices as a single line, perhaps with a start or end marker, as well as a marker to show the highest or lowest points during a day's trading. For the data shown to the right, they also make an excellent way of showing the result of the election over time. Each sparkline shows 25 years of election victories. The shape of the line shows the trend, and a coloured marker indicates which party won each year.

Sparklines are an extremely condensed form of visual communication. As well as being used on a chartmap, they commonly are used inline as a component of text— **For instance, to explain how California has voted from 1920** ·····~········~········· **to 2016 by embedding the sparkline in a passage of text.** It's a high-resolution way of explaining a pattern graphically, and in a context that doesn't necessarily require additional space for a full graph or chart. Such mechanisms can be used as part of general communication wherever a word or numerical data might be. They can be embedded easily in sentences, tables, spreadsheets or, as to the right, a map.

Sparkline compactness makes it a good option for positioning on a map. Combined with a geographical base that also has a minimalist design, they provide a way of showing a lot of detail and the complexity of the relationship over time but without the overhead of positioning graphs with full axes and labels. The only concession here is a small y-axis indicator to anchor the trailing lines and give them numerical context.

Alternatives to the linear sparkline are also possible, such as a column or bar chart. **For instance, a bar chart could be used to show California from 1920** ▪▫▪▫▪▫▪▪▫▪▪▫▪▫▪▫▫▫▫▪▪▪▪ **to 2016 using blue and red columns to illustrate the share of the vote and the winner for each year.** These also could be used on a chartmap.

The vertical scale on each sparkline acts as a way of seeing the share of the vote. Without it it's difficult to assess the relative position of dots vertically and between states.

It also acts as a leader line for the state label in most instances.

Congestion in the Northeast means modifying the location of some of the sparklines to avoid overlap. It's a bit ugly but with long sparklines and small narrow states there's not much of an alternative.

Winning share of vote (%)

100
50
0

No data

Republican win
Democratic win
Other win

1920 ⟶ 2016

Dialing it in

Dials as a Dorling cartogram

Dials are used for all sorts of visual display of information, from the odometer in a car to the gauge on a pressure cooker to the simple clockface of a wristwatch. This might seem to be an argument for analog, mechanically driven displays, but even the most digital of modern displays often revert to showing a digital version of an analog dial. They are so prevalent that it makes sense to draw on their utility to present data that has moved from one position in time to another, such as the shift in voting from 2012 to 2016. **Dials are effectively a form of multivariate chart.**

To the right, multivariate dials are used to show the swing in the vote at a state level. The map uses a Dorling cartogram as its base because the circular shape makes graphical sense. The circles are scaled by Electoral College vote for context, so the dials make a quantifiable statement against the backdrop of what the candidate wins by, perhaps a relatively small swing. The circles could have been presented across a geographical map as a backdrop, but the decision here was to keep the overall appearance detached from geography, as if looking at a series of dials on a dashboard.

In fact, the dashboard metaphor is useful here because web-delivered data visualisation content often makes use of digital dashboards that comprise separate components. Each dial relays different but related data, often from live feeds so the dashboard updates in real time.

Dials can cause anxiety, though, because the needle moves. Small moves can lead to big changes. On election night in 2016, the *New York Times* used a dial to predict the result based on forecast models. As results trickled in, the needle, which had started the night predicting an 80 percent chance of a Clinton win, began to move right. Social media had a meltdown as people followed the needle's swing that illustrated how badly wrong the polls and forecast models were. But more than that, the needle fed anxiety because of its animated jitter. It was supposed to suggest uncertainty. But people disliked that visual metaphor, and even began mocking up their own dials with labels that expressed their own tension waiting for the result.

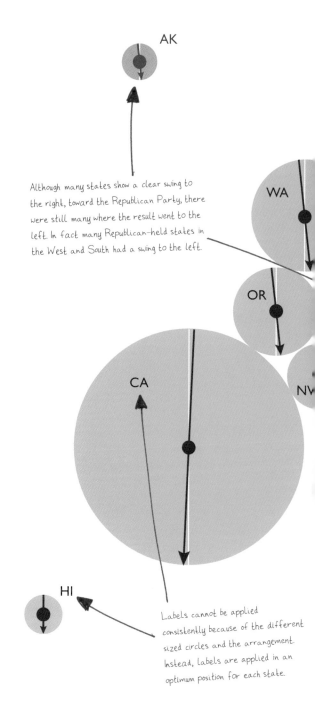

Although many states show a clear swing to the right, toward the Republican Party, there were still many where the result went to the left. In fact many Republican-held states in the West and South had a swing to the left.

Labels cannot be applied consistently because of the different sized circles and the arrangement. Instead, labels are applied in an optimum position for each state.

Win Presidency

> 95% Trump

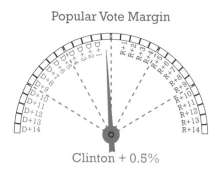

Popular Vote Margin

Clinton + 0.5%

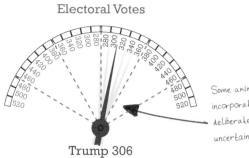

Electoral Votes

Trump 306

Some animated dials incorporate a jitter to deliberately illustrate uncertainty in the pred●

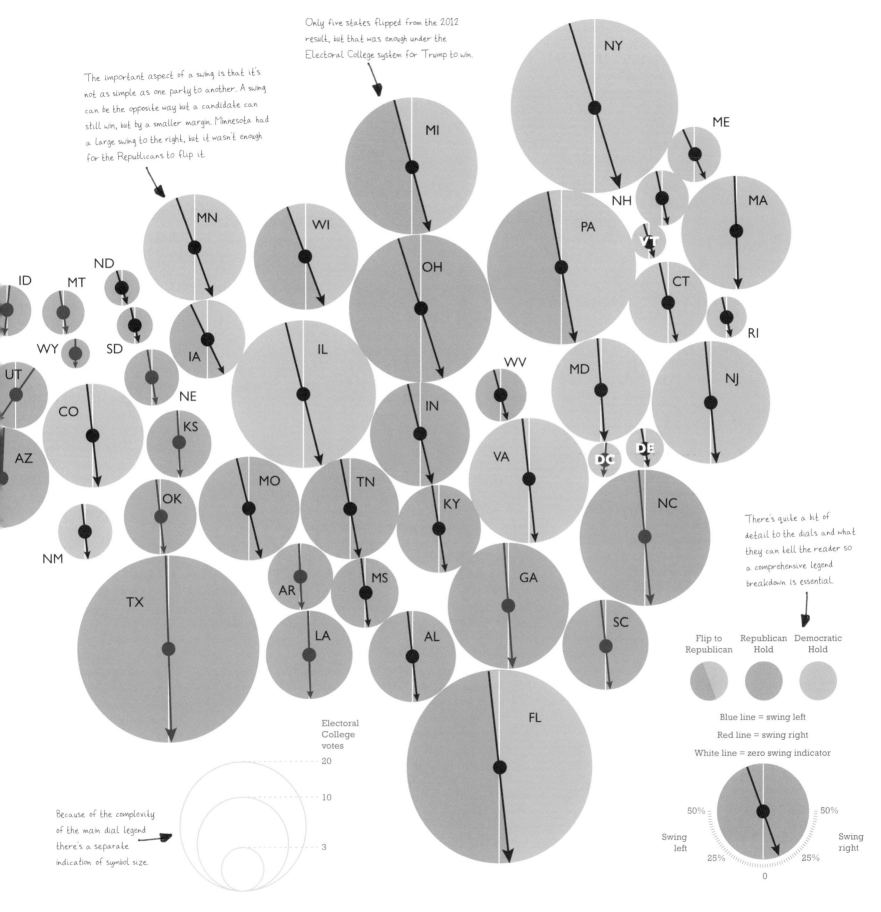

The important aspect of a swing is that it's not as simple as one party to another. A swing can be the opposite way but a candidate can still win, but by a smaller margin. Minnesota had a large swing to the right, but it wasn't enough for the Republicans to flip it.

Only five states flipped from the 2012 result, but that was enough under the Electoral College system for Trump to win.

There's quite a bit of detail to the dials and what they can tell the reader so a comprehensive legend breakdown is essential.

Flip to Republican   Republican Hold   Democratic Hold

Blue line = swing left
Red line = swing right
White line = zero swing indicator

Electoral College votes

20
10
3

Because of the complexity of the main dial legend there's a separate indication of symbol size.

50%   50%

Swing left   Swing right

25%   25%

0

8. 3D maps

Although 3D maps are not new, they have become more popular as a vehicle to display data, particularly on the web. Humans naturally see the world in three dimensions via perspective views at an oblique angle to the world so it makes sense to think that data can be presented in a similar way. Except that's not how 3D works when dealing with thematic maps which encode a variable in the z-dimension. It is this characteristic that requires careful handling when presenting data in 3D, either statically or via an interactive web environment.

3D maps can be static but are rarely effective for thematics either in print or on screen. This chapter will explore some of the reasons this is so and how to mitigate against the issues. When delivered digitally, with other interactive components, 3D can provide compelling views of data. Some of the same limitations exist in an interactive environment, and there are new ones to consider that do not similarly impact planimetric maps.

The key to an effective 3D map is to determine when it's appropriate to take advantage of the z-dimension to bring something meaningful to the map. This chapter illustrates best practice and makes clear the additional perceptual and cognitive loads placed on the map reader when you decide to take a map into the third dimension.

Round and round

Extruded prisms on a digital globe

Upon turning to this page, at first glance the 3D map likely will appear as a visual delight, which is the message conveyed to the brain. Globes and 3D representations do this to people. Humans have a natural fascination with the planet as an orb-shaped object, and with the knowledge that the map isn't forcing geographies onto flat land through mathematical projections. So if it looks good and avoids projections, it must be the perfect representation of data, right?

It's true that using 3D to encode an additional variable provides an expanded palette for cartographic endeavours. **The globe to the right reproduces the diverging-hue choropleth seen in chapter 2, but now total votes are encoded in the z-dimension, and height shows the differences in total vote (turnout) across the country.** It creates an attractive image, but how is the image mediating the impression of the data compared with, say, the 2D version of the same map?

Whenever a globe is viewed from a single static angle, the view of the data is impacted in several ways. **The perspective view creates foreshortening such that the foreground is larger in the field of view than the background.** This creates difficulties for accurately seeing relative sizes and amounts across the whole map. **Whenever objects are extruded or 3D symbology is positioned on a 3D surface, there's inevitably occluded detail that sits behind.** Even when the viewing zenith and azimuth are carefully selected to minimise this, it's impossible to deal with completely. There's always the possibility that extruding data points will create a spiky data ball with prisms pointing in all directions. This problem becomes more acute as more of the globe is mapped. For a relatively small area, it's less of a problem.

Geography always creates difficulties. The United States is a large place covering a large extent on the curved globe and comes with the added complexity of the non-contiguous states of Alaska and Hawaii. **To show the whole country in its true position on the globe, a viewpoint must be chosen that allows the whole country to be seen.** That has the impact of positioning the contiguous states at an odd viewing angle while positioning Alaska and Hawaii at the globe's edge. If this was an atlas of the United Kingdom, the issue would be almost irrelevant.

In short, using a digital globe for thematic data is fraught with difficulties. **3D globes certainly provide ample opportunity to create an interesting visual but not one that is going to lead to meaningful interpretation.** There are far more useful 3D maps to be made that minimise some of the problems associated with digital globes.

Avoiding placing a prism at the globe's nadir (with the viewer looking straight down on top) at least avoids seeing the flat top of a prism with no way of seeing any height.

Most of the maps in this book reposition Alaska and Hawaii as insets which works because there's no connecting, underlying geography. Many thematic maps float their geographies against a neutral background which makes insets sit comfortably in an incorrect geographical position.

However, on a virtual globe the same approach doesn't make visual sense so real geographies remain intact. For the United States this obviously creates difficulties. For smaller maps with complete contiguity this drawback would not be as apparent.

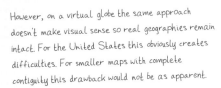

The same symbolisation scheme is used here as the diverging choropleth map so the same legend can be used. It identifies the classification of data, and the way in which values of red and blue go from light to dark across the range.

The additional encoding of total votes as prism height is noted, but because perspective varies across the globe a graduated scale would be meaningless.

Legend diagram warns reader to be aware of foreshortening.

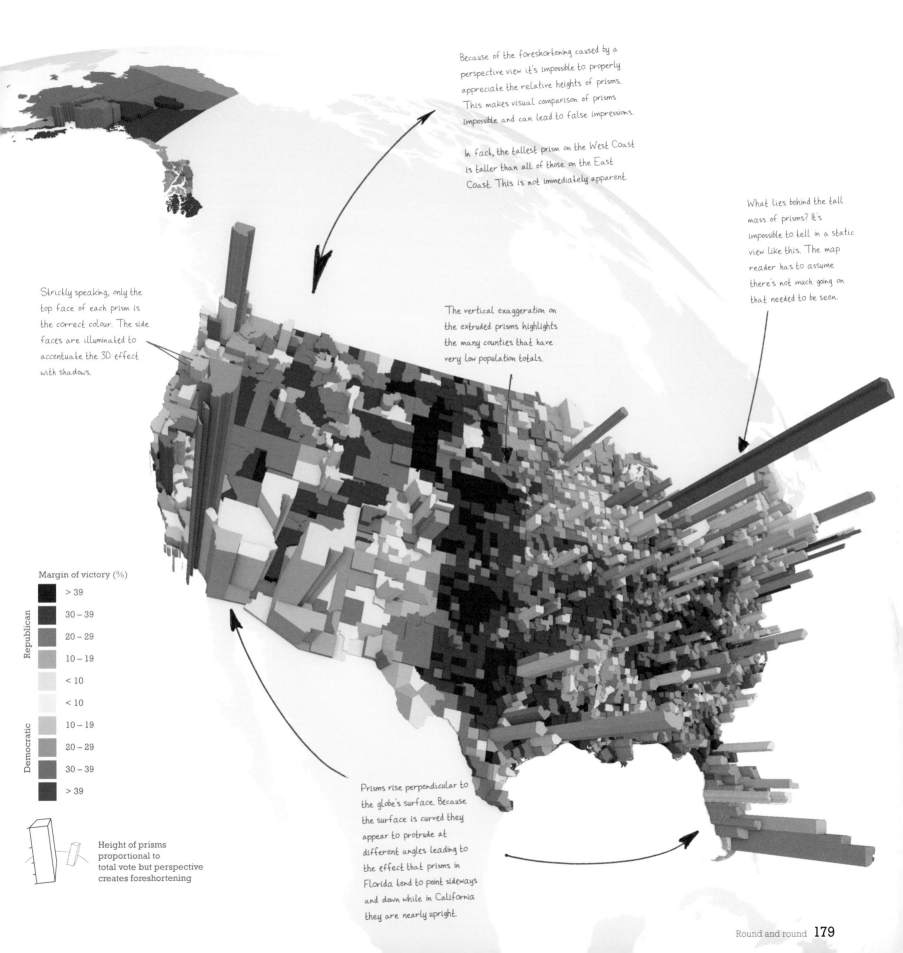

Because of the foreshortening caused by a perspective view it's impossible to properly appreciate the relative heights of prisms. This makes visual comparison of prisms impossible and can lead to false impressions.

In fact, the tallest prism on the West Coast is taller than all of those on the East Coast. This is not immediately apparent.

What lies behind the tall mass of prisms? It's impossible to tell in a static view like this. The map reader has to assume there's not much going on that needed to be seen.

Strictly speaking, only the top face of each prism is the correct colour. The side faces are illuminated to accentuate the 3D effect with shadows.

The vertical exaggeration on the extruded prisms highlights the many counties that have very low population totals.

Margin of victory (%)

Republican

> 39
30 – 39
20 – 29
10 – 19
< 10

Democratic

< 10
10 – 19
20 – 29
30 – 39
> 39

Height of prisms proportional to total vote but perspective creates foreshortening

Prisms rise perpendicular to the globe's surface. Because the surface is curved they appear to protrude at different angles leading to the effect that prisms in Florida tend to point sideways and down while in California they are nearly upright.

Reach for the sky

Extruded prisms on a flat surface

Although globes remain alluring, many applications of 3D in thematic cartography benefit from using a planar approach. **Here, to the right, the same data used on the curved digital globe on the previous page is positioned on a flat surface rather than a curved surface.** Conventionally, and in static cartography, 3D models of data are viewed from a single viewpoint and often in perspective. The terms *viewpoint* and *perspective* are not synonymous though.

Setting a viewpoint describes the relationship between the position of the 3D model and the viewer. This becomes a set point in 3D space governed by the distance from the 3D model and the angle of the model to the eye. This is a different way of viewing a map compared with a planimetric map viewed from above, which is effectively orthogonal to the eye and at the nadir at all viewing angles. Every 3D map incorporates a zenith, the angle at which the viewpoint is positioned with respect to the horizon. Even looking directly down on a 3D model, nadir is correct only at the viewpoint.

Perspective refers to the perception of depth in the image and is, in part, a function of where the viewpoint is set. Depending on the viewpoint, perspective can modify the impression of extruded areal prisms in profound ways. Consider the following diagram. A square block can take on many visual forms, depending on whether the viewpoint is set below, parallel to, or above the block. Furthermore, if the block is positioned in the foreground of the model (closer to the viewpoint), it will appear much larger than if it is set in the distance. This creates foreshortening of images.

These issues have a major impact when using extrusion to encode the height of objects with a data variable. **Encoding total votes as the height of the counties to create a prism map, viewed in perspective, appears to be more logical than the version presented on the curved globe surface, except it still suffers similar flaws.** Prisms in Florida appear much larger than those in Washington toward the rear of the model. This can limit the ability to estimate properly relative heights, and therefore data values. With irregularly shaped objects, this problem is exaggerated because there is no consistent basis for comparing like for like across the map. Occlusion is also a perennial problem in 3D static maps.

Los Angeles County, CA, had a total vote count of 2,464,419 and is the tallest prism on the map. Scaling the prisms is difficult when you have a large data range. They could all have been scaled down but that would have resulted in a great many very flat prisms. Alternatively, a log scale could have been used but that arguably creates greater difficulties in interpretation because people do not naturally see lengths as anything other than linearly scaled.

LA County has three areas, hence three components to the prism. The two smaller prisms are Catalina Island, and San Clemente Island, just off the Pacific coast.

View from below

Foreground

Vanishing point

View from side

Horizon

Background

View from above

Even on a planar surface the prisms still point in slightly different angles. The surface may not be curved, but perspective is still in play.

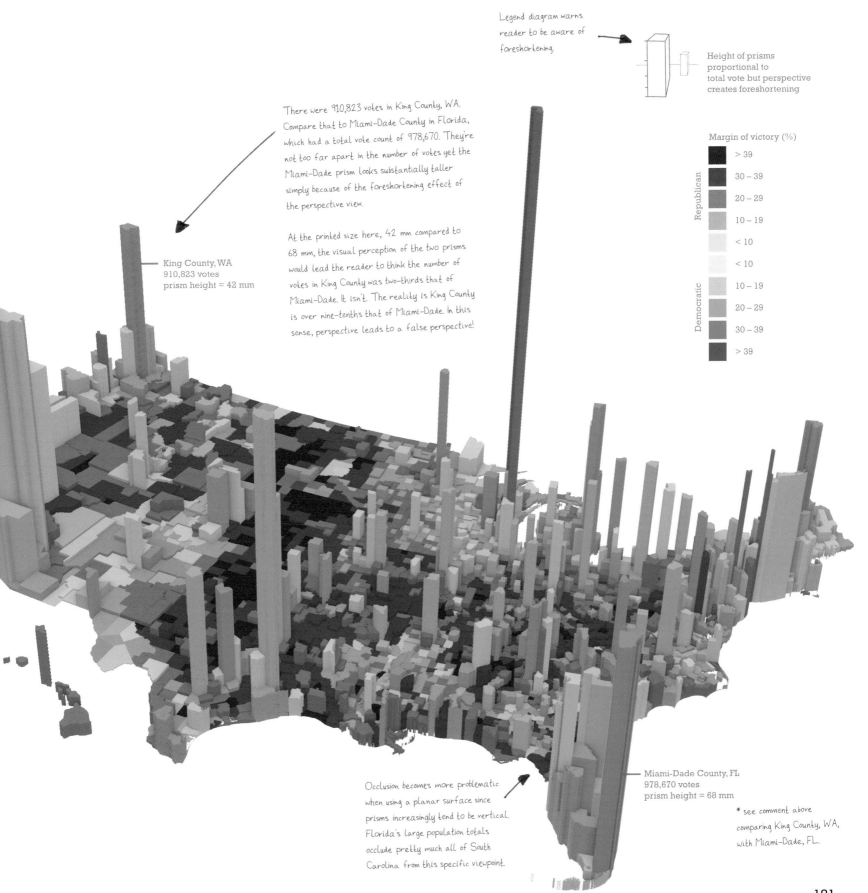

Legend diagram warns reader to be aware of foreshortening.

Height of prisms proportional to total vote but perspective creates foreshortening

There were 910,823 votes in King County, WA. Compare that to Miami-Dade County in Florida, which had a total vote count of 978,670. They're not too far apart in the number of votes yet the Miami-Dade prism looks substantially taller simply because of the foreshortening effect of the perspective view.

At the printed size here, 42 mm compared to 68 mm, the visual perception of the two prisms would lead the reader to think the number of votes in King County was two-thirds that of Miami-Dade. It isn't. The reality is King County is over nine-tenths that of Miami-Dade. In this sense, perspective leads to a false perspective!

King County, WA
910,823 votes
prism height = 42 mm

Margin of victory (%)

Republican
> 39
30 – 39
20 – 29
10 – 19
< 10

Democratic
< 10
10 – 19
20 – 29
30 – 39
> 39

Miami-Dade County, FL
978,670 votes
prism height = 68 mm

Occlusion becomes more problematic when using a planar surface since prisms increasingly tend to be vertical. Florida's large population totals occlude pretty much all of South Carolina from this specific viewpoint.

* see comment above comparing King County, WA, with Miami-Dade, FL.

Proper perspective

Extruded prisms in an axonometric view

If globes, with their curved surfaces, and perspective models, with their foreshortening, are not optimal for 3D thematic cartography, the question remains: Is there a way of creating a 3D model that supports effective 3D cartography? As with any map, the short answer is that there always will be compromise in what can be shown, and how it is shown. **Some of the drawbacks of globes and perspective models can be mitigated by using an axonometric view.**

Axonometric **refers to a form of parallel, or orthographic, projection in which sight lines are perpendicular to the plane of projection.** An object then can be rotated around one of its axes but the objects remain parallel to one another. The most common form of axonometric projection is isometric, in which the rotation and tilt are constrained so that angles between each pair of axes are 30 degrees and sum to 120 degrees. Consider the diagram to the right of a cube drawn in perspective and its counterpart drawn in an isometric projection. Using a pure isometric form does not always translate well to maps because the appearance is distorted. But the map can be rotated and tilted to find a suitable viewpoint while maintaining axonometric properties.

When extrusion encodes data values in the z-dimension, the key cognitive process for the reader is the ability to estimate not only the height in an absolute sense but also relatively across the map. As noted, globes and perspective views do not support that cognitive task. Axonometric views do, which is its primary utility when making 3D thematic maps.

The practical result is that all parts of the 3D model are equally foreshortened, meaning that scale is consistent across the entire map—not only from left to right but also from top to bottom or background to foreground. The key decision is what viewpoint gives the most useful view and limits the impact of occlusion. The small inset version below is still projected in an axonometric view, but the tilt angle is lower and occlusion much greater, rendering it less useful.

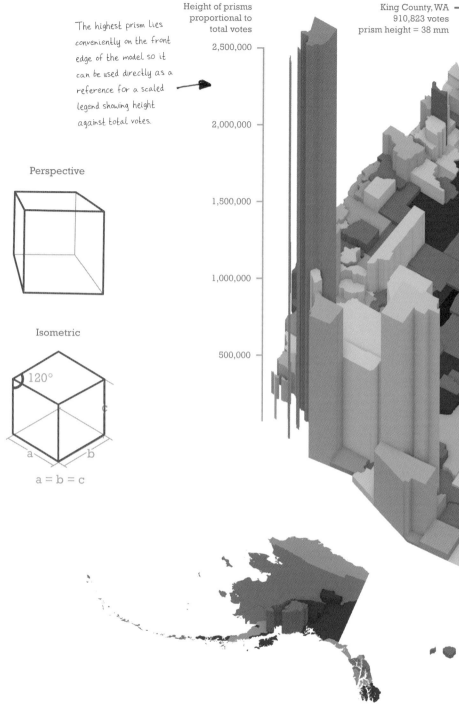

The highest prism lies conveniently on the front edge of the model so it can be used directly as a reference for a scaled legend showing height against total votes.

Height of prisms proportional to total votes

2,500,000

2,000,000

1,500,000

1,000,000

500,000

King County, WA
910,823 votes
prism height = 38 mm

Perspective

Isometric

120°

a = b = c

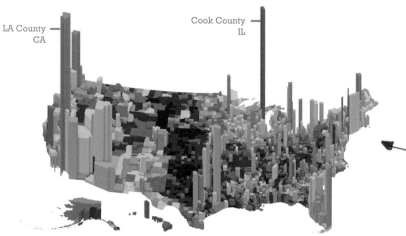

LA County
CA

Cook County
IL

Assessing height is really only possible if the base of the prism can be seen, and account is taken of the relative position front to back.

For instance, Cook County, IL, might at first glance appear to be the tallest prism, yet its base is toward the back so it's still relatively smaller than LA County.

If the same measurement is performed on this map as the previous perspective view of the same prisms then both King County, WA, and Miami-Dade County, FL, have prisms that are identical at 38 mm in height. This is an accurate reflection that at this scale their total votes are broadly the same.

Legend diagram explains prisms are comparable in height.

Height of prisms proportional to total vote and equally scaled across the map

Margin of victory (%)

Republican
> 39
30 – 39
20 – 29
10 – 19
< 10

Democratic
< 10
10 – 19
20 – 29
30 – 39
> 39

Occlusion is still problematic, as with any 3D rendering with solid vertical prisms and a fixed viewpoint. Compared to the previous perspective view there's less occlusion of South Carolina due to Florida's prisms.

The map appears distorted to the human eye precisely because it's not seen in perspective. The vertical character of all the prisms immediately shows that the map is presented in a parallel, axonometric projection.

Miami-Dade County, FL
978,670 votes
prism height = 38mm

* see comment above comparing King County, WA, with Miami-Dade, FL.

Peaks and valleys

Extruded filled contours

In the same way that isolines provide a mechanism to break away from rigid administrative boundaries in planimetric maps, the same can be achieved in 3D. Rather than using county boundaries and extruding the area to a height relative to the total vote, the map to the right applies the same idea to an isopleth map to create a 3D version.

For the planimetric isopleth, map contours were generated to represent the margin of victory across the United States. These contours created a more organic landscape of peaks and valleys. The contours were plotted at 5 percent intervals representing the margin of victory so the area between each contour is at some point between adjacent contour values. **The 2D statistical surface demarcated by contours can be converted into discrete classes (filled contours), and then extruded as a 3D surface.** This creates a stepped contour 3D model in which each step on the map represents areas of values binned into discrete classes.

Because contours are nested, the result is a 3D terrain of the data, albeit a somewhat abstract representation. The extrusion is essentially classified so there's less reliance on the need to use an axonometric projection because the classified colours are tinted by elevation, and readers can count the number of steps from the baseline to identify a broad elevation value and, by inference, data value.

Occlusion is still a potential problem so the value used to extrude each step becomes a fine balance between creating enough variation in the z-dimension and not creating such large mountains and deep valleys that will obscure too much. An alternative approach (below) shows Republicans above and Democrats below, effectively pushing the Democratic steps below a mean surface and having them appear as lakes with contours mimicking bathymetry considering their blue hue.

The main map encodes margin of victory as height above a baseline of zero with the same vertical exaggeration for both the Democratic and Republican parties. It creates valleys of marginal territory meandering between peaks and ranges of more partisan territory.

Coastal cliff face

Some of the Aleutian Islands have been omitted (a form of cartographic generalisation) simply because they form an overbearing set of promontories for this design at this scale.

Above

Below

This map positions the Democratic data below the zero baseline which accentuates the winning peaks.

In the same way as an isopleth map benefits from some labels to aid people's orientation then so does this 3D version. The difference here is that labels do not sit flat on the map, they are billboarded above the location with a small vertical leader line that descends to the surface of the 3D model.

There's a little transparency applied to a halo that sits behind the label just to help lift it away from some of the darker areas of the map to maintain visibility, and hence legibility.

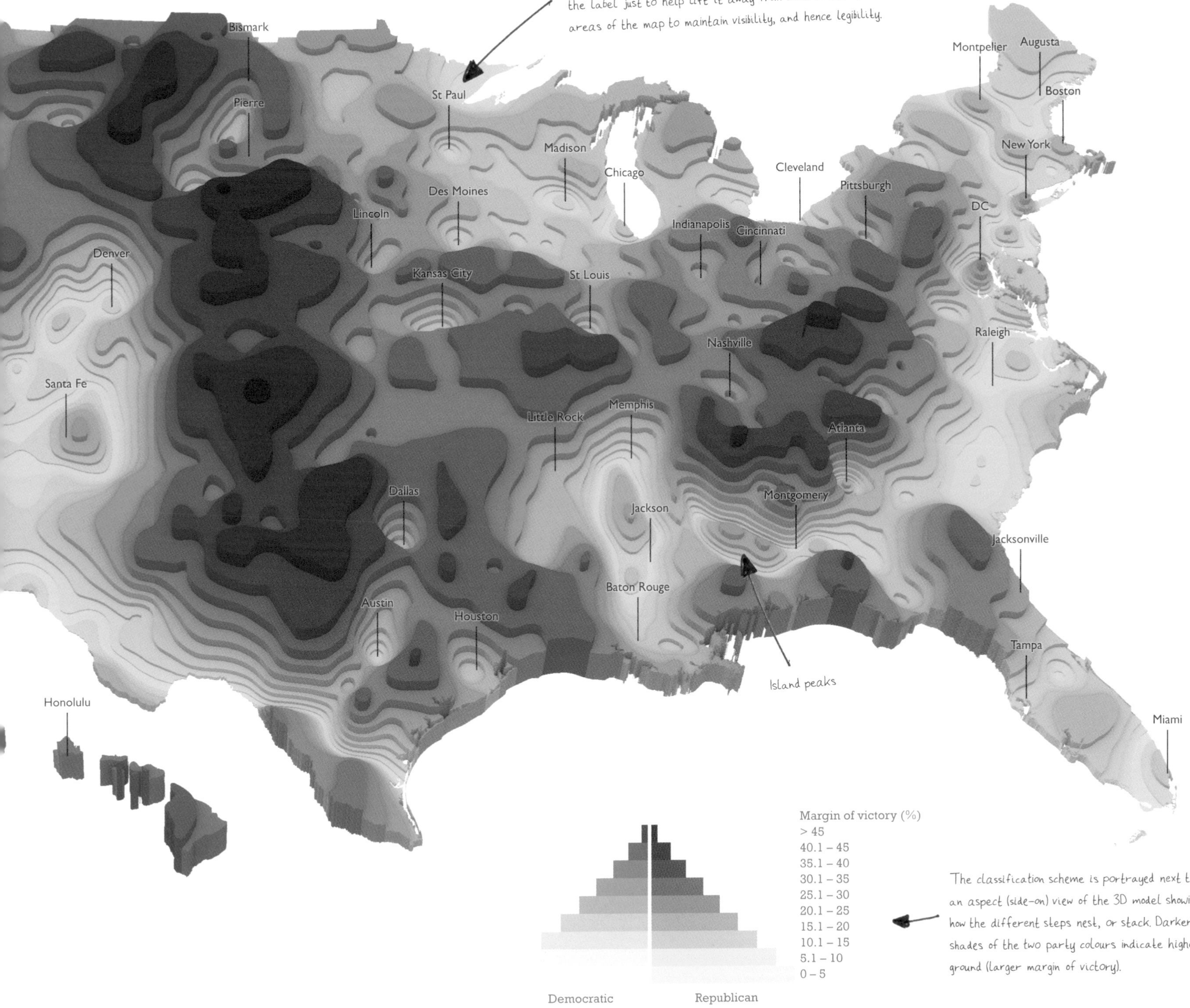

Bismark

St Paul

Pierre

Madison

Chicago

Cleveland

Montpelier   Augusta

Boston

Des Moines

Lincoln

New York

Pittsburgh

Denver

Kansas City

St Louis

Indianapolis Cincinnati

DC

Santa Fe

Nashville

Raleigh

Little Rock   Memphis

Atlanta

Dallas

Jackson

Montgomery

Jacksonville

Austin

Houston

Baton Rouge

Island peaks

Honolulu

Tampa

Miami

Margin of victory (%)
> 45
40.1 – 45
35.1 – 40
30.1 – 35
25.1 – 30
20.1 – 25
15.1 – 20
10.1 – 15
5.1 – 10
0 – 5

The classification scheme is portrayed next to an aspect (side-on) view of the 3D model showing how the different steps nest, or stack. Darker shades of the two party colours indicate higher ground (larger margin of victory).

Democratic   Republican

View from the hill

3D block diagram with a draped surface

The basic purpose of a 3D map is to show something useful in the z-dimension. But this doesn't have to be something different from what is shown in the planar dimension. On the map to the right, the interpolated surface planimetric map illustrated earlier in the book is used as a starting point. In planimetric view, the dark red and dark blue are peaks among a landscape of lighter hues. Because darker hues are seen as more prominent and interpreted as more, the impression is of a naturally undulating statistical landscape that shows the peaks of where voters gave either Donald J. Trump or Hillary Clinton a large margin of victory, and where the margin was particularly slim in the in-between valleys.

To display this interpolated surface map in 3D, the process requires draping the empirical surface across the top of another dataset. The dataset used to define elevation is referred to as a *digital elevation model* (DEM). Conventionally, DEMs are datasets that model real-world topographic elevation but they can model anything that encodes a value in lieu of height. They are a dataset that defines height values across the extent of the map.

In this case, the same data is being used as an elevation surface as well as the draped interpolated surface of election results. The margin of victory ranges from 0 percent to just below 50 percent, and these values are stored in a raster dataset as individual cell values. **When the interpolated surface is draped across the DEM, the peaks in the symbolised drape match the undulations in the DEM, and a 3D impression forms that emphasises the interpolated surface.** In some respects, it helps visualise more clearly where the peaks and troughs are in the data. The choice of viewing angle is designed not only to fit the page dimensions sensibly but to avoid as much occlusion as possible.

One other characteristic that this map uses is vertical exaggeration. Data values vary 0–50 in the DEM, and a map that uses those values as real-world units (metres, in this case) wouldn't show much variation in height at a national scale. **Vertical exaggeration is used to increase the height of the DEM, providing a view that uses elevation more suited to the scale at which the map is seen.** For this map, a vertical exaggeration of 5,000× might sound like a lot but it turns the maximum peak in the interpolated surface into a mountain that is roughly 50 km high in real-world units. It's an arbitrary number, though. It's defined by trial and error to give a surface with enough variation to make it suitable, but not so much as to create bizarre peaks, cliffs, and occlusion.

Finally, is it 3D? Not really. It has no volumetric properties. It's more accurately referred to as a *2.5D draped surface*.

There are many ways this sort of 3D model can be styled. The surface could be completely smoothed and devoid of contours, or a triangulated irregular network could have been used instead of a DEM. Instead, this is a gridded surface that captures a little of the abstract nature of interpolating surfaces across space. By using lighting effects, the shadows accentuate the gridded effect.

A uniform dropped border is added to the whole map to create the block appearance, and provide a sense of relative height, particularly at the map edges.

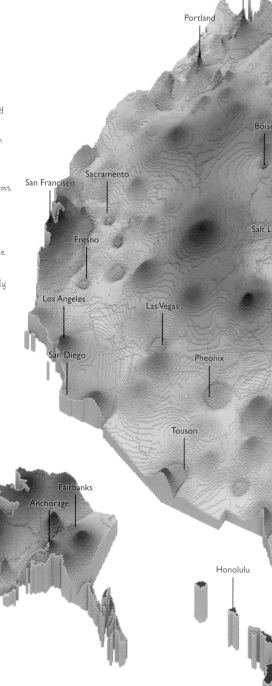

Compared with the impression the stepped contour map on the previous page gives, of a very generalised landscape, this version shows the data to a much finer degree of detail.

It shows far more variation across space rather than having the data smoothed out by averaging it across an area. Given it's a representation of the 3,141 counties, the nuance shown in the highly detailed contours seems valid.

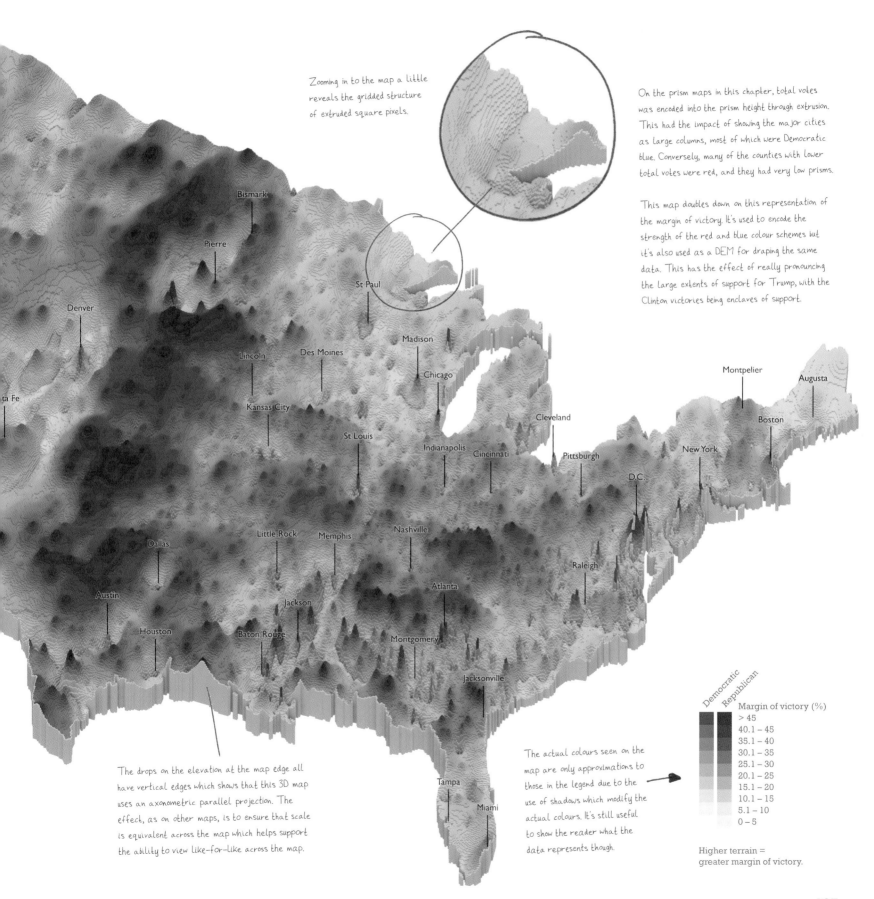

Zooming in to the map a little reveals the gridded structure of extruded square pixels.

On the prism maps in this chapter, total votes was encoded into the prism height through extrusion. This had the impact of showing the major cities as large columns, most of which were Democratic blue. Conversely, many of the counties with lower total votes were red, and they had very low prisms.

This map doubles down on this representation of the margin of victory. It's used to encode the strength of the red and blue colour schemes but it's also used as a DEM for draping the same data. This has the effect of really pronouncing the large extents of support for Trump, with the Clinton victories being enclaves of support.

Bismark

Pierre

St Paul

Madison

Denver

Lincoln

Des Moines

Chicago

Montpelier

Augusta

ta Fe

Kansas City

Cleveland

Boston

St Louis

Indianapolis

Cincinnati

Pittsburgh

New York

D.C.

Little Rock

Memphis

Nashville

Raleigh

Dallas

Austin

Jackson

Atlanta

Houston

Baton Rouge

Montgomery

Jacksonville

The drops on the elevation at the map edge all have vertical edges which shows that this 3D map uses an axonometric parallel projection. The effect, as on other maps, is to ensure that scale is equivalent across the map which helps support the ability to view like-for-like across the map.

Tampa

Miami

The actual colours seen on the map are only approximations to those in the legend due to the use of shadows which modify the actual colours. It's still useful to show the reader what the data represents though.

Democratic
Republican

Margin of victory (%)
> 45
40.1 – 45
35.1 – 40
30.1 – 35
25.1 – 30
20.1 – 25
15.1 – 20
10.1 – 15
5.1 – 10
0 – 5

Higher terrain =
greater margin of victory.

Us and them

Triangulated irregular networks

There are many ways to build a surface from data. The example to the right uses a triangulated irregular network (TIN) as a surface with the results at state level draped over the top. **A TIN is a continuous surface made of triangular facets.** Each facet is defined by the values of the three vertices that define it. The slope and aspect of each facet are a function of the relationship between the three vertices. It's often a highly generalised surface with the planar facets approximating the general slope and aspect of the surface they represent.

TINs are based on Delaunay triangulation in which, for a given set of discrete points in a plane, there is a triangulation in which none of the points is in the circumcircle of the triangle (see below). The circumcentres of Delaunay triangles become the vertices of a different surface type called Voronoi tessellation.

Delaunay triangulation with circumcircles and their centres

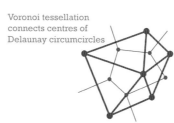

Voronoi tessellation connects centres of Delaunay circumcircles

For this map, a gridded square cartogram is used as the base, so each state is represented by a square of the same size, and all are organised in regular rows and columns. Each square is defined by four corner vertices, which are given a value of zero. **A vertex at the centre of each square is given a value of the percentage margin of victory that the winning candidate gained, and connecting each of the corner vertices with the central vertex gives four triangular facets for each state.** This basic TIN is modified by the connections of the centre vertices to one another that introduces facets which do not necessarily include vertices at zero. It makes little sense for an empirical surface to pass through data values of zero when they share a similar value. Where Democratic states neighbour Republican states, the TIN passes through lines of zero, and a metaphorical wall is built.

This example is more illustrative than analytical but it's possible to get a sense of the relative margins of victory through a visual comparison of heights. **Where there is little variation in the data values between neighbours, there tend to be larger facets, which are shallower and have more widely spaced vertices.** Where data values between neighbouring states are greater, steeper facets appear. A perspective view is preferred over axonometric because the intent is not to measure the heights of the pyramids, and mixing squares and isometric views creates an awkward view.

This map is shown in perspective so the foreground is more prominent than the background. The flags help cement that scale is different across the map.

There's no vertical scale provided because measuring pyramid height is difficult anyway, and that's before the problems of foreshortening are considered.

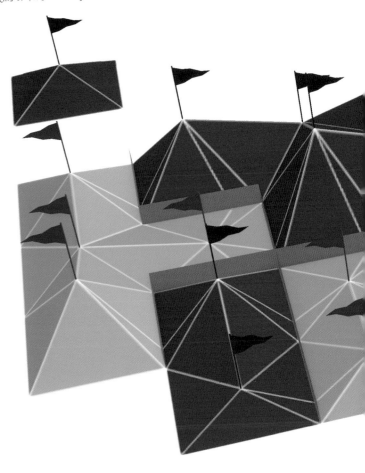

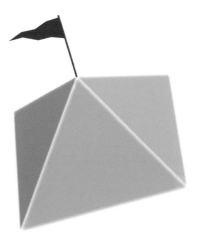

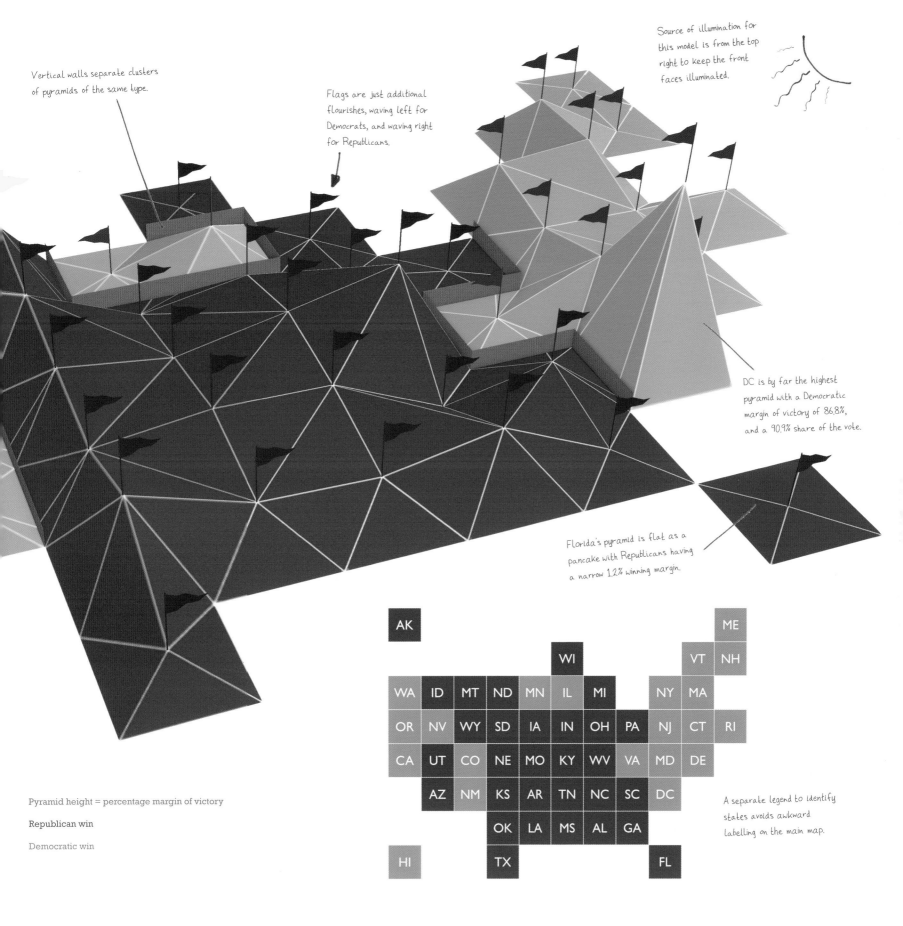

Vertical walls separate clusters of pyramids of the same type.

Flags are just additional flourishes, waving left for Democrats, and waving right for Republicans.

Source of illumination for this model is from the top right to keep the front faces illuminated.

DC is by far the highest pyramid with a Democratic margin of victory of 86.8%, and a 90.9% share of the vote.

Florida's pyramid is flat as a pancake with Republicans having a narrow 1.2% winning margin.

Pyramid height = percentage margin of victory

Republican win

Democratic win

A separate legend to identify states avoids awkward labelling on the main map.

AK										ME
				WI					VT	NH
WA	ID	MT	ND	MN	IL	MI		NY	MA	
OR	NV	WY	SD	IA	IN	OH	PA	NJ	CT	RI
CA	UT	CO	NE	MO	KY	WV	VA	MD	DE	
	AZ	NM	KS	AR	TN	NC	SC	DC		
			OK	LA	MS	AL	GA			
HI			TX				FL			

Chromastereoscopic colour encoding

3D views of data don't have to be rendered by turning planimetric into a 3D model. **It's possible to modify the way in which a standard 2D planimetric map is symbolised to reveal information in 3D or, more precisely, fake 3D.** Several techniques achieve this state, such as the creation of anaglyphs in which the image is split into two components that then are read using glasses with different-coloured filters in each lens that force each eye to see two slightly different images. The brain then reassembles those two images in stereo to perceive a 3D image.

An alternative is the chromastereoscopic approach, which uses the same trick of forcing the eyes to see aspects of the image in different ways, but instead of coloured filters, the glasses contain small prisms. As light hits the prisms, the different wavelengths are refracted at different angles, so the eyes then see colour in an ordered way. Red wavelengths are refracted at a shallower angle than blue, and the eyes bend slightly more inward to 'see' red. This has the effect of leading the brain to interpret the image as red being closer than blue. Different hues are ordered between these two ends of the electromagnetic spectrum. A dark background improves depth perception for chromastereoscopic colour schemes.

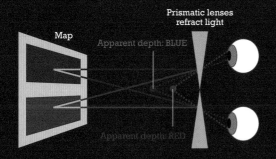

Armed with this knowledge, you can design the map with objects to appear closer to the eye being shown in the red spectrum and objects to appear farther away being shown in the blue spectrum. **For political maps in which red and blue are key colours and in which red won, the map orders the results by promoting reds as closer to the eye and blues as farther away.** Grab a pair of chromastereoscopic glasses, and for most, this page will jump out in full fake holographic 3D.

The benefit of this approach over other stereoscopic techniques is that the map itself appears readable without chromastereoscopic glasses, albeit using some odd colours. As with any map, those who have some form of colour deficiency might not be able to see the map in 3D considering its specific

of the electromagnetic spectrum an unmodified chromastereoscopic colour ramp would simply be the colours of the rainbow in order. This version omits many hues to re-create the red to blue continuum.

Different parts of the colour ramp have been tweaked to include some subtle visual steps to give a hint at contoured elevation across the empirical data surface.

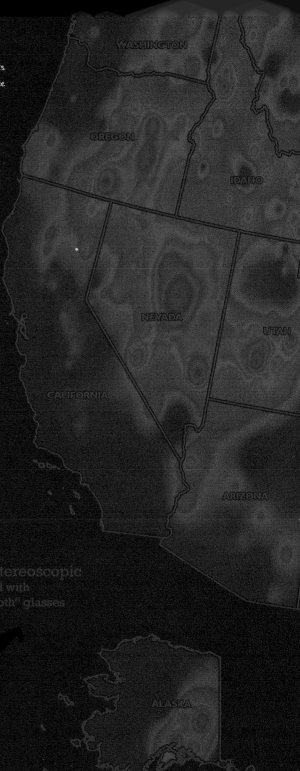

All the labels and linework are red because red sits in the highest viewing plane in chromastereoscopic 3D. Light labels would look better in 2D, but wouldn't work in 3D.

The chromastereoscopic colour scheme is mapped onto the continuous interpolated surface that represents the data.

Margin of victory (%)

Republican

> 45

20

0

Democratic

20

> 40

Whenever bold choices are made about colours and other design elements there are often consequences for other map detail. Labels are in red so they sit across all other detail when viewed chromastereoscopically. With such saturated map colours and a dark background they have to be larger, and also use a dark halo to distinguish them from the map colours.

From the ground up

Illuminated 3D columns

3D maps, in general, exhibit the same sorts of strengths and limitations as their 2D equivalents. The prism, stepped surface, and draped surface maps all suffer from the same problem of a map that uses arbitrary area boundaries that exhaust space. Area-based maps tend to overemphasise large geographical areas in comparison with smaller, often more densely populated areas. By adding volume to those areas, the exaggeration is both propagated and magnified. But as with planimetric maps, options exist to use a different approach which mitigates this concern.

Instead of using areal symbols as the base, points can represent administrative areas with one point in each area. **3D maps can be made from the same data as a point rather than an area.** The map therefore becomes a reference and a canvas upon which you position the symbols. In the case of a 3D map, they can be 3D symbols instead of, or in addition to, flat 2D versions.

On the map to the right, the United States works merely as a plinth. **Each county is represented by a single point, but for many symbols, the point is represented by a 3D column symbol.** Each column is scaled to a consistent diameter so that each county is treated similarly, and none are over- or underemphasised purely by virtue of the different-sized geographies they encompass. The height of each column is proportional to the number of votes that the winning candidate won. Total votes have been used as a variable on many maps in this book but this one uses a subset. The columns are red and blue as expected, to indicate a Republican or Democratic victory, respectively.

Occlusion and congestion still exist in parts of the map, but to counter it, flat symbols are used for any county with less than 20,000 winning votes. **The diameter of each symbol and column is neither too narrow, which would leave the columns looking more like poles, nor too wide, which would leave them coalescing into some sort of vertical blob.** The other important part of presenting a map like this in a static mode is to select a viewpoint that helps avoid as much occlusion as possible, and using a parallel axonometric projection ensures columns can be assessed relative to one another in height. The most congested area is to the east and northeast, so this area is positioned toward the rear of the map so it doesn't occlude too much.

The advantage of a map like this is that it accounts for geographical differences and reduces the visual impact of large, sparsely populated areas. A drawback is the problem of comparing heights from different bases by virtue of the position of columns across the map, and map readers may not find that intuitive.

Because this map uses an axonometric parallel projection to equalise scale across the map, the vertical scale can be used to assess the height of the columns related to the number of votes the winning candidate received.

All areas that had a winning vote of < 20,000 votes are symbolised with the same height of column to avoid difficulties in having imperceptibly small cylinders in some areas.

Winning votes

1 million
900,000
800,000
700,000
600,000
500,000
400,000
300,000
200,000
100,000
0

* LA County – see note for New York City

There's a hint of transparency used to vary the appearance of the columns. Columns are increasingly transparent where the percentage share of the vote was low, which illustrates a different variable than the number of votes the winner received. It's perfectly possible to see a large number of winning votes and a small margin of victory simply by virtue of the population distribution.

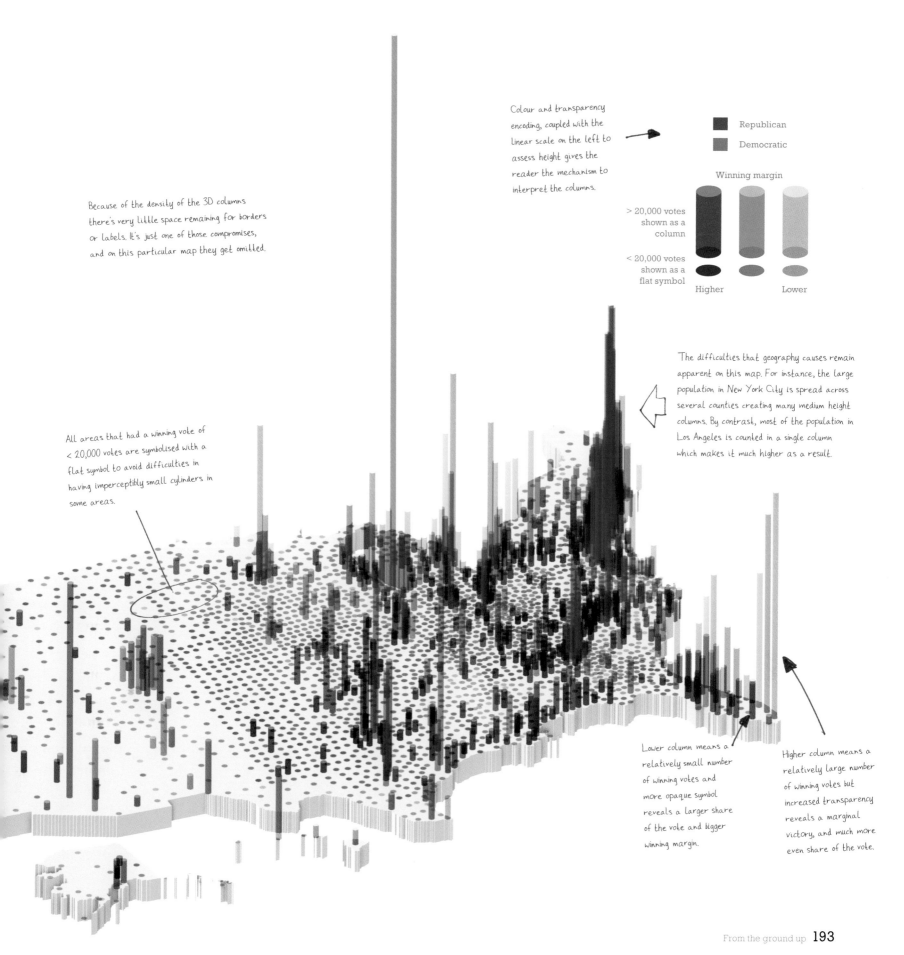

Colour and transparency encoding, coupled with the linear scale on the left to assess height gives the reader the mechanism to interpret the columns.

Republican

Democratic

Winning margin

> 20,000 votes shown as a column

< 20,000 votes shown as a flat symbol

Higher   Lower

Because of the density of the 3D columns there's very little space remaining for borders or labels. It's just one of those compromises, and on this particular map they get omitted.

The difficulties that geography causes remain apparent on this map. For instance, the large population in New York City is spread across several counties creating many medium height columns. By contrast, most of the population in Los Angeles is counted in a single column which makes it much higher as a result.

All areas that had a winning vote of < 20,000 votes are symbolised with a flat symbol to avoid difficulties in having imperceptibly small cylinders in some areas.

Lower column means a relatively small number of winning votes and more opaque symbol reveals a larger share of the vote and bigger winning margin.

Higher column means a relatively large number of winning votes but increased transparency reveals a marginal victory, and much more even share of the vote.

Raising the stakes

Stacked chips

Compromises exist in every map made, and arguably, more so when creating a 3D map. Effective design relies on the ability to understand the constraints of the technique and to work with them or around them in a way that helps the map reader. In planimetric cartography, there are only two dimensions, and many best practices are understood. Adding a third dimension adds complexity and additional constraints and considerations. As the maps in this chapter already have demonstrated, one of the difficulties lies in appreciating that extruding symbols in the z-dimension—and expecting readers to assess how height relates to a variable—is not an easy perceptual or cognitive task. Often, it is compounded by the aspect or projection used and people's natural tendency to see a 3D image as volumetric.

How can the information encoded vertically make sensible use of the z-dimension, mitigating for the increased cognitive load on the map reader? The trick lies in a simple perceptual and cognitive task—counting, which some of the maps in chapters 5 and 7 also employed. Human beings are extremely good at counting objects as opposed to estimating lengths, areas, or volumes.

This map uses a stacked chips approach to illustrate the distribution of Electoral College votes across the United States according to who won the state. **Each chip represents one Electoral College vote, and seeing all the chips facilitates counting.** This sort of map allows perfect data recovery if that's one of the goals that the map reader wants from the map.

If counting is so useful, why aren't all maps made this way? Well, as ever, there are compromises, and with this map, the choice of dataset has major consequences. The data on Electoral College votes is basic. There are only 538 chips to distribute. It would be impossible to show the data for total votes on a 1:1 basis though chips could be classified into, say, 100,000 votes per chip, and then stacked. Precision could be lost with this approach because the sum of stacked chips always will round to the nearest 100,000. Chips even could be colour-coded like poker chips by different values, though for this data, the loss of the blue and red encoding might be a compromise that is a step too far.

Stacked chips make for an attractive solution to showing a basic dataset for a 3D map. This map is effectively a 3D version of the point-based compromise cartogram from chapter 5. Instead of relying on abstract symbology, the geographical base retains the visual link to known places to help people interpret the map's data with reference to places they identify with.

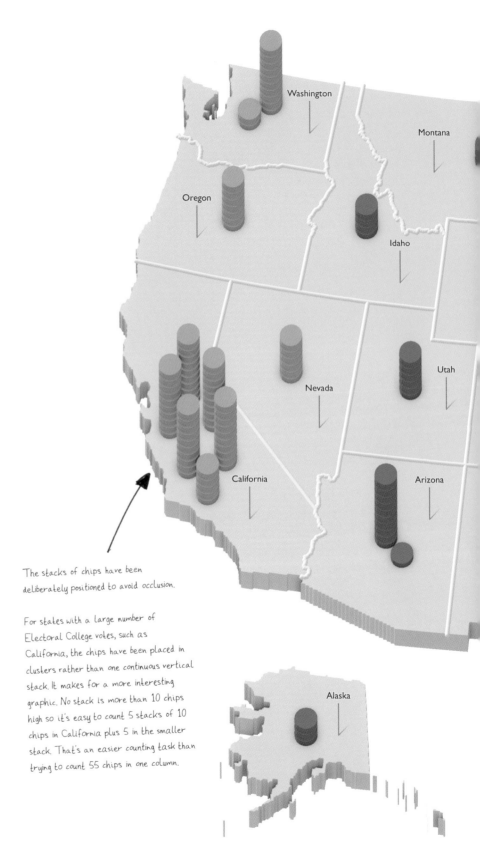

The stacks of chips have been deliberately positioned to avoid occlusion.

For states with a large number of Electoral College votes, such as California, the chips have been placed in clusters rather than one continuous vertical stack. It makes for a more interesting graphic. No stack is more than 10 chips high so it's easy to count 5 stacks of 10 chips in California plus 5 in the smaller stack. That's an easier counting task than trying to count 55 chips in one column.

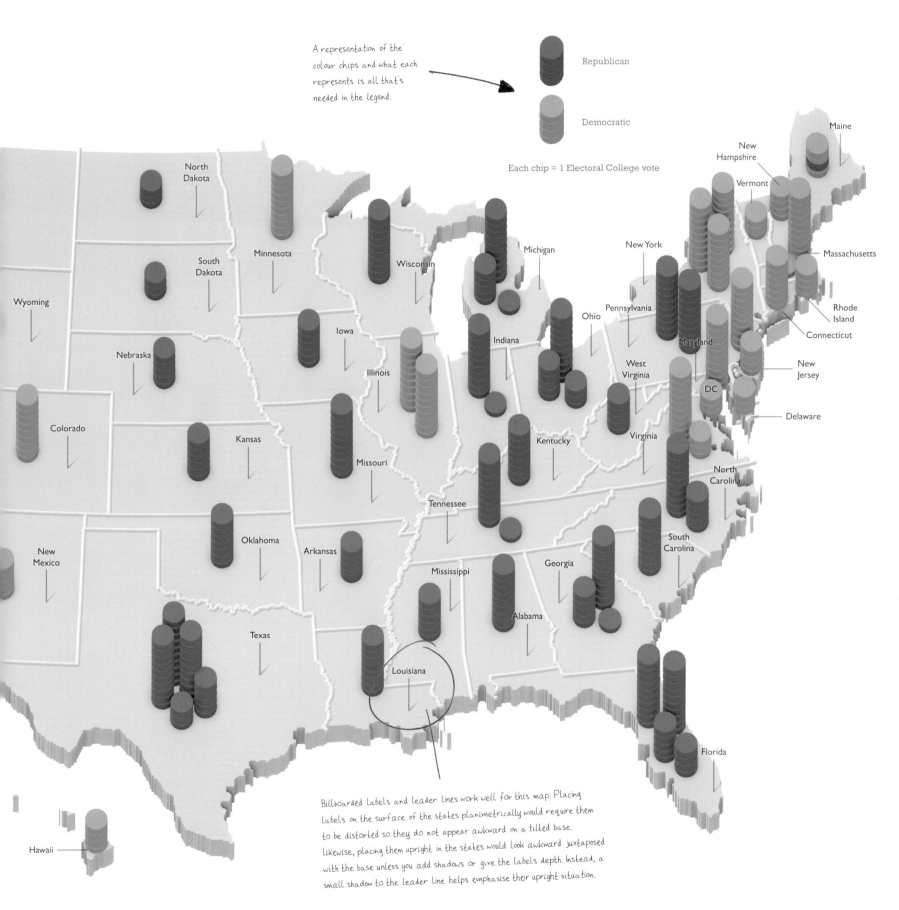

A representation of the colour chips and what each represents is all that's needed in the legend.

Republican

Democratic

Each chip = 1 Electoral College vote

Maine

New Hampshire

Vermont

Massachusetts

New York

Rhode Island

Pennsylvania

Connecticut

North Dakota

Ohio

New Jersey

Minnesota

Wisconsin

Michigan

Maryland

Wyoming

South Dakota

Iowa

Illinois

Indiana

West Virginia

DC

Delaware

Nebraska

Virginia

Colorado

Kansas

Missouri

Kentucky

North Carolina

New Mexico

Oklahoma

Arkansas

Tennessee

South Carolina

Texas

Mississippi

Alabama

Georgia

Louisiana

Florida

Hawaii

Billboarded labels and leader lines work well for this map. Placing labels on the surface of the states planimetrically would require them to be distorted so they do not appear awkward on a tilted base. Likewise, placing them upright in the states would look awkward juxtaposed with the base unless you add shadows or give the labels depth. Instead, a small shadow to the leader line helps emphasise their upright situation.

Power to the people

Dasymetrically distributed 3D people

The irony of making maps of population data is that it normally means turning information about people into an abstract form of symbology. This may be a point, line, or area and can be at many different levels of spatial or statistical aggregation. But it's still mapping something very human. **Instead, 3D allows you the opportunity to put people on the map.**

The map to the right shows 538 people positioned across the United States. Although the locations of each figure are random, there is a logic. The data shows the US Electoral College electors, coloured in either the red of Republicans or the blue of Democrats. The figures obviously conform to state boundaries, so the count of Electoral College electors is distributed within state boundaries. **Position of the people is broadly based on a dasymetric approach, which locates them to reflect the pattern of population distribution.**

The use of 3D models that are non-geometric is well supported by 3D cartography. More scope is given to use different approaches to symbology when using a basemap as a plinth, on top of which you can have people wandering or any number of other types of 3D symbology. **Using models that represent real people is a little more playful than using simple geometric symbols.** It also might be argued that they are easier to identify with. The symbols here are obviously people, and there's less of a barrier to trying to understand what, for instance, an extruded 3D geometric shape means.

But what of the symbols themselves? **This map deliberately uses a gender-neutral architectural model symbol.** It would be possible to populate the map with mini Donald J. Trump or mini Hillary Clinton figures, but the more lifelike the symbols seem, the closer the map gets to being a cartoon rather than an objective representation of data. Neutral symbols, coloured red and blue, do enough of a job without making them caricatures. There's some random rotation so the figures face different directions for visual interest. They could be forced to face left or right to keep in line with their politics but the map becomes visually jarring when they are oriented in a rigid rather than organic pattern.

This map, then, develops the use of a more human approach to symbology with the ability for readers to count the data points on the map to recover the information. Like the stacked-chip approach, it works well for relatively small datasets. Trying to populate this map with many more figures, or even scaling them in size by some other data variable, would make the map's utility quickly deteriorate into an illegible mess. As ever, getting the balance between representing some useful aspect of the data and doing so with a sensible approach to symbology is the goal.

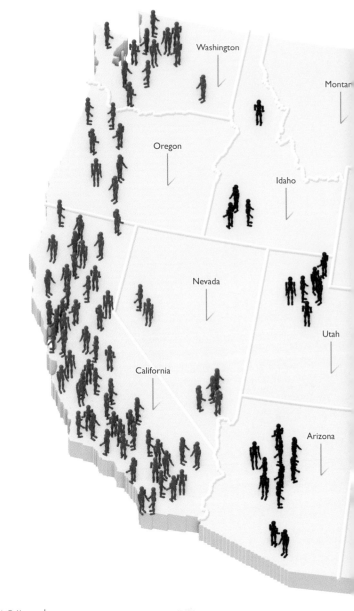

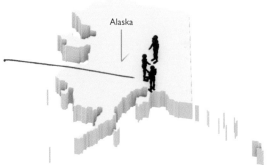

Alaska has 3 Electoral College votes and given the vast expanse of wilderness they are located to broadly reflect Anchorage and Fairbanks. The same sort of positioning is seen in other states where the population is largely located in small, concentrated areas.

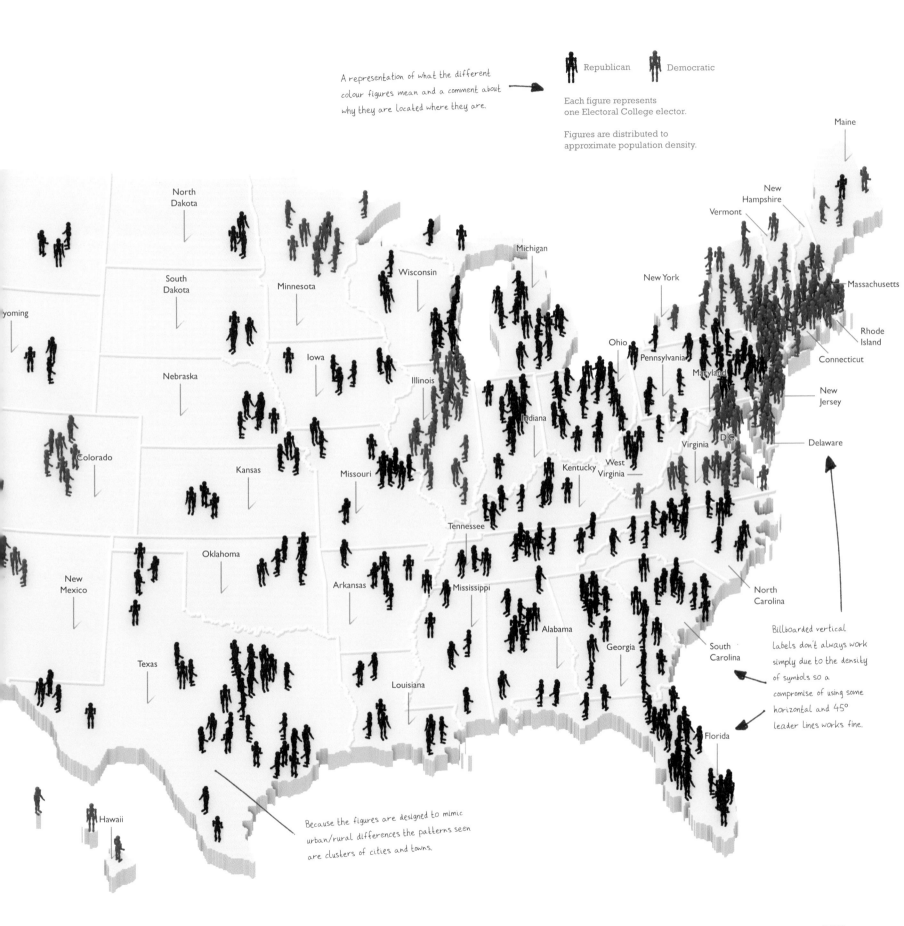

A representation of what the different colour figures mean and a comment about why they are located where they are.

Republican   Democratic

Each figure represents one Electoral College elector.

Figures are distributed to approximate population density.

Maine

New Hampshire

Vermont

Massachusetts

North Dakota

Michigan

New York

Rhode Island

Wisconsin

Connecticut

Minnesota

Ohio

South Dakota

Pennsylvania

Maryland

Wyoming

Iowa

New Jersey

Nebraska

Illinois

Indiana

D.C.

Delaware

Virginia

Colorado

West Virginia

Kentucky

Kansas

Missouri

Tennessee

Oklahoma

North Carolina

New Mexico

Arkansas

Mississippi

Alabama

South Carolina

Texas

Georgia

Louisiana

Florida

Hawaii

Billboarded vertical labels don't always work simply due to the density of symbols so a compromise of using some horizontal and 45° leader lines works fine.

Because the figures are designed to mimic urban/rural differences the patterns seen are clusters of cities and towns.

Q*Map

Extruded waffle grid

Whether it's prisms or a surface or any other type of area feature, encoding something additional in the z-dimension is the key to 3D thematic mapping. The title of this map is a play on the classic arcade game Q*BERT® because it extrudes a grid of squares and takes on the same isometric block appearance.

The basis of this map is the waffle grid cartogram of historic election wins from 1920 to 2016 illustrated earlier but now modified into 3D. The planimetric version in chapter 7 used a gridded cartogram in which each state was represented by a square, and each square was then subdivided into a 5 × 5 grid of 25 smaller areas. Each of the squares represents a single election cycle beginning with 1920 in the bottom left, running left to right, then bottom to top, and finishing with 2016 in the top right. In the planimetric version, each square is symbolised by the winner of the election in that state in that year. Because the pattern is repeated for each state, there's a consistent basis for visual comparison on the basis of familiarity. The same information is in the same place for each state, and the only change is in colour.

This map encodes share of the vote, as a percentage, into the z-dimension, which provides additional detail to give the map reader more than only the binary winner. It becomes a visual statement of the extent of the win in a year. The z-dimension could have shown total votes (used on other extruded maps in this chapter), except that would show the large variation in population distribution between states, and perhaps detract from the idea of a first-past-the-post system in which gaining 50 percent of the votes is the critical cut-off regardless of how many voters turn out. It also provides an interesting landscape that supports comparison of like for like because it's percentages and not absolute values.

The map also uses a pure isometric form of the axonometric parallel projection in which internal angles are 120 degrees. It gives a regular grid with scale consistent throughout, from front to back. The top and two sides of each rectangular column can be seen at the same time and in the same proportions. None of the pairs of axes (i.e. x-z, y-z, or x-y) are parallel to the plane of the page itself, and the main function is to support a general overview of the data and the relative values and trends from place to place.

The drawback of using tightly packed extruded columns, of course, is that you cannot see all of every column, and occlusion is a problem. By adding a frame above the columns that demarcates the edges of the squares that represent each state and extending vertices at each corner, you at least can estimate how far from the 100 percent hypothetical maximum each column sits. By inference, it's possible then to estimate the value of the height of the column and the percentage share of the vote.

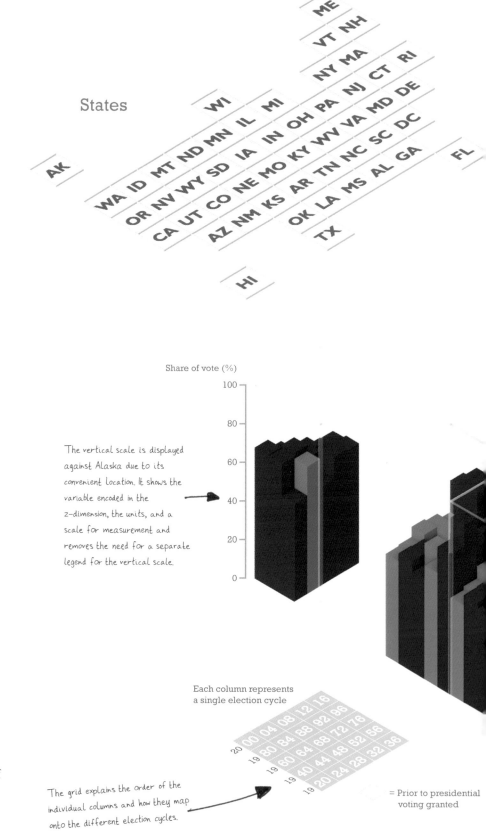

States

The vertical scale is displayed against Alaska due to its convenient location. It shows the variable encoded in the z-dimension, the units, and a scale for measurement and removes the need for a separate legend for the vertical scale.

Share of vote (%)

Each column represents a single election cycle

The grid explains the order of the individual columns and how they map onto the different election cycles.

= Prior to presidential voting granted

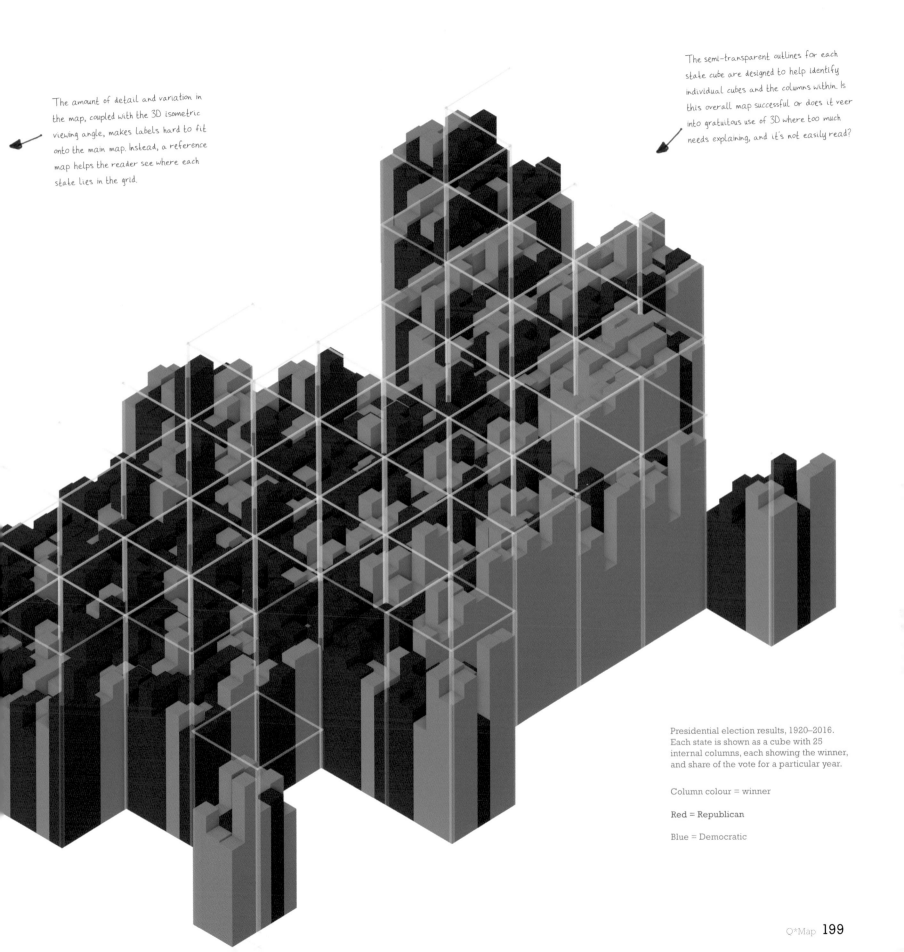

The amount of detail and variation in the map, coupled with the 3D isometric viewing angle, makes labels hard to fit onto the main map. Instead, a reference map helps the reader see where each state lies in the grid.

The semi-transparent outlines for each state cube are designed to help identify individual cubes and the columns within. Is this overall map successful or does it veer into gratuitous use of 3D where too much needs explaining, and it's not easily read?

Presidential election results, 1920–2016. Each state is shown as a cube with 25 internal columns, each showing the winner, and share of the vote for a particular year.

Column colour = winner

Red = Republican

Blue = Democratic

Electoral causeway

Stratified areal space-time cube

A single map has graphical limits, which can be impacted by the size of the final product as well as the complexity of the phenomenon being shown. And for many maps, it's important to show the temporal dimension to illustrate how phenomena change from one period to another or across multiple periods. Animated maps and small multiples are common ways of achieving this, and modelling time in the z-dimension is another.

Time can be plotted on the z-axis in a 3D map, which supports stacking maps one on top of another, creating a form of space-time cube. In the example to the right are 25 separate maps, each mapping the winner of an election from 1920 to 2016 by state as a gridded hexagonal cartogram.

The earliest dated map is at the bottom of the stack, and each subsequent map is positioned offset vertically in the z-dimension with equal spacing until the result of the 2016 election is placed atop. **Each map is extruded to a consistent height giving a space-time block of hexagons.** The design takes its lead from nature by re-creating the appearance of a grid of geological columns of basalt such as those found at Devils Postpile National Monument in California or the Giant's Causeway in Ireland.

The same approach could have been taken with a geographical map but the complexity of a space-time cube is sometimes hampered further by using complex real-world shapes. The simpler the shapes, the easier it is to disentangle the map, so arguably, a gridded cartogram works better than a geographical map. **The regular shapes help keep the map simple in structure, and therefore easier to understand.** Occlusion is inevitable because some parts of the map are stacked in front of others, and the ability to see the internal structure of each stack is lost. This is overcome in an interactive environment with the ability to cut away layers in both the horizontal and vertical planes.

Plotting time vertically provides a good way to see how the election result changes from one cycle to another by state. A quick glance at an individual stack shows immediately whether a state has been predominantly blue or red or flipped, or even had the odd year when neither of the candidates for the two main parties won.

You could make plenty of alternative choices in modifying this map. Different geometric shapes are one option, as are nested symbology for each state, perhaps to show the breakdown of the result per state per year by party. But the overall appearance quickly deteriorates if symbols are overloaded. 3D already complicates the view. Keeping the rest simple is a good tenet to follow.

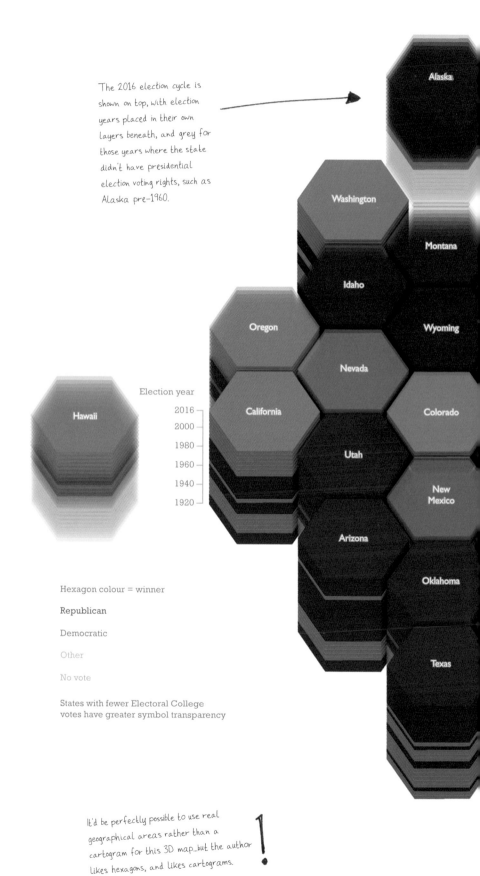

The 2016 election cycle is shown on top, with election years placed in their own layers beneath, and grey for those years where the state didn't have presidential election voting rights, such as Alaska pre-1960.

Election year

2016
2000
1980
1960
1940
1920

Hexagon colour = winner

Republican

Democratic

Other

No vote

States with fewer Electoral College votes have greater symbol transparency

It'd be perfectly possible to use real geographical areas rather than a cartogram for this 3D map...but the author likes hexagons, and likes cartograms.

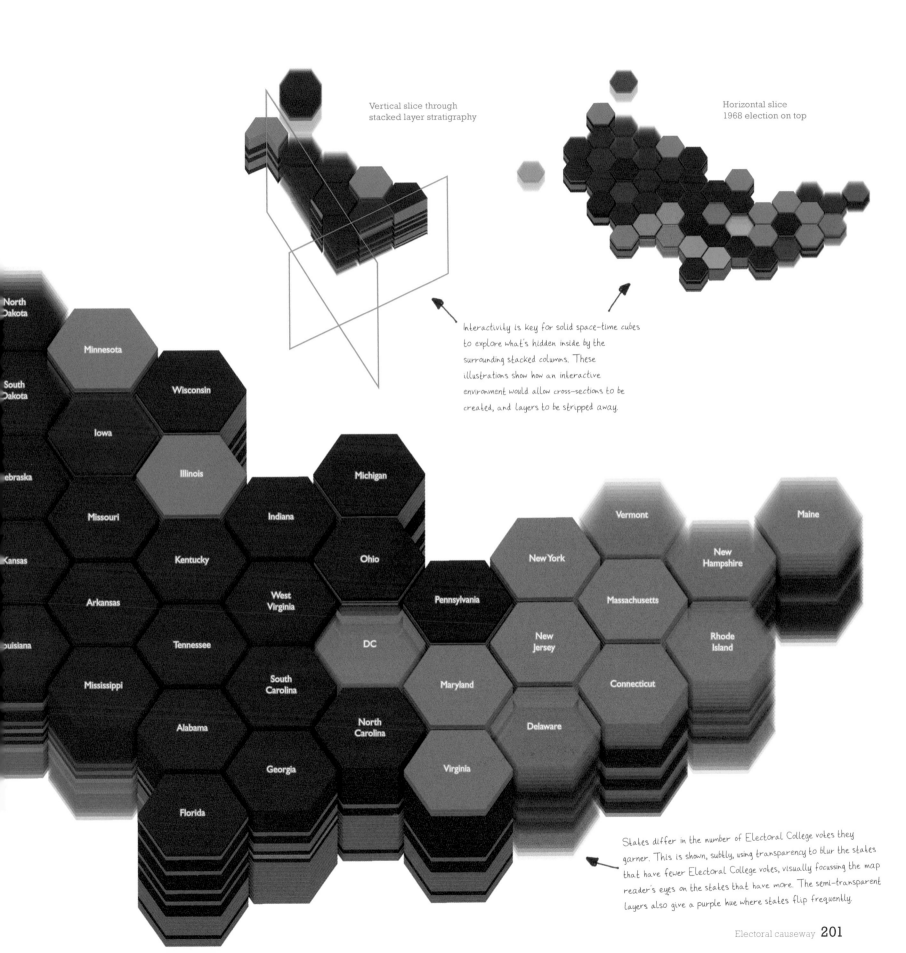

Vertical slice through
stacked layer stratigraphy

Horizontal slice
1968 election on top

North
Dakota

South
Dakota

Minnesota

Wisconsin

Iowa

ebraska

Illinois

Michigan

Missouri

Indiana

Kansas

Kentucky

Ohio

Vermont

Maine

New York

New
Hampshire

Arkansas

West
Virginia

Pennsylvania

Massachusetts

ouisiana

Tennessee

DC

New
Jersey

Rhode
Island

Mississippi

South
Carolina

Maryland

Connecticut

Alabama

North
Carolina

Delaware

Georgia

Virginia

Florida

Interactivity is key for solid space-time cubes
to explore what's hidden inside by the
surrounding stacked columns. These
illustrations show how an interactive
environment would allow cross-sections to be
created, and layers to be stripped away.

States differ in the number of Electoral College votes they
garner. This is shown, subtly, using transparency to blur the states
that have fewer Electoral College votes, visually focussing the map
reader's eyes on the states that have more. The semi-transparent
layers also give a purple hue where states flip frequently.

Can't see the forest for the trees

Data spikes

A lot of maps sit between what is conventional and what stretches into artistic data visualisation. To the right is a series of data spikes created from an interpolated grid of the election data. Is the image pleasing? Is it possible to hate a map? Does one care enough about what they're seeing to want to engage with it more? Is it a map? Is it a #cartofail?

In an objective sense, this data visualisation does not allow the recovery of much meaning, although deliberately so because it's created with a more artistic than analytical purpose. The projection used is a perspective one, and that means assessing the relative height of the data spikes is problematic, so a vertical scale is omitted. Red hues are Republican and blue are Democratic. For data that does not have clear colour associations, interpreting even this sort of basic information can be difficult. There isn't a map base to locate the spikes or labels to aid interpretation, but because the density of spikes creates a familiar shape, it perhaps proves a base isn't always necessary.

Multiple variables are encoded into each data spike but there's no graphical legend. To include one would suggest a more analytical purpose. **Simple statements that make clear what variables the spikes represent are enough for the map reader to form an idea of what they are seeing.** But the complex interplay of variables would make trying to deconstruct each spike into its constituent parts impossible.

What is compelling and useful about this sort of data visualisation is that it's arguably attractive and eye-catching, though that, of course, is also a subjective statement. **Often, such visuals veer into becoming data art because their intent is to turn data into something visual, for the simple purpose to visually delight.** And if that's the goal, it's fit for the purpose. If it includes legends and explanatory statements, it begins to suggest it supports a more analytical set of functions, which it cannot achieve. So the downside to this sort of data visualisation is that, all too often, they are made in an attempt to make an objective product which mischaracterises what they are capable of doing.

The whole purpose of this book is to showcase the good—and ignore the bad and the ugly. Along the way, it explains what works and what isn't so good for different maps and techniques. It would have been easy to ignore these sorts of data spike maps (or data spike balls if they're on a globe) because they have hardly any analytical function. But they do support a basic function in that they can create a visually different piece of data art. And that has worth, or not, you be the judge.

Note, not calling this a map.

Data spike encoding

Red = Republican

Blue = Democratic

More transparent = fewer total votes for winning candidate

Darker colour = larger share of the vote for winning candidate

Height = total votes (turnout)

How would you use this data visualisation? The point is that you don't. These sorts of dataviz serve one purpose, and that's to imagine a dataset in an often unconventional, more artistic form. You like it to look at or you don't. If you don't expect any more from such forms of artistic data visualisation you won't be disappointed.

Sometimes it's worth making a map in order to appreciate that actually, it isn't a particularly good idea (or possibly even a map). Many a map never sees the light of day due to the author ending up not wanting to share it, or peer review and critique suggesting it might not be worthy. This is one such example, in the opinion of its author but hey, you might like it, so it got included.

It's a sign!

3D gridded chartmap

In data visualisation, not many approaches are so heavily derided as turning perfectly good 2D graphs, charts, or plots into 3D. The 3D pie chart causes all sorts of perceptual problems because of the inability of the eyes and brain to assess relative angles and surface area when presented in perspective. Even 3D bar graphs, or almost any graph in which depth is added, comes up for criticism. But depth can be added that reflects data itself, and not simply a gratuitous graphical accoutrement. Using 3D on graphs to add an additional variable can extend its information potential.

The same might be said for 3D maps because the first question when considering 3D over planimetric approaches is simple: **What is gained by going into 3D, and can the additional issues and constraints be handled satisfactorily?** Answering that question might differ depending on subjectivity and the demands of the intended audience as much as determining the best way to show the data. So it's with a leap of faith that the map to the right, which plots graphs in 3D and sits them on top of a 3D map, is created. It's a 3D chartmap combining elements from the two preceding chapters.

Because of the geography of the United States, the use of a gridded cartogram for the base makes more sense because it creates a uniform amount of space for each state's graph. The graphs themselves show the share of the vote for each candidate with red Republican columns extending right, blue Democratic columns extending left, and orange-coloured other candidates extending forward. The vertical axis represents time with layers of graphs from 1920 to 2016. The winner for each election cycle is shown without transparency, whereas the losing candidate's share of the vote is shown in 80 percent transparency so there's a visual cue for more easily identifying the winner.

The map is presented using a parallel axonometric projection in an isometric configuration to ensure that the graphs stand upright and maintain the property of being scaled consistently across the map. The result also might be considered a form of space-time cube in which time is the z-dimension. Because the individual columns are fairly narrow, there's little overlap, and you can see the temporal dimension on each graph without the usual occlusion that comes with volumetric space-time cubes (such as the stacked hexagons earlier in the chapter).

Ultimately, venturing into the world of 3D for either graphs or maps can be problematic, but with a cautious approach and an attempt to keep it simple, they can work separately or in combination as a 3D chartmap.

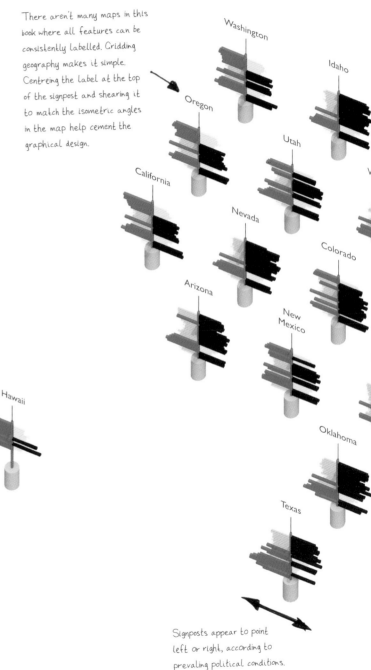

There aren't many maps in this book where all features can be consistently labelled. Gridding geography makes it simple. Centreing the label at the top of the signpost and shearing it to match the isometric angles in the map help cement the graphical design.

Signposts appear to point left or right, according to prevailing political conditions.

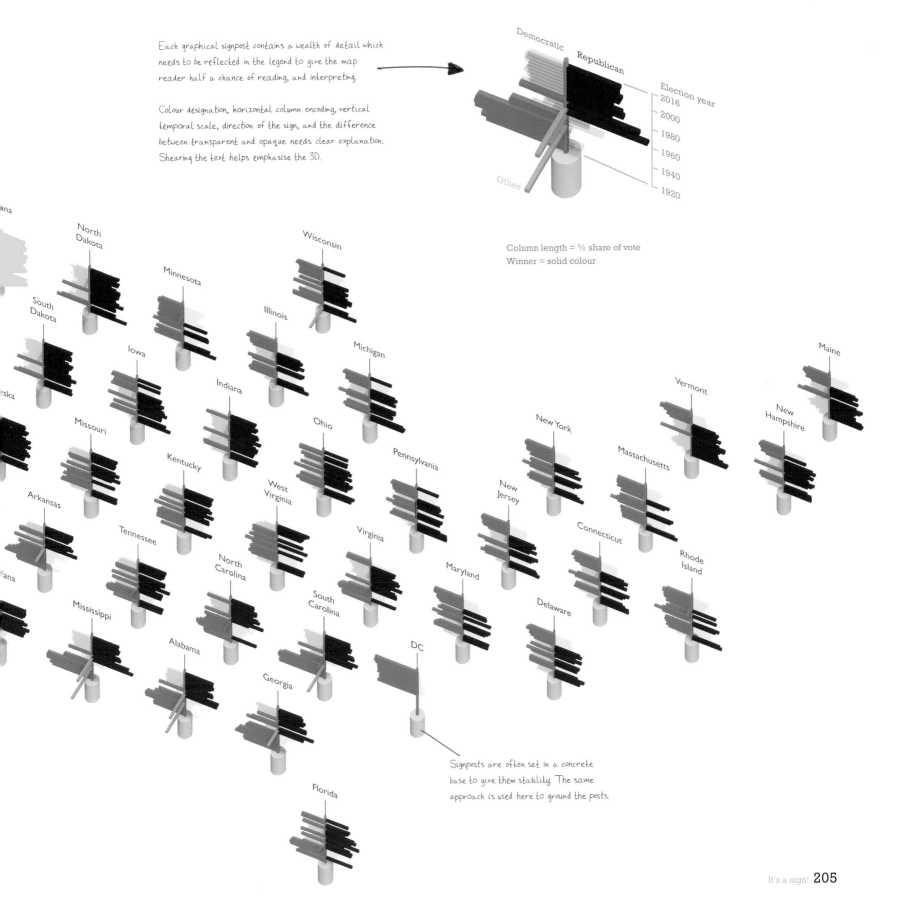

Each graphical signpost contains a wealth of detail which needs to be reflected in the legend to give the map reader half a chance of reading, and interpreting.

Colour designation, horizontal column encoding, vertical temporal scale, direction of the sign, and the difference between transparent and opaque needs clear explanation. Shearing the text helps emphasise the 3D.

Democratic

Republican

Election year

2016
2000
1980
1960
1940
1920

Other

Column length = % share of vote
Winner = solid colour

North Dakota

Minnesota

Wisconsin

South Dakota

Illinois

Iowa

Michigan

Indiana

Missouri

Ohio

Pennsylvania

Vermont

New Hampshire

Maine

Kentucky

New York

Massachusetts

Arkansas

West Virginia

New Jersey

Connecticut

Rhode Island

Tennessee

Virginia

North Carolina

Maryland

Mississippi

South Carolina

Delaware

Alabama

DC

Georgia

Florida

Signposts are often set in a concrete base to give them stability. The same approach is used here to ground the posts.

9. Curiosities

Most of this book has paid attention to an exploration of conventional forms of mapping of empirical data using the data from the 2016 US presidential election as a case study. Along the way, the intent has been to vary the data processing and analysis to support alternative visual treatments that tease out information in different ways. Some of the maps, graphics, charts, and plots have been more visually challenging than others, but they've all kept within sensible limits to act as exemplars for best practice. But what happens when mapping goes up to 11?

This chapter presents a few examples of how the same election data can be presented in artistic, experimental, and often bizarre ways. The utility of some of these techniques is deeply questionable for everyday needs but maps do not have to be confined to convention. Rules (such as they are, considering that they're more best practices) are there to be broken, and there are opportunities to unshackle maps and take them beyond the defaults.

The intent in this chapter is not to be comprehensive or exhaustive. Rather, the intent is to encourage experimentation and perhaps inspire tangential thinking about how to reimagine the map should the opportunity arise.

You have been warned.

Winner takes all

Single fill for the win

For many, the stark red and blue map showing the result on a state or county basis gives a false impression of reality. Accusations of bending the truth are common because there's no nuance in those maps. President Donald J. Trump used those maps because there is a predominance of red, and as a victorious Republican president (by whatever margin doesn't matter), it's precisely the map any winner would use to reinforce the result and demonstrate the extent of victory, particularly to their base.

But perhaps Trump missed a trick?

Results are aggregated from precincts, which define a set of addresses assigned to polling places, to larger units such as counties, and then totalled at state level. The state result determines who takes the Electoral College votes assigned to the state. But what about the country? **The winner takes all and governs the country.** The result of the popular vote doesn't count in a presidential election unless it coincides with winning enough states to win the Electoral College. Proportional representation is not the system in place. So perhaps the maps on this spread are the real result?

Winner

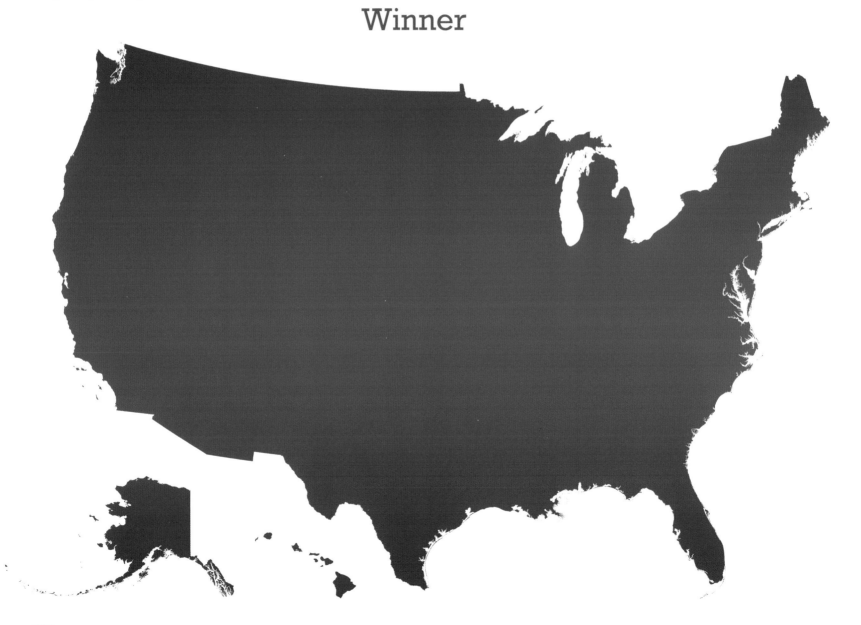

...and the loser falls.

Loser

A textured finish

Pattern fills

It's only a few decades since making maps was a manual process involving drawing pens and the photomechanical techniques of production. **Computers have so revolutionised the practice of mapmaking, and design more generally, that it's possible to forget what maps used to look like.**

A map of election results in a newspaper or a book much more likely would be in black and white than full colour, and making distinctions between classes of data across a map required the use of pattern fills rather than solid fills because of the limitations of design and production technology. Pens may have been used to draw consistent parallel lines to create a hatched fill. Pre-printed sheets of adhesive-backed patterns might have been used to build fills out of dots. The same basic tenets applied that remain in use today—namely, for qualitative differences, symbolisation is used that does not visually promote one class over another. For quantitative data, it's important to show areas as darker or lighter according to the corresponding empirical value. The following two maps illustrate these approaches using black ink on a white background with line and dot fills. Even without a legend, gaining a sense of the data and message is an intuitive map-reading process.

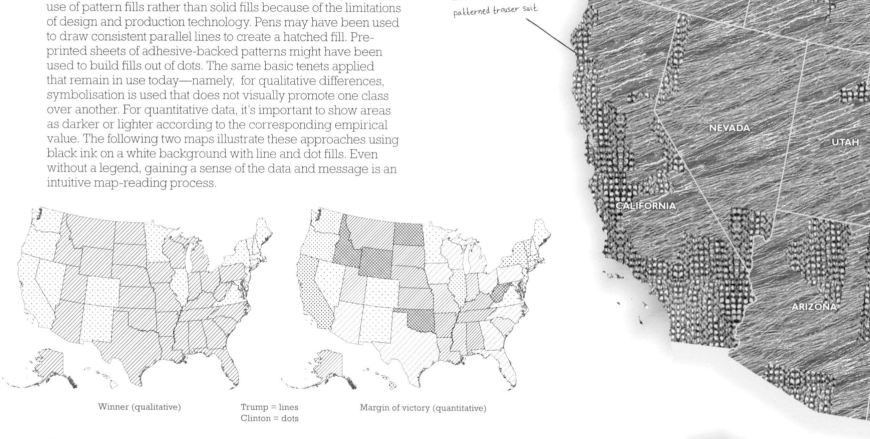

A shock of golden hair...

...and a swatch of a patterned trouser suit.

Winner (qualitative)

Trump = lines
Clinton = dots

Margin of victory (quantitative)

For the main map to the right, a pattern fill is used to show the qualitative difference between the counties that Donald J. Trump won and the counties that Hillary Clinton won. The only difference between the line and dot fills above and the main map is the use of modern technology to generate the pattern. The pattern fills are small clipped bitmaps that are used as a recurring pattern fill. Personality counts for a lot in elections. People arguably vote as much for the person and who they are based on aspects of their demeanor, dress, and character. This map goes down that route by riffing off an aspect of each candidate's public persona that became as much a part of the narrative of their candidacy as their ideas and policies.

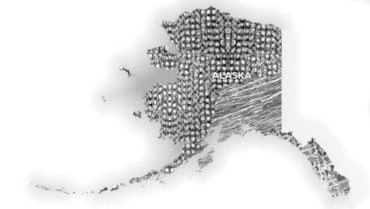

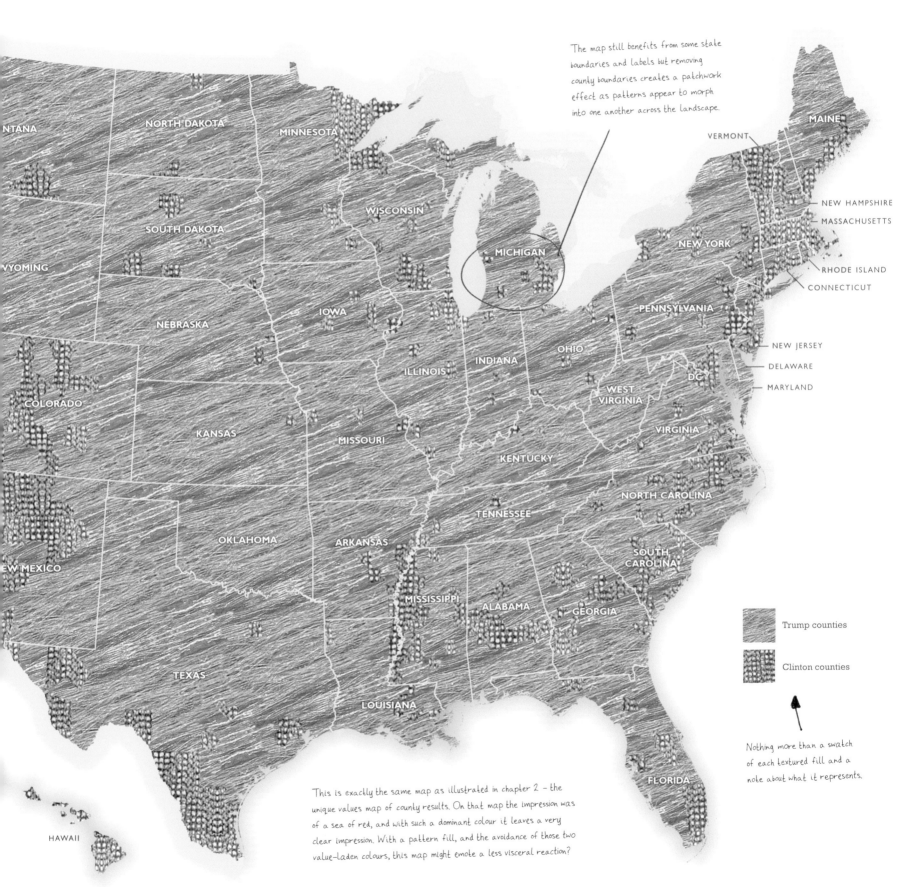

The map still benefits from some state boundaries and labels but removing county boundaries creates a patchwork effect as patterns appear to morph into one another across the landscape.

MONTANA
NORTH DAKOTA
MINNESOTA
MAINE
VERMONT
WISCONSIN
NEW HAMPSHIRE
SOUTH DAKOTA
MASSACHUSETTS
WYOMING
MICHIGAN
NEW YORK
RHODE ISLAND
CONNECTICUT
IOWA
PENNSYLVANIA
NEBRASKA
OHIO
INDIANA
NEW JERSEY
ILLINOIS
DC
DELAWARE
COLORADO
WEST VIRGINIA
MARYLAND
KANSAS
MISSOURI
VIRGINIA
KENTUCKY
NORTH CAROLINA
TENNESSEE
NEW MEXICO
OKLAHOMA
ARKANSAS
SOUTH CAROLINA
MISSISSIPPI
ALABAMA
GEORGIA
TEXAS
LOUISIANA
FLORIDA
HAWAII

Trump counties

Clinton counties

Nothing more than a swatch of each textured fill and a note about what it represents.

This is exactly the same map as illustrated in chapter 2 – the unique values map of county results. On that map the impression was of a sea of red, and with such a dominant colour it leaves a very clear impression. With a pattern fill, and the avoidance of those two value–laden colours, this map might emote a less visceral reaction?

Islands in a storm

Imagined landscape

The process of choosing what to put on a map or, conversely, what to leave off often is borne out of the desire to clarify some aspect of the data. Some maps are used to help inform a narrative, or a story, particularly in the visual journalism domain. Maps are used to bring clarity to aspects of the story that the overall piece is trying to tell. For instance, on one of the maps in chapter 2, only the flipped counties were shown, and all others were omitted from the map to reinforce that dimension in the data. Equally, only those counties that voted Republican or only those counties that voted Democratic could be selected and displayed as separate maps.

In fact, this approach was taken by Tim Wallace in an article in the *New York Times* ('The Two Americas of 2016,' November 16, 2016) when he created two maps displayed as complementary pieces. Titled *Trump's America* and *Clinton's America*, the former was a map of the US that covered most of the land area, with holes and a ragged coastline where Hillary Clinton's counties would have been. The latter was the inverse and took on the appearance of numerous small islands reflecting the location of isolated cities. By land area, Donald J. Trump won more than 3 million square miles (85 percent) compared with Clinton's 530,000 square miles, yet most of the population of the US lives in urban areas with Clinton's map representing 174 million people (54 percent), and Trump's map representing 148 million (46 percent). Excluding other candidates, Clinton won the popular vote with 62.1 million votes (50.5 percent) compared with Trump's 61.0 million votes (49.5 percent).

Although the results of elections often are thought of as divided, with Republican or Democratic, red or blue, us or them, and winners or losers, often votes are much closer than you might think considering the binary nature of the first-past-the-post system. **The map to the right selects those areas that are remarkably bipartisan and far more similar in their political expression at the ballot box than not.** These areas would be the most purple on a blended-hue choropleth map but instead they are represented using imagery to paint a picture of the topographic landscape on the basis of a selection criterion. **The map is intended to support a narrative that emphasises similarities as more profound than differences.**

This sort of map represents a more artistic approach. Selected counties could be shown in the same mid-purple but that would have been a little boring in graphic terms. By imagining the counties as islands, adding a faint coastline vignette around the extracted satellite imagery, and adding some real and playful names for the resulting land and water features, the map comes to life as an imagined landscape.

It's a fake map with fake islands and a fake sea but a fun way to take a different look at the election data.

Is this a map, or is it data art? There are certainly elements that are maplike, and the data is obviously of a map, but this sort of artistic representation eschews much of what's normally seen on a map.

It's designed to be seen, enjoyed, and not taken too seriously. It belongs more to the artistic end of the spectrum of cartography. It's relevant, and accurate, but not a product that is designed to support more serious commentary on the results.

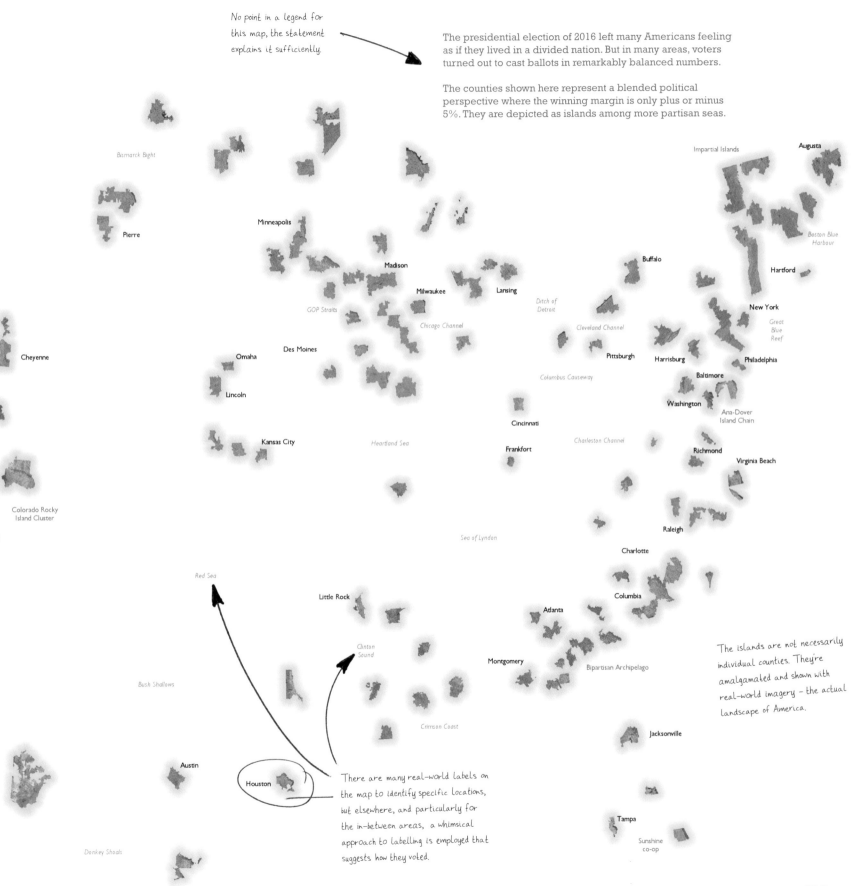

No point in a legend for this map, the statement explains it sufficiently.

The presidential election of 2016 left many Americans feeling as if they lived in a divided nation. But in many areas, voters turned out to cast ballots in remarkably balanced numbers.

The counties shown here represent a blended political perspective where the winning margin is only plus or minus 5%. They are depicted as islands among more partisan seas.

Bismarck Bight

Impartial Islands

Augusta

Minneapolis

Boston Blue Harbour

Pierre

Madison

Buffalo

Hartford

Milwaukee

Lansing

New York

GOP Straits

Ditch of Detroit

Great Blue Reef

Chicago Channel

Cleveland Channel

Cheyenne

Omaha

Des Moines

Pittsburgh

Harrisburg

Philadelphia

Columbus Causeway

Baltimore

Lincoln

Washington

Ana-Dover Island Chain

Cincinnati

Kansas City

Heartland Sea

Charleston Channel

Richmond

Frankfort

Virginia Beach

Colorado Rocky Island Cluster

Raleigh

Sea of Lyndon

ove

Charlotte

Red Sea

Columbia

Little Rock

Atlanta

The islands are not necessarily individual counties. They're amalgamated and shown with real-world imagery – the actual landscape of America.

Clinton Sound

Montgomery

Bipartisan Archipelago

Bush Shallows

Crimson Coast

Jacksonville

Austin

Houston

There are many real-world labels on the map to identify specific locations, but elsewhere, and particularly for the in-between areas, a whimsical approach to labelling is employed that suggests how they voted.

Tampa

Donkey Shoals

Sunshine co-op

Pop art carte

Halftone symbology

A lot of maps in this book approach the shading of a county through the generation of a fill or pattern. And many are a variation on using dots as part of that design. For instance, the dot density area map demonstrated how randomly placed dots can overcome some of the limitations that solid fills have. Dasymetric dot density maps go one step further. The map to the right is an extension of these ideas and uses a fairly simple technique that borrows from an old-school approach to printing: the Ben Day process, a form of halftone.

Halftones are a reprographic technique that uses dots of different sizes and spacing, but with a consistent colour and value, to create the illusion of a smooth fill or gradient. If dots are small enough or close enough together, the human eye can't see them individually, and instead you see a grey tone or a specific colour based on the mix of the individual dot colours.

The halftone has a long history stemming from the mid-1800s and was used heavily in map production until the late 1900s. Four-colour offset lithography also relies on the use of four printing inks, cyan (C), magenta (M), yellow (Y), and black (key, or K). Different screens are used to lay down ink in different amounts. So, for instance, a 10 percent magenta screen printed in combination with a 50 percent cyan and 20 percent yellow would create a midblue colour. The inks don't mix. The perception of a specific colour is a function of the 'subtractive' colour model. The CMYK printing inks absorb certain wavelengths of colour and reflect others. Their properties, the percentage of ink placed on the page via dots, and the pattern in which they are laid down by virtue of the screen angles gives the perception of a particular colour.

Comic book art in the mid-1900s also used halftone to good effect. To keep comics cheap, paper stock and printing were kept to a minimum in terms of cost, which led to a style that has become synonymous with comic art. **The dot patterns borne from the accidental four-colour separations create an almost unique aesthetic.** In fact, this style also was appropriated in the 1960s and 1970s by fine art and became a characteristic of pop art by Andy Warhol, Roy Lichtenstein, and others. Today, the style is considered retro and is what this map attempts to evoke.

This map uses a regular grid of dots for each candidate, positioned so that the location of blue dots is the inverse of red. They are sized by total vote and coloured according to the candidate's affiliation. The effect borrows from halftone (though it doesn't employ different screens at different angles) and gives a pop art–inspired representation of the pattern of voting. Reds appear dominant in some places, blue in others, and an element of mixing occurs to give the impression of bipartisan purple in some places.

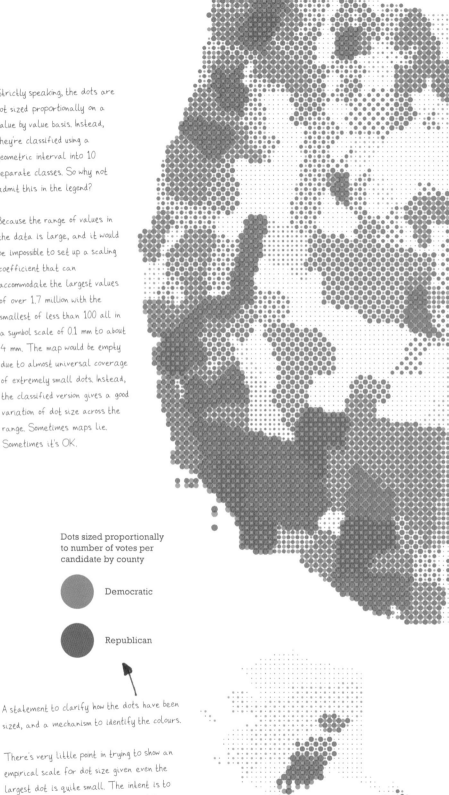

Strictly speaking, the dots are not sized proportionally on a value by value basis. Instead, they're classified using a geometric interval into 10 separate classes. So why not admit this in the legend?

Because the range of values in the data is large, and it would be impossible to set up a scaling coefficient that can accommodate the largest values of over 1.7 million with the smallest of less than 100 all in a symbol scale of 0.1 mm to about 4 mm. The map would be empty due to almost universal coverage of extremely small dots. Instead, the classified version gives a good variation of dot size across the range. Sometimes maps lie. Sometimes it's OK.

Dots sized proportionally to number of votes per candidate by county

Democratic

Republican

A statement to clarify how the dots have been sized, and a mechanism to identify the colours.

There's very little point in trying to show an empirical scale for dot size given even the largest dot is quite small. The intent is to show a relative pattern, not an absolute one.

The Ben Day process differs from the halftone process in that Ben Day dots are always of equal size and distribution in a specific area.

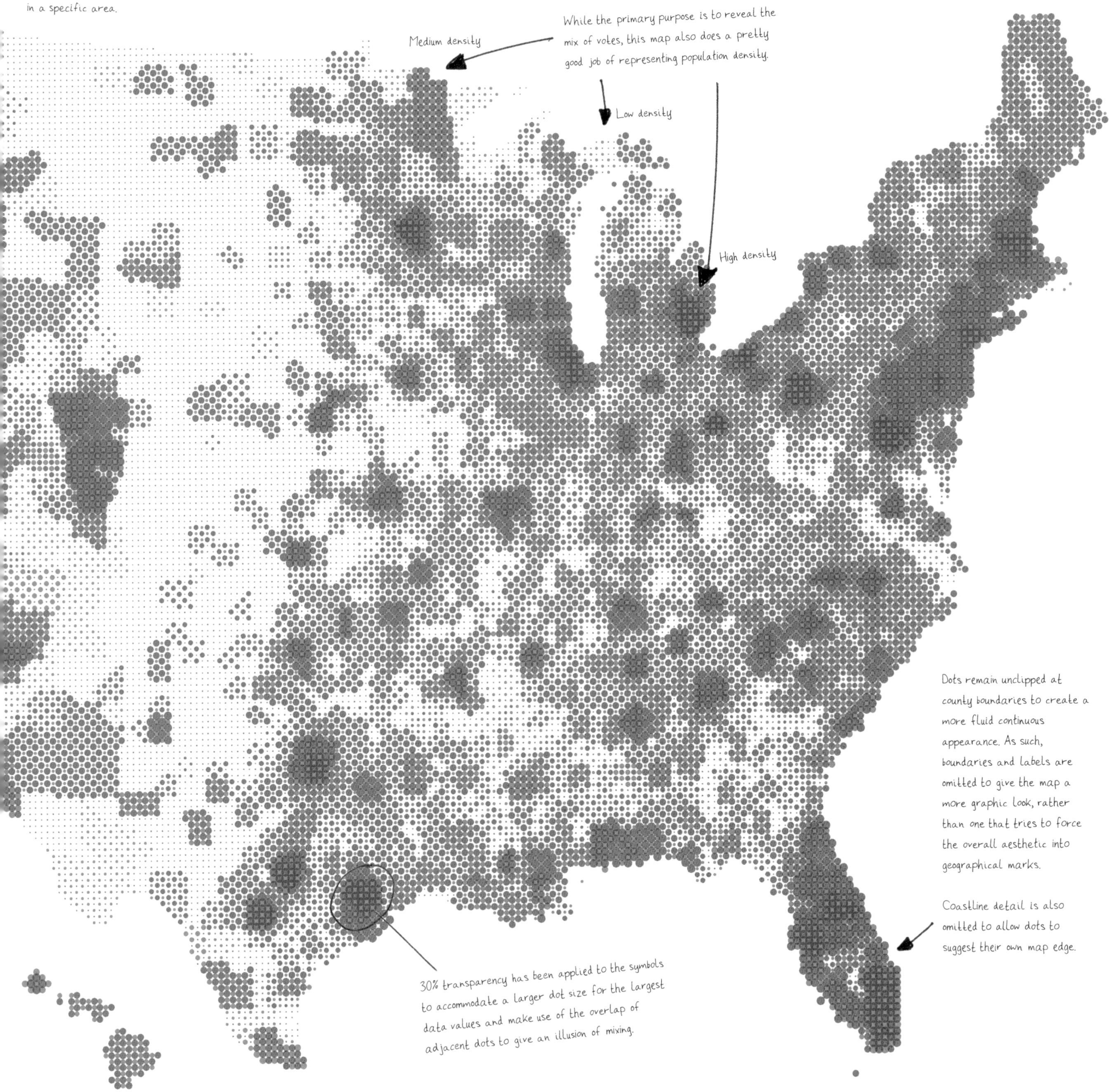

Medium density

While the primary purpose is to reveal the mix of votes, this map also does a pretty good job of representing population density.

Low density

High density

Dots remain unclipped at county boundaries to create a more fluid continuous appearance. As such, boundaries and labels are omitted to give the map a more graphic look, rather than one that tries to force the overall aesthetic into geographical marks.

Coastline detail is also omitted to allow dots to suggest their own map edge.

30% transparency has been applied to the symbols to accommodate a larger dot size for the largest data values and make use of the overlap of adjacent dots to give an illusion of mixing.

Mountains and molehills

Multivariate symbol landscape

Multivariate symbology comes in many forms. Some of the more inventive ways to design symbology come from thinking about the data in imaginative ways. The map to the right was inspired by an example published in the *Washington Post* that envisaged the electoral results as a mountainscape. **Although the results of elections are simple in many respects, reporting only the outcome would lead to an empty newspaper (or website).** The nuance in how the result is achieved is much more of an interesting story because it shows how different parts of the country voted. Local similarities are borne out, large urban/rural disparities are seen, and identifying broad patterns can begin to help unravel how the election was won or lost.

Here, the county results are displayed as a triangular symbol that has numerous pieces of information encoded. **The height of each symbol encodes the total number of votes cast, which helps give a sense of the distribution of the population.** Larger mountains appear as single peaks for large urban centres and major cities or along ranges, particularly on the Eastern Seaboard where a string of major cities exist in close proximity. By contrast, more sparsely populated areas tend to appear more like a series of small molehills with much smaller symbols.

The width of each mountain changes according to the margin of victory. Indeed, there's a more subtle metaphor at work with a large broad base given to the mountains where the margin of victory was larger, reflecting a larger base for the winning party in that county. The mountain narrows when there's a more bipartisan electorate.

Colour is used in its usual way to encode the winning party. The main symbol elements are accentuated by a few graphical flourishes (transparency, bold line weights, and white peaks), and the mix of these different elements creates a rich set of different symbols. No two symbols on the map are the same because of the way in which different characteristics of the vote are mixed.

The result is an abstract statistical landscape. **Each mountain is positioned so that its base is centred in the county, but that means for many of the larger mountains, the peak is likely well outside the area itself.** It's as if the map is planimetric but with a landscape seen in aspect across the top. The fact that the mountains are in outline only and have no fill helps avoid the usual problems of occlusion when using solid fill symbols. There's clutter where there are many overlapping symbols, but the overall effect is the aim, and the compromise is losing the ability to identify every single county or unique result and being able to quantify it.

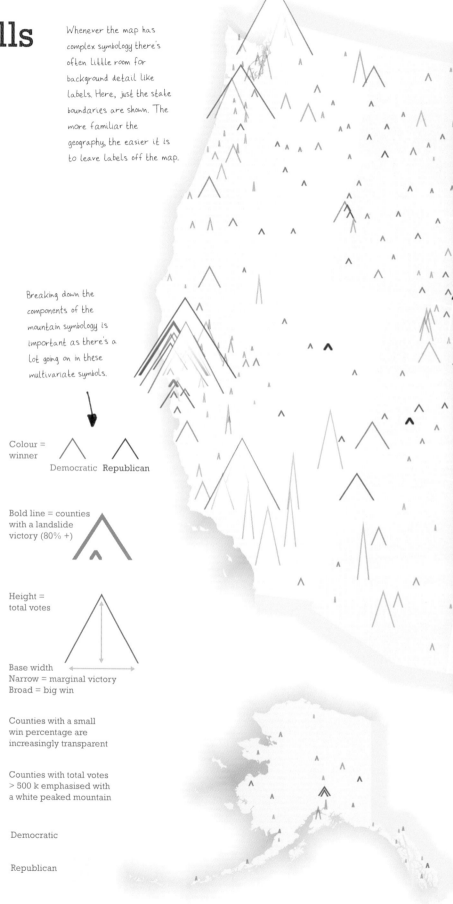

Whenever the map has complex symbology there's often little room for background detail like labels. Here, just the state boundaries are shown. The more familiar the geography, the easier it is to leave labels off the map.

Breaking down the components of the mountain symbology is important as there's a lot going on in these multivariate symbols.

Colour = winner

Democratic   Republican

Bold line = counties with a landslide victory (80% +)

Height = total votes

Base width
Narrow = marginal victory
Broad = big win

Counties with a small win percentage are increasingly transparent

Counties with total votes > 500 k emphasised with a white peaked mountain

Democratic

Republican

Some states, like North Dakota, contain symbols which each have similar characteristics across all variables. Others, like Minnesota, have much greater variation.

States are coloured according to the winning party just to give the background a bit more context. Context provides the map reader the ability to see how counties are organised within the overall result.

Because of the vast difference between heavily populated areas and sparsely populated areas this symbology creates a landscape characterised by many small, rounded molehills punctuated by large, sharply peaked mountains.

Each landform tends to be surrounded by similar landforms, creating ranges. Changes in the landscape tend to occur along lines, creating ridges and valleys.

The hint of transparency applied to symbols with a lower win percentage gives the mountainscape a touch of depth. Mountains with a larger win percentage appear more distinct.

Politics will tear us apart*

Joy plot

Profiles or transects long have been used to show change in some phenomenon by illustrating the changes in height along a line drawn on a map. For instance, elevation profiles are a side-view of the changes in elevation along the profile line as if the terrain has been sliced open and can be seen in aspect. Conceptually (or, if making them by hand), straight-line profiles can be constructed from a contour map along any line drawn atop. The profile is constructed by plotting the position of contours along the line, and then interpolating a curve between the contour lines to represent the surface between these known points. Transects add to the profile line by illustrating features that the line bisects, such as geological strata.

The same approach can show change in any empirical value across a surface, with stacked horizontal lines displaying a value of data vertically. Such maps have been referred to as *joy plots* (or *ridgeline plots*). They take their name from the seminal debut album *Unknown Pleasures* by Joy Division, from Salford, England, released in 1979 by Factory Records. The reference isn't to any of the music, but rather, the iconic album cover for the record. Designed by graphic designer and Factory Records cofounder Peter Saville, the cover shows a reinterpretation of the image of radio waves from the first discovered pulsar. It creates a series of vertically stacked lines that vary in height. **The term 'joy plot' often is used to reference a set of stacked lines that takes on the appearance of that classic Joy Division album cover.**

When applied to an empirical surface, the technique produces ridges and valleys but with intriguing overlaps that allow you to see behind an area that ordinarily would be occluded with a solid surface. The spacing of the lines is vital. Too few, and you lose a great deal of character from the underlying data. Too many, and the outcome will resemble a mess of spaghetti.

Vertical exaggeration almost always is required to translate many similar low values to something that gives a more mountainous appearance. This is a commonly used method in cartography to emphasise a surface to bring the salient characteristics into view. Without vertical exaggeration, the raw percentage share of the vote would appear flat at this scale. This map has 10,000 × vertical exaggeration, so that at this scale, the peaks of the highest share of the vote appear as mountains approximately 45 km in height.

Information retrieval is not the aim of this map. It's an abstract representation of the results, more usefully used to capture attention and invite curiosity.

* A nod to Joy Division's classic track 'Love Will Tear Us Apart'.

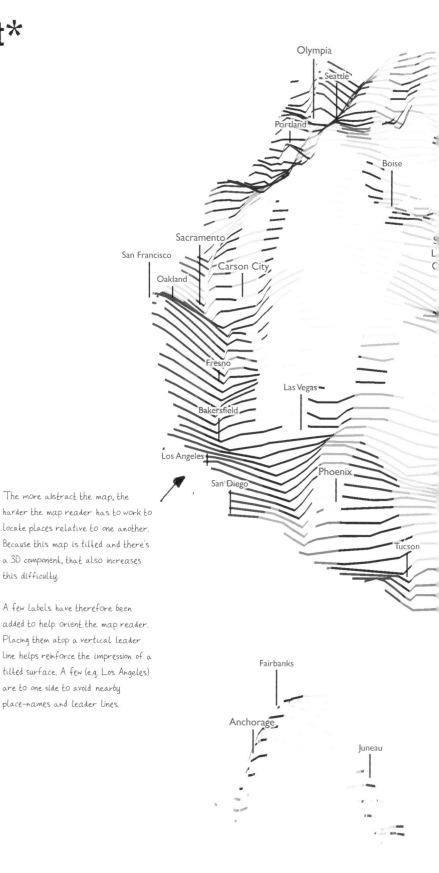

The more abstract the map, the harder the map reader has to work to locate places relative to one another. Because this map is tilted and there's a 3D component, that also increases this difficulty.

A few labels have therefore been added to help orient the map reader. Placing them atop a vertical leader line helps reinforce the impression of a tilted surface. A few (e.g. Los Angeles) are to one side to avoid nearby place-names and leader lines.

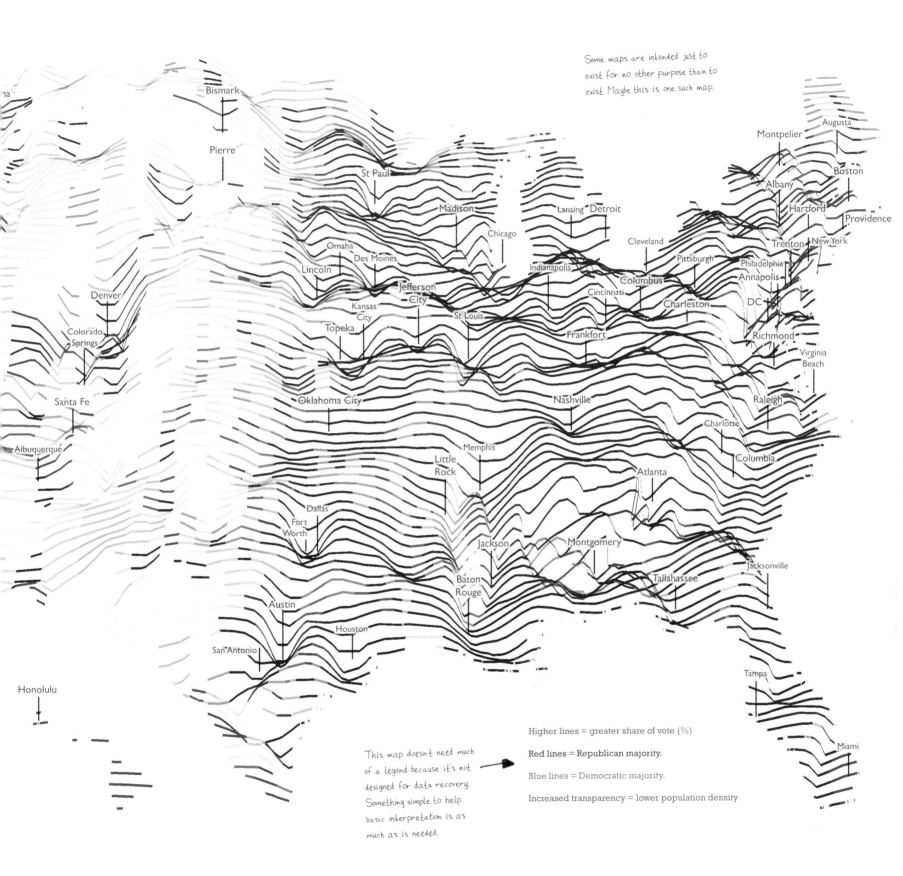

Some maps are intended just to exist for no other purpose than to exist. Maybe this is one such map.

Bismark

Pierre

St Paul

Madison

Lansing   Detroit

Chicago

Montpelier   Augusta

Cleveland   Boston

Albany   Hartford

Omaha   Indianapolis   Columbus   Pittsburgh   Philadelphia   Trenton   New York

Des Moines   Annapolis

Lincoln   Cincinnati   DC

Denver   Jefferson   Frankfort   Charleston   Richmond

Kansas   City   Virginia

Colorado   Topeka   St Louis   Beach

Springs

Santa Fe   Nashville   Raleigh

Oklahoma City   Charlotte

Albuquerque   Memphis   Columbia

Little

Rock   Atlanta

Dallas

Fort   Jackson   Montgomery   Jacksonville

Worth   Baton   Tallahassee

Austin   Rouge

Houston

San Antonio   Tampa

Honolulu   Miami

This map doesn't need much
of a legend because it's not   ➤   Higher lines = greater share of vote (%)
designed for data recovery.
Something simple to help   Red lines = Republican majority.
basic interpretation is as
much as is needed.   Blue lines = Democratic majority.

                                             Increased transparency = lower population density

Old skool pen plot

Horizontal proportional lines

The map to the right is inspired by 1980s pen plotter technology, which itself supported a form of printing that had its roots in hand-drawn techniques. You don't have to go back too far to a time when many maps were printed in black and white, and because printing block colours in any shade was extremely expensive, pattern fills were commonplace. These fills may have been hand-drawn with a ruling pen and so were limited to lines of different shapes (dashed, solid) and widths (thin, thick) or regularly placed dots, as shown earlier in this chapter. By varying the combination of shapes, line angles, and so forth, it was possible to create patterns that mimicked light to dark and build a shading scheme that allowed the map reader to see less ink as lower data values and more ink as higher data values.

Then pen plotters arrived. **Pen plotters were one of the earliest forms of computer peripheral printing devices, which can be traced back to the mid-1950s.** They were mechanical and consisted of a pen holder to hold a pen and a bed or drum that held the paper. Paper would move in the y-axis, and the pen holder would move along rods on the x-axis. The pen was lowered to apply ink to the paper, and so the combination of movement in three planes (x, y, and z) allowed the plotting of a map. In terms of style, the maps were rudimentary and paralleled hand-drawn techniques using combinations of line shape, thickness, and orientation to create vector-drawn pattern fills for early computer-drawn thematic maps. Plotted maps could incorporate some degree of colour encoding simply by changing the colour of the pen.

Computer graphics and printing technology long have surpassed the pen plotter era but the style of precisely drawn repeatable lines remains a distinctive approach to designing a map. **In the example here, a set of evenly spaced horizontal lines clipped to the shape of the United States is the basic graphic primitive.** The lines are symbolised to encode three pieces of information. Line thickness encodes the percentage share of the vote and is effectively a proportional line symbol. Thicker lines represent a larger share of the vote. Colour shows the majority winner in the same way as two pens, red and blue, might have been interchanged on a pen plotter. **The final piece of information, population density, is encoded using transparency, and although this certainly wasn't possible with pen plotters, anyone who used one will recall the effect of when the pen began running out of ink and colour was applied erratically and with less saturation.** It's a nod to that moment of having to change the pen for a new one and therefore having to begin plotting the map again.

The patterns here are only one configuration. Vertical lines, cross-hatching, dashed lines, and fishnet grids all could yield interesting but different effects.

Thicker lines = greater share of the vote (%)

Red lines = Republican majority.

Blue lines = Democratic majority.

Increased transparency = lower population density

Virtually the same legend as the map on the previous page because it's the same variables but line thickness replaces height as a way of encoding the percentage share of the vote.

This map is stark. The horizontal lines are simply indicative of a cross-section running across the empirical landscape. And yet this causes issues. Because they are sample lines the map misses an enormous amount of detail from in between the lines. For instance, the spacing of the lines misses most of St Paul and Minneapolis in Minnesota. These would be indicated by blue elements in the line. So the lines only pick out general trends. They can easily miss out on differences at a more detailed level.

On a map that tends toward the abstract, it's often unnecessary to use the trappings associated with 'normal' maps. There's no coastline or other external borders here. The impact of the regular horizontal lines allows a sharp external shape to be seen even though it's not directly used.

Proportional line symbols are graduated with thicker lines showing a greater share of the vote and thinner lines a lower share. The lines are coloured by the majority as red or blue.

Transparency is applied to modulate the appearance of the map by population density, making the technique akin to a value-by-alpha map.

Ishihara election test

Dot density Dorling cartogram

If a Dorling cartogram isn't abstract enough, even more abstraction can be built into the symbology. To the right, a Dorling cartogram is scaled by the total votes cast (turnout). Inside each circle are many more circles (dots), each one equivalent to 1,000 votes, and coloured according to the candidates. **It's a dot density Dorling cartogram and looks a little like the Ishihara colour deficiency test diagrams developed by Dr. Shinobu Ishihara in 1917.**

The effect is dramatic and invites curiosity, principally because it's different. Because of the proximity of the relatively small dots, perceptually the reader sees some mixing. The 18 states, plus DC, that tend toward blue are those that Hillary Clinton won and which have a blue state label to indicate the result. The extent of victory by share of the vote can be seen in the strength of the blue, so DC is almost completely blue although it is relatively small because of its much smaller population size. By contrast, Donald J. Trump gained 26 states in a convincing manner, which tend toward red because of the internal mixing of the coloured dots. These states have a red state label. **Without the labels, it'd be hard to discern which states went which way in many cases because the mix of coloured dots is almost even there.**

Elections often are won or lost on fine margins, and visually in a few states it's difficult to identify whether there is a tendency toward red or blue. That's because there isn't much of a tendency at all. In fact, they appear a little more purple because of the perceptual mixing of colour, reminiscent of a blended-hue colour scheme that goes from red, through purple, to blue, which was used on a choropleth map example in chapter 2. The purple dashed outline indicates a marginal result of less than 2 percent. Clinton won New Hampshire and Minnesota by a thin margin. Trump narrowly won Michigan, Florida, Wisconsin, and Pennsylvania, all states he managed to flip from Obama's victory in 2012 and which were crucial to him passing the magic 270 Electoral College votes required to win the presidency.

What's perhaps even more interesting is that Trump ended up not needing Florida to win. Trump gained Pennsylvania, Michigan, and Wisconsin, with a total of 107,330 more people voting Republican than Democratic, or approximately 0.09 percent of all people who voted. But what are numbers without comparison with something more relatable? Put another way, that's roughly the number of people who would fit in a major sports stadium. **If 36,000 Republican voters had been invited to each of Beaver Stadium (Pennsylvania, capacity 106,572), Michigan Stadium (capacity 107,601), and Lambeau Field (Wisconsin, capacity 81,441) and were persuaded to vote differently, Hillary Clinton would have been the 45th president of the United States.** Even if this is how it had played out, the maps in this book still would look broadly the same.

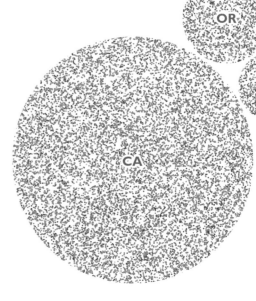

This map originally used a red or blue coloured outline to signify the winner for each state and labels were white. Two reviewers commented that the outlines seemed to dominate and led the eye to see dots as more red, or more blue by virtue of the outline. Instead just encode the winner with differently coloured text. So here is that version, and peer review and critique made the map much better as a result. Most maps in this book have been improved by others making critical comments, and suggesting small, or sometimes large, changes that have led to a better product. Shun critique at your peril.

Legend identifies dot value, and symbol sizing

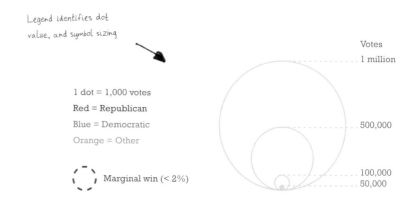

1 dot = 1,000 votes

Red = Republican

Blue = Democratic

Orange = Other

Marginal win (< 2%)

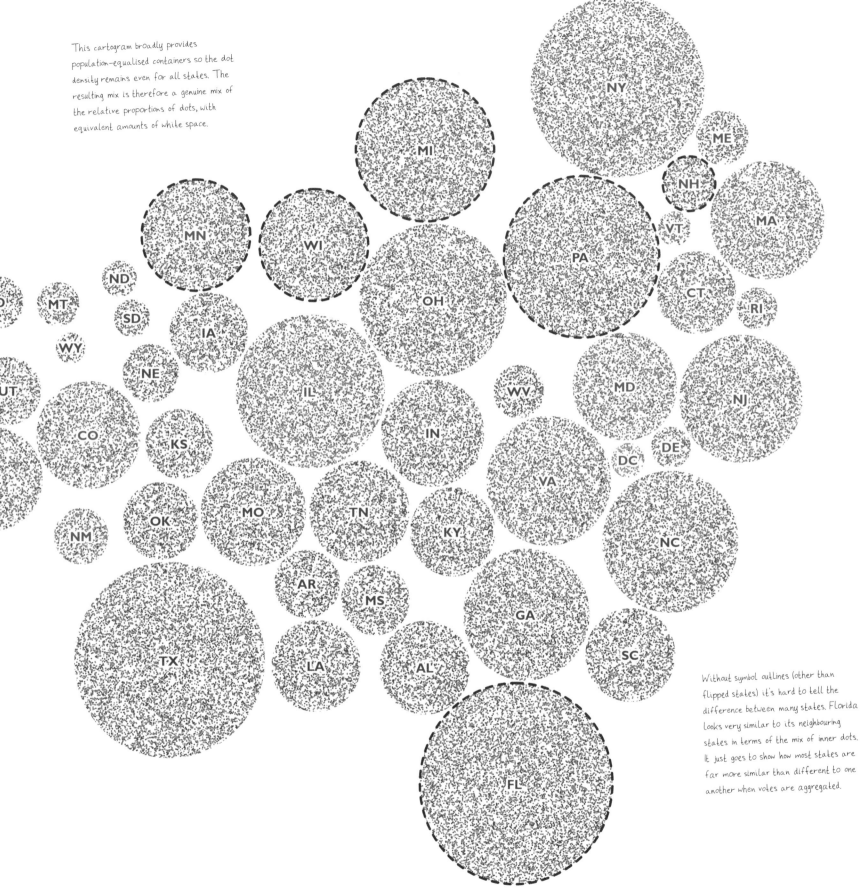

This cartogram broadly provides population-equalised containers so the dot density remains even for all states. The resulting mix is therefore a genuine mix of the relative proportions of dots, with equivalent amounts of white space.

Without symbol outlines (other than flipped states) it's hard to tell the difference between many states. Florida looks very similar to its neighbouring states in terms of the mix of inner dots. It just goes to show how most states are far more similar than different to one another when votes are aggregated.

Presidential puzzle

Abstract tessellated cartogram

The outcome of the 2016 presidential election was puzzling to many. If the intent of a map is to riff off that narrative, the map might need a graphical style to match. **A cartogram immediately transforms geography into a graphical rather than geographical style so it's a good starting point.** But instead of using a regular approach, you can go beyond geometrical shapes and do something more artistic.

The map to the right uses pictorial symbology to create something different. **It's a puzzle based on the mathematics of tessellation made famous by the Dutch artist M. C. Escher.** Escher's famed graphical work was influenced heavily by recurring shapes and so-called impossible constructions. All of Escher's tessellations begin with one of three regular shapes that can tessellate on a plane—namely, a triangle, square, or hexagon. There are three ways to deform the shapes: reflections (about a line of reflection), translations (deformation of an edge so that it is compensated on the opposing edge), and rotations (in which two opposite edges are shaped differently, and then rotated by 90 degrees about one of the vertices).

This map was inspired by the 1943 lithograph 'Reptiles' since the lizard on Escher's work has the same basic shape as a human of four limbs and a head. It is based on a square shape and requires translations of two opposite edges, and then rotated copies for the two remaining edges. Once the basic shape, illustrated below, is created, internal design is applied.

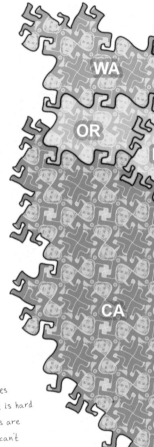

Different people will see different things in this sort of map. For some, it's just the fun of seeing intertwined presidential candidates, particularly where they share a common state boundary. The idea that these borders demarcate territories or political battlegrounds might be one way of interpreting the arrangement of the figures.

For others, the pattern that might emerge gives rise to different imagery. Because the figures are tessellated, other regular repeated shapes come to the fore. You might see a cross appearing as a repeated pattern, or maybe a flower head. The candidates might appear to be holding hands in some areas, or punching each other elsewhere.

But remember, it's just a playful, fun image. It's not to be taken literally, or seriously, as reinforced in the headline. It's just riffing off an artistic style.

1. Begin with a square	2. Reshape two opposite sides	3. Copy the two sides and rotate about a vertex	4. Design the figures	5. Tessellate

The mini Trump and Clinton figurines create a complex visual image that is hard to disentangle. Because light strokes are used as part of the figurines they can't really be used for other map boundaries. Instead, a dark stroke is used for state boundaries.

Abbreviated labels are also larger and emboldened across a busy graphic to enable them to be more visible, and legible.

The figurines are extended beyond the map in outline only, just to provide a background for the map, soften the transition between the map and the page, and because why not!

And Escher's original tessellated reptiles add to the whole page design.

The map shows 538 Electoral College votes, each as either Donald J. Trump or Hillary Clinton, tessellated into a pattern that resembles the general shape of the United States and subdivided into the number of Electoral College votes allocated to each state. These Electoral College votes sum to determine who wins the presidency. More populated states such as California (CA) have more Electoral College votes. Less populated states, such as North Dakota (ND), have fewer. There are 306 Donald J. Trump figures on the map and 232 Hillary Clinton figures. The visual ratio of red to blue is therefore 290 to 228, giving a visual representation of victory by the amount of colour.

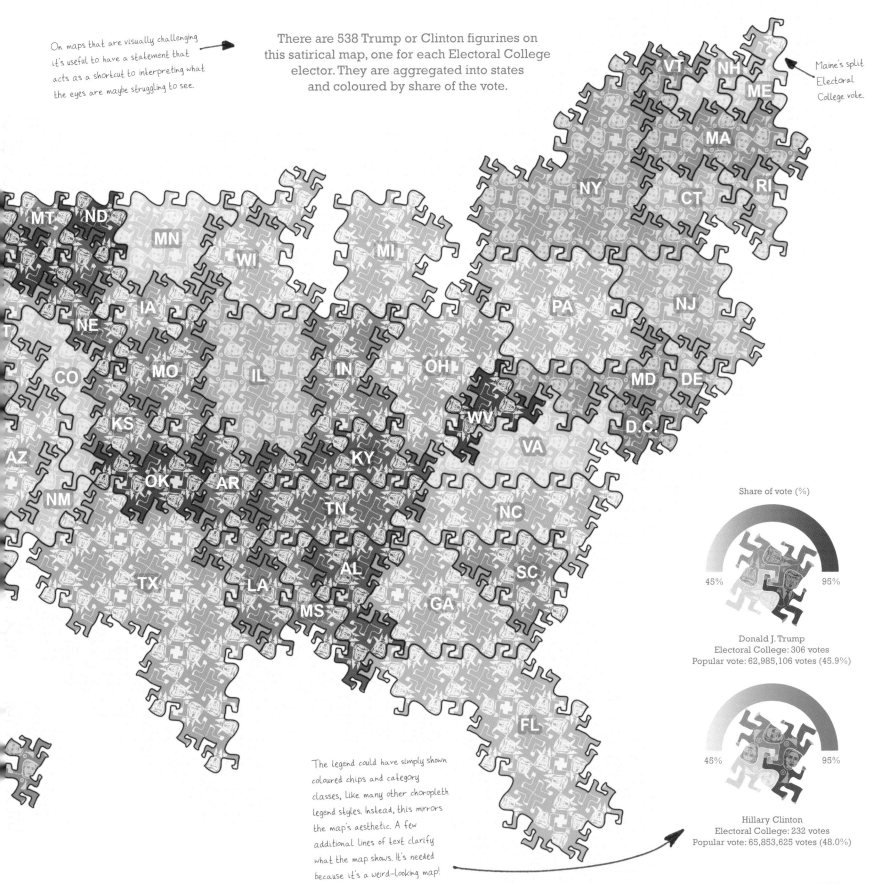

On maps that are visually challenging it's useful to have a statement that acts as a shortcut to interpreting what the eyes are maybe struggling to see.

There are 538 Trump or Clinton figurines on this satirical map, one for each Electoral College elector. They are aggregated into states and coloured by share of the vote.

Maine's split Electoral College vote.

Share of vote (%)

45%   95%

Donald J. Trump
Electoral College: 306 votes
Popular vote: 62,985,106 votes (45.9%)

45%   95%

Hillary Clinton
Electoral College: 232 votes
Popular vote: 65,853,625 votes (48.0%)

The legend could have simply shown coloured chips and category classes, like many other choropleth legend styles. Instead, this mirrors the map's aesthetic. A few additional lines of text clarify what the map shows. It's needed because it's a weird-looking map!

Go ahead, make my day

Steampunk'd gridded cartogram

Most maps in this book belong to a collection. That collection is partly defined by the topic chosen as a dataset for the maps. The initial work in chapter 1 explained the development of a consistent visual language. In essence, this is how an atlas is designed to ensure that, as each page is turned, there's a consistency and harmony to what is seen. Colours, layout, typographical components, and many other dimensions are generally similar throughout, even though the maps might be very different. **Alternatively, deliberately do something entirely different, and the choices are endless.**

Restraint is almost always a key tenet to cartographic design because it leads to cleaner, more useful maps, but that doesn't mean it's a hard-and-fast rule. It's just a best practice. So why not marry the representation of the data with a visual aesthetic commonly seen in other artistic representations? **A map can be mid-century modern. It also can be science fiction.** These aesthetics might evoke particular impressions of the data or tie them to a specific visual metaphor. The look and feel that is created is more than just the data; it's the whole map, and different emotional responses can be generated by different visual styles.

Instead of a map designed to show the data in a neutral way, brave stylistic treatments provoke the map reader to accept or reject the visual style as much as what it is trying to show. So how do you react to this steampunk aesthetic? **Inevitably, the further the visual style is taken from what might be considered the norm, the more likely the readership will be divided.**

Steampunk fiction is inspired by 19th-century industrial steam-powered machinery. It harks back to British Victorian or the American Wild West periods, or perhaps fantastical machines from the minds of authors like H. G. Wells or Jules Verne. Dials are often an important component, and to the right, a hexagonal gridded cartogram has been given the steampunk treatment. It's over the top and excessive. Each of the states has a dial that has all sorts of graphical embellishments that would be rejected immediately on a more restrained map. The same dials could have been shown without the visual adornment. **The data is almost secondary to the striking visuals.** That's perhaps also part of the appeal.

Maps therefore can be fashionable and riff off design ideas seen in art, literature, film, architecture, or any other creative endeavor. But fashions ebb and flow. **The more flourish is applied to the map to engineer a different look and feel, the more it can be left looking like the equivalent of a pair of flared jeans.** Everyone's got a pair in the cupboard, but who dares to be seen wearing them (until they come back into fashion, obviously)?

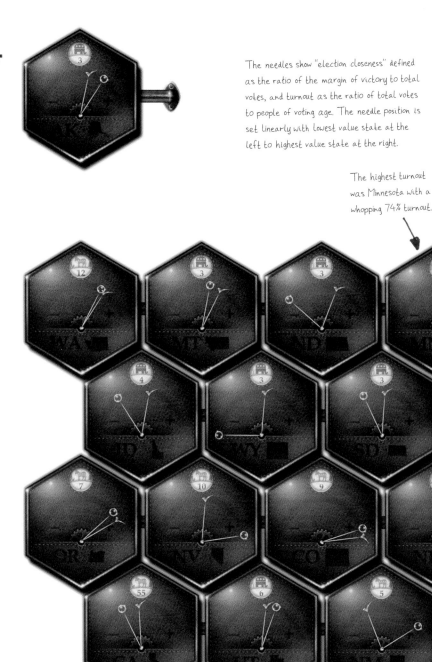

The needles show "election closeness" defined as the ratio of the margin of victory to total votes, and turnout as the ratio of total votes to people of voting age. The needle position is set linearly with lowest value state at the left to highest value state at the right.

The highest turnout was Minnesota with a whopping 74% turnout.

The lowest turnout (so the needle is bottomed out to the left) was Hawaii at 42.6% of voting aged people.

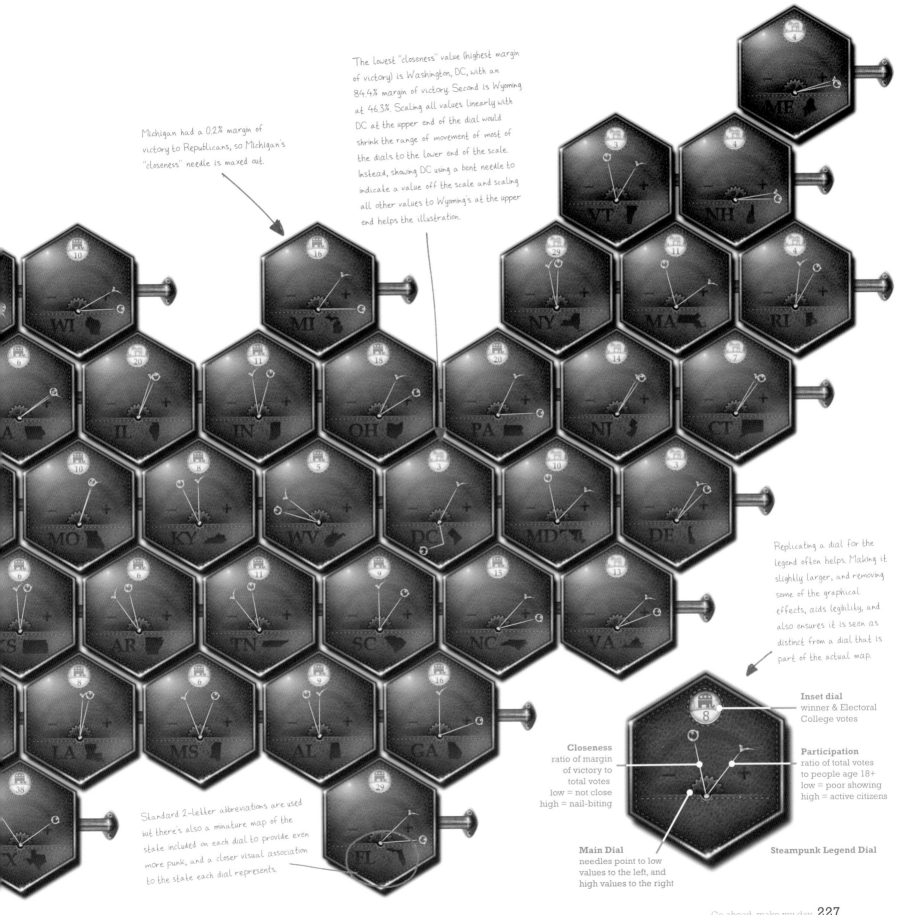

Michigan had a 0.2% margin of victory to Republicans, so Michigan's "closeness" needle is maxed out.

The lowest "closeness" value (highest margin of victory) is Washington, DC, with an 84.4% margin of victory. Second is Wyoming at 46.3%. Scaling all values linearly with DC at the upper end of the dial would shrink the range of movement of most of the dials to the lower end of the scale. Instead, showing DC using a bent needle to indicate a value off the scale and scaling all other values to Wyoming's at the upper end helps the illustration.

Replicating a dial for the legend often helps. Making it slightly larger, and removing some of the graphical effects, aids legibility, and also ensures it is seen as distinct from a dial that is part of the actual map.

Standard 2-letter abbreviations are used but there's also a miniature map of the state included on each dial to provide even more punk, and a closer visual association to the state each dial represents.

Inset dial
winner & Electoral College votes

Closeness
ratio of margin of victory to total votes
low = not close
high = nail-biting

Participation
ratio of total votes to people age 18+
low = poor showing
high = active citizens

Main Dial
needles point to low values to the left, and high values to the right

Steampunk Legend Dial

The force is strong in this one

Value-by-alpha dasymetrically equalised repetition

Introducing the *vader*: a hybrid thematic map combining a value-by-alpha layer across hex-binned areas derived from dasymetrically processed data. The hexagon's destiny is now complete. Combined with the other techniques, it can rule the galaxy! Well, that's unlikely, but it's an example of mashing up different mapping techniques to create something new.

Many of the maps in this book are based on what might be called *pure* cartographic techniques. That is, they are established techniques that have a long history of use, and for which best practice largely is understood and can be applied readily. But making maps doesn't begin or end with single, well-trodden techniques. **Mapping is ripe for experimentation, and armed with a little understanding of the opportunities and constraints of different techniques, you can mash up several and create something new, innovative, or just plain crazy. The map to the right falls into the latter of these categories.**

As seen in previous chapters, the dasymetric technique processes data from one set of areal units and reapportions it to new areas on the basis of the spatial distribution of an additional variable. In this example, data from election results at county level was reapportioned on the basis of land classification of only populated areas. A dot density technique then symbolised the data, and this subsequently was used as a framework to bin the point-based data into hexagons of a uniform size. **It dispenses with some of the problems of a straightforward choropleth by equalising areas.** This provides a visual consistency across the map with no one area distorting visual perception by virtue of being a different size. A value-by-alpha treatment then modifies the appearance of the hexagons to vary the saturation of colour by population density. Hexagons with a low overall number of voters tend to dissolve into the background and disappear from the map (turning away from the dark side maybe?).

This map has gone through such considerable processing that the data apportioned to each hexagon likely has a high degree of uncertainty. Except the patterns still broadly fit what other maps in this book have shown, and as a stylised treatment, it creates an interesting aesthetic. It demonstrates invention— that you can blend a range of cartographic techniques for a visually compelling result. Accuracy? Not so much. **It also demonstrates how you can take a range of techniques and shoehorn them into an acronym that can be used to name the map value-by-alpha dasymetrically equalised repetition (vader).**

Impressive. Most impressive. The journey to the dark side is complete.

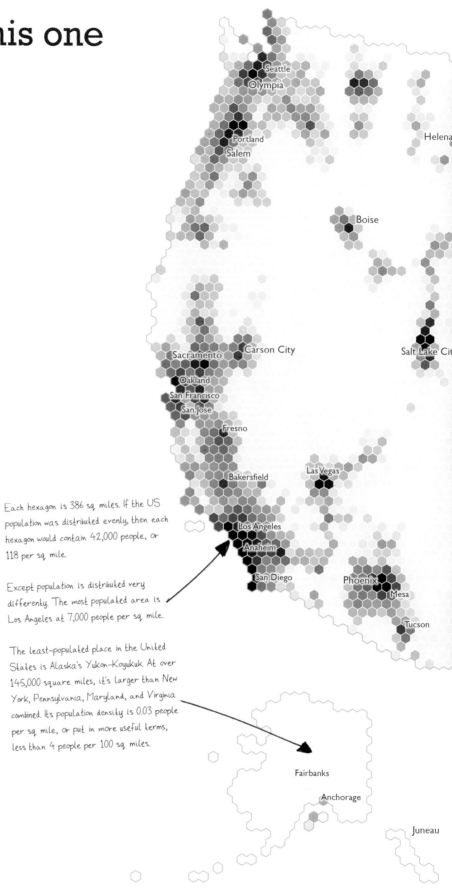

Each hexagon is 386 sq. miles. If the US population was distributed evenly, then each hexagon would contain 42,000 people, or 118 per sq. mile.

Except population is distributed very differently. The most populated area is Los Angeles at 7,000 people per sq. mile.

The least-populated place in the United States is Alaska's Yukon-Koyukuk. At over 145,000 square miles, it's larger than New York, Pennsylvania, Maryland, and Virginia combined. Its population density is 0.03 people per sq. mile, or put in more useful terms, less than 4 people per 100 sq. miles.

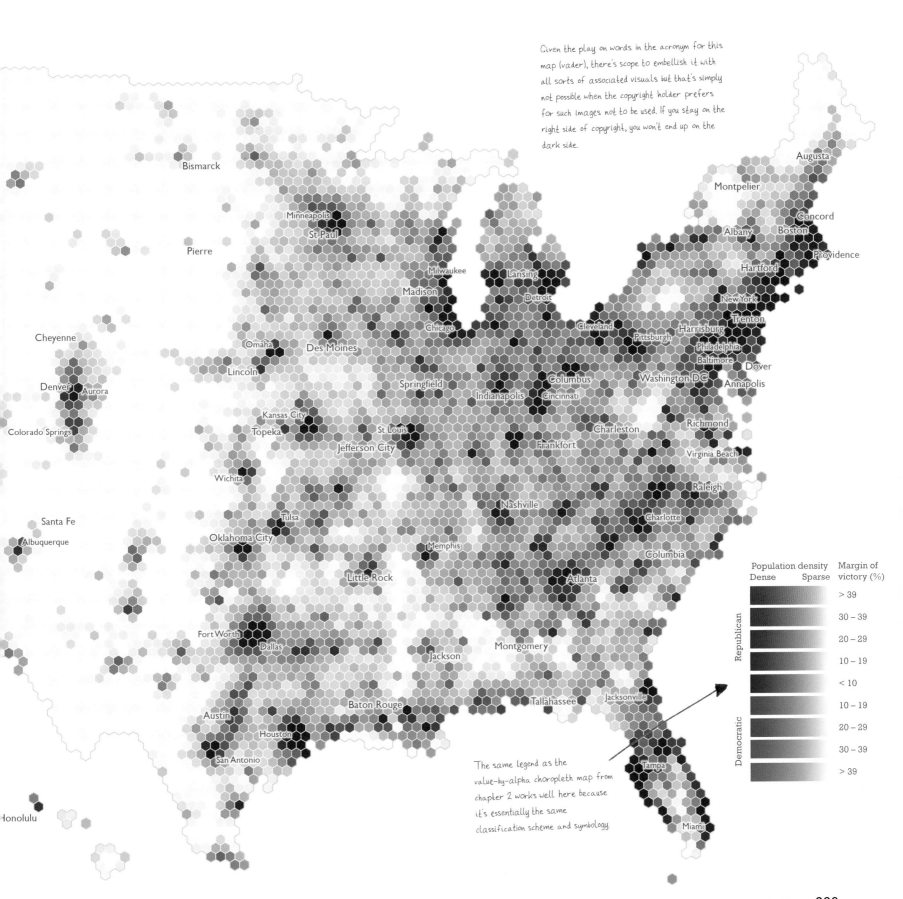

Given the play on words in the acronym for this map (vader), there's scope to embellish it with all sorts of associated visuals but that's simply not possible when the copyright holder prefers for such images not to be used. If you stay on the right side of copyright, you won't end up on the dark side.

Bismarck

Montpelier

Augusta

Concord

Minneapolis
St Paul

Pierre

Albany
Boston
Providence

Milwaukee
Lansing
Hartford

Madison
Detroit

Cheyenne

Omaha
Des Moines
Chicago
Cleveland
Harrisburg
Trenton
New York

Pittsburgh
Philadelphia

Lincoln
Columbus
Washington DC
Baltimore
Dover

Denver
Aurora

Springfield
Indianapolis
Cincinnati
Annapolis

Colorado Springs

Kansas City
Topeka
St Louis
Charleston
Richmond

Jefferson City
Frankfort
Virginia Beach

Wichita

Santa Fe
Raleigh

Albuquerque
Nashville
Charlotte

Tulsa

Oklahoma City
Memphis
Columbia

Little Rock
Atlanta

Fort Worth
Dallas
Montgomery
Jackson

Tallahassee
Jacksonville

Austin
Baton Rouge

Houston

San Antonio
Tampa

Honolulu
Miami

Population density
Dense   Sparse

Margin of
victory (%)

> 39

30 – 39

20 – 29

Republican
10 – 19

< 10

10 – 19

20 – 29

Democratic
30 – 39

> 39

The same legend as the
value-by-alpha choropleth map from
chapter 2 works well here because
it's essentially the same
classification scheme and symbology.

A picture is worth 2,638 counties

Modified pictorial map

This entire book has been focussed on how to take a dataset and massage it into different representations of maps, graphs, charts, or plots. There's an almost endless number of ways in which data can be manipulated and symbolised to generate many types, styles, and variations of visual representation. **The map to the right isn't predominantly of data. It's a map that uses a pictorial style, modified by data.**

The image clips a famous photograph taken by Carlos Barria during a conversation with *Reuters* journalists nearly 100 days into Donald J. Trump's presidency. The photograph itself shows Trump at the Resolute desk in the Oval Office of the White House as he handed out copies of a map of the results of the 2016 presidential election. *Reuters* quotes Trump as saying, **'Here, you can take that, that's the final map of the numbers. It's pretty good, right? The red is obviously us.'**

Maps long have been used as shapes to be turned into other forms or used as a framework for an illustration. When clipped to the shape of the US, the maps on the desk appear neatly along the Gulf Coast, and in the Alaska inset, and make for an interesting juxtaposition of image and map shape. Combining it with an overlay that uses transparency to show the strength of Republican support brings a different perspective. The more the underlying photograph shines through, the more staunchly Republican the area. Opacity increases the more marginal and Democratic the area.

The map Trump is sharing in the photograph showed his version of the truth. Commentators proclaimed it warped reality. Yes, the map shows a lot of red but it isn't wrong. It appears earlier in this book as an example of a diverging-hue choropleth. If Hillary Clinton had won, she'd have used a map that was much bluer and which spoke to her version of the truth.

Lining up the photograph inside the map results in some fortuitous delights such as the map on the desk sitting squarely in the Alaska inset. Alaska in Alaska – meta!

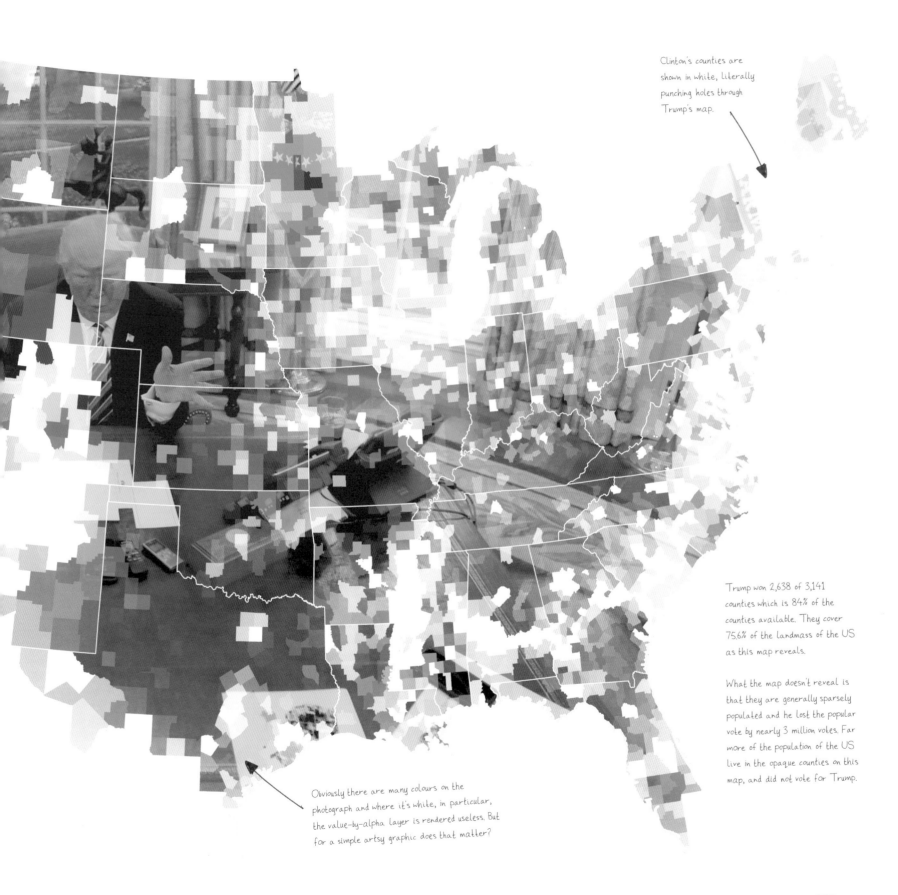

Clinton's counties are shown in white, literally punching holes through Trump's map.

Trump won 2,638 of 3,141 counties which is 84% of the counties available. They cover 75.6% of the landmass of the US as this map reveals.

What the map doesn't reveal is that they are generally sparsely populated and he lost the popular vote by nearly 3 million votes. Far more of the population of the US live in the opaque counties on this map, and did not vote for Trump.

Obviously there are many colours on the photograph and where it's white, in particular, the value-by-alpha layer is rendered useless. But for a simple artsy graphic does that matter?

Epilogue

The making of a viral map

The maps you have seen in this book have been made over time. Many are updated versions of maps I made for the 2012 presidential election result. The collection has been slowly built, and when I've hit on something I've felt is worthy of sharing, I generally post it on Twitter or write a blog on the technique. Occasionally, one or two get a bit of traction online. On one occasion, the map became a mini viral hit: the dasymetric dot density map in chapter 3 (p. 68, to the right).

In 2012, I made a similar dasymetric map for the Barack Obama/Mitt Romney election. It was a product of the mapping technology of the time, made using ArcMap™ and published as a web map using an early version of ArcGIS® Online. At the smallest scale, 1 dot is 1,000 votes. At the largest, 1 dot is 10 votes, and if you'd printed the map, it would be as large as a football field. It took three months to cajole the largest scale map onto the web. I wanted to update the map for 2016, and the four intervening years brought significant new software capabilities. For 2012, I had to generate up to 12 million points and position them because there was no way to use transparency at the symbol level in the dot density renderer. Recall, cartography is an art, science, and technology, and although the art and science of dasymetric dot density had been well understood for many decades, the technology for building a map with more than 120 million dots wasn't there yet.

With the release of ArcGIS® Pro, I instead could use the dot density renderer and let the software take the strain, and if I were to go all out, why not try and make a map where 1 dot was 1 vote? So, for me, the map was a technical challenge, and part of what I do in my work at Esri is to push the software to see what it is capable of, to test it, and to show others what capabilities it affords.

The map I made is a product of many decisions, each one of which propagates into the map. As discussed on the spread in chapter 3, a dasymetric map takes data held at one spatial unit (in this case, counties) and reapportions it to different (usually smaller) areas. It uses a technique developed by the late Waldo Tobler called *pycnophylactic reallocation modelling*. Those different areas are, broadly, urban. The point of the map is to show where people live and vote rather than painting an entire county with a colour which creates a map that often misleads. (I actually learnt of Waldo Tobler's passing while my computer was running his model to make the map.)

I used the National Land Cover Database to extract urban areas. It's a raster dataset at 30 m resolution. I used the impervious surface categories and created a polygon dataset with three classes, broadly dense urban, urban, and rural. I then did some data wrangling in ArcGIS Pro to reapportion the Democratic and Republican total votes at county level into the new polygons. There's some weighting involved so the dense urban polygons get (in total) 50 percent of the data. The urban polygons get 35 percent of the data, and the rural polygons get 15 percent of the data. Then I used the dot density renderer in ArcGIS Pro to draw the dots, one for each vote, resulting in a map with nearly 130 million dots.

The result is a map that pushes the data into areas in which people live. It leaves areas in which no one lives devoid of data. It reveals the structure of the US population surface in addition to showing the variable being mapped with blue dots representing Democratic votes and red dots representing Republican votes. Most maps that take a dasymetric approach will end up like this but I think there's value in the approach. To me, it presents a better visual comparison of the amount of red and blue than the standard county-level map that maps geography, not people, and overemphasises relatively sparsely populated large geographical areas.

It's also worth noting at this point that the dasymetric technique is not simply masking geography. Masking areas that have low population densities is a quick method to approximate a dasymetric technique but it has several drawbacks. Firstly, you lose data. You are not reapportioning the data reflected in the areas masked. If every vote in an election counts, it ought to be on the map. Secondly, the unmasked areas remain unaltered, and the geographies still play an important part in how the map is read. Even a large county which has one large city might avoid being masked by virtue of the population density being skewed by that city's population total.

The map finished rendering on my computer late one Tuesday afternoon in early March 2018. It had over 128 million dots and took 35 minutes for my computer to draw. Technical challenge achieved, and ArcGIS Pro nailed it with dots that had about 95 percent transparency so those in dense urban areas coalesced and created wonderful shades of purple. This is a map that I couldn't have made in the previous election cycle. I was excited, and so I took a quick screen grab, sent out a tweet, and went home to walk Wisley the dog with my partner, Linda.

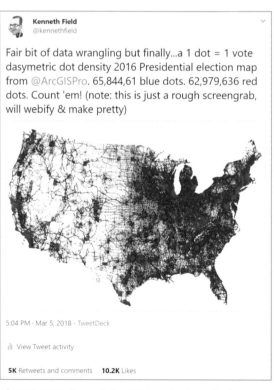

Kenneth Field
@kennethfield

Fair bit of data wrangling but finally...a 1 dot = 1 vote dasymetric dot density 2016 Presidential election map from @ArcGISPro. 65,844,61 blue dots. 62,979,636 red dots. Count 'em! (note: this is just a rough screengrab, will webify & make pretty)

5:04 PM · Mar 5, 2018 · TweetDeck

View Tweet activity

5K Retweets and comments   **10.2K** Likes

According to twitter analytics, this tweet has been liked over 10,000 times, retweeted 5,000 times. It has been seen by nearly 2 million people on the internet, and over 250,000 have had some engagement with it.

And that, I thought, was that. I'd put the map on the back burner and returned to doing layout reviews for my first book. But then something unexpected happened. My cell phone started pinging. Slowly at first but then a little more during the evening as people began to see the map on Twitter and liked or retweeted it. That's nice, I thought. I went to bed. On Wednesday morning, I woke to a relative avalanche of likes and retweets. I spent the day in Palm Springs at the Esri Developer Summit, and my phone never stopped making beeping sounds. By the end of the day, the tweet with the map had received around 3,000 likes and had been retweeted 2,000 times. By Thursday morning, it had received 7,000 likes and a little over 3,000 retweets. The side-effect of this 15 minutes of map fame is I also picked up an extra 1,000 followers (a 25 percent increase) in the same 24 hours on my nearly 10-year-old Twitter habit.

The map had gained such attention because no one had seen one like it for the election before. I hadn't appreciated that, and there was a problem. The screen grab was quick and dirty, and I'd got ahead of myself releasing the map for public view. It was by no means the finished article, and as the tweet itself explains, my intention always was to polish it, create a high-resolution version, and publish a web map. I didn't have time to do that work in the weeks immediately after the tweet as I was deep into final editing work on *Cartography.*

On the plus side, the map was well-reviewed. It was critiqued on a massive scale. Questions arose on specific patterns based on individual knowledge. Anomalies were pointed out. Queries on the maths involved and the result were posed. Many people offered fascinating insights into the gaps and the patterns through Twitter replies. And yes, many people quite correctly pointed out that Hawaii and Alaska were missing. Getting eyes on the map was fantastic and helped shape the final published map. The irony, of course, is the finished map, and the associated tweet to publicise its release a few weeks later, received barely a blip on the internet. To this day, the original tweet still gets far more interest than the finished map.

More than anything, I was blown away by the nice things that have been said about the map. It showed the election result in a different way than people had been aware of or used to. It told a different story from one that many had been used to.

My only regret is that I excitedly tweeted a rough version. I should have waited until I made the map properly. At the very least, it ignited a discussion. It brought different cartographic eyes to the dataset and illustrated to the public that there are different ways to represent the data. Many referred to it as a 'more truthful' map, and that was the angle that reporters were keen to take when interviewing me for various articles over the ensuing months. Those interviews also were enlightening because the critical question many wanted an answer to was 'What is the best map?' My answer, as always, was 'It depends'. So with that in mind, of those in this book, what do I consider the best map for different circumstances, and what are my personal preferences?

In terms of showing the actual result of the election, it comes down to how many Electoral College votes the winner secured, and that is a function of the aggregated sum of votes won at a state level. In these terms, the red and blue state map (p. 16) is hard to beat. It explains the result simply, using a familiar geography that most would find easy to interpret. That said, when you're trying to find who got to the magic 270 Electoral College votes, a graphic showing that detail (p. 134) would suffice. Sometimes, the best map is no map.

It's only when people want to get a little more analytical that other maps become more useful. Identifying smaller or larger shares of the vote or establishing margins of victory are both interesting metrics. A value-by-alpha choropleth (p. 30) goes a long way to ensuring that an area-based map can be used without the huge distortions that sparsely populated geography often has on the map. A well-scaled proportional symbol map (p. 56) or the aforementioned dasymetric dot density map (p. 68) also provide a way of dealing with the issue of different-sized geographies and population densities at the same time.

Perhaps more than any of these maps, displaying the data itself as a label, symbolised with different reds and blues and sized according to votes (p. 82), arguably provides all the benefits of more traditional techniques, without the drawbacks. Whether it's share of the vote or margin of victory or some other metric, the benefit of putting the data on the map means it's the perfect mechanism for information retrieval.

Maybe the story is not about the whole dataset, though. For 2016, the key issue was the counties (and, in aggregate, the states) that Donald J. Trump was able to flip. Most states tend to vote the same way, so the campaign always comes down to key battlegrounds. By showing only the counties that flipped (p. 36), you can tune out every county that voted the same in 2012 and 2016 and focus only on the ones where it made a difference. The same flipped counties can be used to highlight the symbols that show the extent of swing from one election cycle to another, and directional arrows (p. 98) do a good job of encoding this additional detail over and above masking counties in which there was no change.

In some of my interviews for the media, one question recurred: 'What map would a candidate be able to show to help their campaign?'. It's a tough one because it differs depending on the area, but let's rephrase the question as 'How can a map be used to encourage a candidate and an electorate?'. Let's take Wisconsin as an example. Trump won Wisconsin by the narrowest of margins, 0.77 percent. The choropleth map of the counties paints a picture of a predominantly red state. Clinton won only 13 of the 72 counties. As a Republican candidate, this is the map you'd use because it would encourage you and your supporters that you're on the winning side, and by quite some margin despite the headline figure of a less than 1 percent margin.

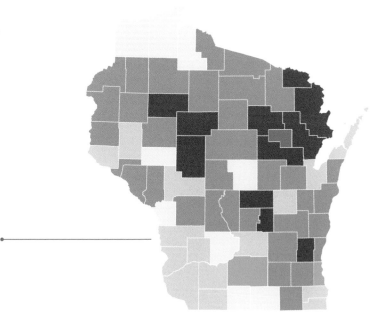

But how might you encourage a Democratic base in the face of such a visually overpowering map? Perhaps the dasymetric dot density map is useful in this scenario. It shows the data constrained to where people live and immediately illustrates the large swathes of sparsely populated rural areas in the state. There's red for sure, but it's spread thinly, and the urban areas are beacons of mid-range purple which show a bipartisan electorate. Small gains in those marginal areas might lead to the gain required at a state level to flip the margin back in favour of the Democratic Party. Madison itself stands out toward the blue end of the spectrum.

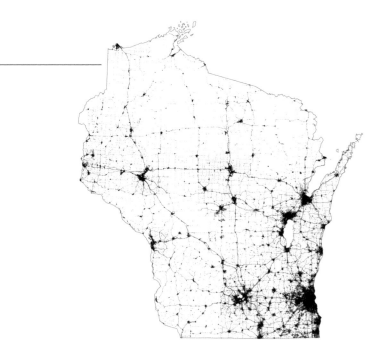

If patterns over time are the phenomenon of interest, there are only two viable options: animation or small multiples. Both have their uses. Small multiples (p. 48) sacrifices detail for an illustration of several maps in the same visual plane and overcomes change blindness on the part of a reader having to look at multiple maps in sequence. Animation takes advantage of a different medium, and being able to replay a sequence of maps or pause and interact is an extremely powerful way of delivering a lot of information unencumbered.

The past decade has seen a shift in the way media handles election data. Cartograms, particularly gridded variations, have become more popular and better understood. On websites, they often are used in conjunction with a geographical equivalent so the map reader can flip to their preference, or even use the geographical version as a way of more easily interpreting the cartogram equivalent.

Cartograms, in general, are efficient graphical devices and because of their innate generalisation can give a clean visual summary. My personal preference is the Dorling cartogram (p. 116), which uses circles as the basic form and which also can be used as a framework for more interesting symbology, such as the dot density version (p. 222). But there are also trends that come and go, and hexagonal gridded cartograms (p. 112) have found favour in many organisations' output. The compromise cartogram (p. 122) provides an excellent way of using a geographical base to situate what is effectively an array of repeated shapes. With any cartogram, there's a cognitive hurdle to overcome in terms of the map reader's ability to interpret, and although examples such as the compromise cartogram are effective, ultimately they can show only a limited dimension of the data.

Combining graphs, charts, and plots with maps is extremely powerful. As a mechanism for conveying detail, it's difficult to beat a well-designed graph, chart, or plot. Positioning them across a geographical backdrop or tessellated into a gridded cartogram adds the spatial dimension. Waffle grids (pp. 150, 154) are good at encoding data in a repeatable way. Stacked bar chart chartmaps (p. 158) and polar area plots (p. 164) are perhaps more visually challenging but provide even more scope for data encoding if bringing clarity to complexity is the purpose of the map.

3D maps are difficult to recommend if the medium of production is static though isometric prisms (p. 182) and using a 3D gridded chartmap (p. 204) with a vertical signpost arguably come as close as possible to providing a useful solution. It's when using an interactive online environment that 3D comes into its own. Design is still crucial, and there's plenty of scope to put too much glitz into a 3D visualisation, but the ability to pan, rotate in different planes, and zoom overcomes one of the main problems of static 3D maps: occlusion. The addition of other interaction such as clicks, rollovers, slicing, and pop-ups often makes 3D an attractive proposition. It's worth remembering, however, that interpreting many planimetric maps requires a significant cognitive load on the part of the map reader, and 3D can add to that if not handled carefully.

Maps often are made to be illustrative or to visually delight rather than have an analytical function at their core. Many of the maps in this book have been made with that in mind. Perhaps the standout is the multivariate symbol landscape (p. 216). Although a multivariate symbol might be used as an analytical map, for me, at least, it's the overall look and feel that makes it somewhat unique. The envisioning of the idea of a mountain landscape of peaks of different sizes and dimensions is an interesting idea and visual motif. Carrying it through to the map creates an interesting, visually striking map that invites the inquisitive reader to spend time with it.

I think that most, if not all, maps in this book are of value in some way or another. This short set of personal observations and insight into my preferences perhaps has shone a light on the ones I feel are of more utility, but perhaps there are many more that I have not even considered or which might be made by combining ideas from those herein.

The map type chosen, or used as a basis for developing an iteration of a technique, is based on a range of key factors that should shape every map made of election or other thematic data. Predominantly, the map should have a clear purpose and defined readership. Account for the constraints and work with them rather than ignore them because they won't go away. Make a clear blueprint, and review examples of what's gone before to learn from what others have found has worked or not worked. Don't be afraid of experimentation, but also be cognisant that if you experiment and it doesn't work, it's fine. There's always another map to be made.

In modified version, I wrote the adjacent basic tenets for my previous book, *Cartography.*, but I felt they were worth repeating here because they are still fundamental to good cartographic practice.

It's worth remembering that making a map is one thing but deciding to publish it is another. Once published, whether formally or via social media, the map becomes public. It becomes fair game for people to comment on and offer opinions. The map may provoke, delight, inspire, frustrate, or anger people. These and many more reactions are

1. Make the map right. There's rarely a short cut. Give yourself the time to work out what you need, and how to do it.

2. When in doubt, follow what's gone before. Best practice has been developed by many people over many decades and serves you well.

3. Always go beyond the defaults. Defaults are inevitable and can be a good starting point but knowing something of cartography helps you adapt them to your needs.

4. Quality trumps quantity. People like maps but they are drawn to great maps. Quality attracts, inspires, sells, and invites a second look.

5. Study great maps. Learn from those maps held in high esteem so you can appreciate the value of them and why they work so well.

6. Give and take critique. Learning how to assess cartographic quality helps you recognise greatness and do more of it. Learn to accept constructive criticism. It's vital.

7. Break the rules (such as they are) when necessary. If you know the best practices, you'll know when you can break them, by how much, and to what effect.

8. Recognise cartographic lineage. Not much is genuinely new. That's OK. Learn from it, use it, and cite it without trying to claim it.

9. Accept that the map is rarely finished. You have to learn when you're done. There will always be tweaks but the old adage of 90% taking 10% of the time and the last 10% taking 90% of the time is often true.

10. Own your work. Put your name on it. Share it and get eyes on it. Build a portfolio.

11. Take it to 11. Making maps can make you incredibly frustrated at times but put your heart into your work and learn to enjoy the ride. The end result is worth it.

commonplace when people look at maps. I strongly encourage experimentation and making maps, whatever your level of experience, but I'd also caution that it behoves any mapmaker to understand that the decisions made in processing the data and designing the map always will form an impression in a map reader's mind. By ensuring you understand a little of how your decisions mediate the message, you can control that to an extent. You also can avoid harsh criticism of the map as an information product by doing exactly what I failed to do when tweeting my dasymetric screen shot: show restraint.

Many a map is stored, unseen, in file drawers or on hard drives. And there's likely a reason for that. Sometimes, making a map is merely for personal development and to see what works or doesn't. Not all output needs to be immediately shared with the world. Many maps I have made never have seen the light of day. There are many maps I made for this book that didn't cut it. Restraint is knowing what an audience will find useful and which gives added value to the subject being mapped. Simply adding to the noise around a subject is not always useful.

It's hard to pick a single map from this book that I categorically would express as a favourite or a go-to technique. A well-designed choropleth is hard to beat in terms of people's familiarity and ability to read it, although the basic pitfalls all too often are managed poorly. Proportional symbols make a lot of sense, particularly if the metric of interest is totals. I have a penchant for cartograms but I also am acutely aware they are often hard to read and divisive because people seem either to love them or hate them. If in doubt, a dot density map is one of the most useful maps that maintains geography and deals with issues of different-sized geographies. Beyond these mainstay techniques, there's literally a world of design opportunity, but the more complex the map, the more care is needed to walk the fine line between something useful and something that leaves people baffled.

But you're going to push me, right? Okay, for me, it's the dasymetric dot density map; my little moment of viral map fame wins. It encodes all the data, where it truly exists, and lays it bare. It's not without fault, because no map is, but it's as close to an objective, truthful display as I think there is, particularly for the dataset that is the subject of this book. You may disagree. That's fine.

As a postscript to my map gaining a little traction on the internet, and like a lot of material placed online, it soon was modified. Data journalist Thomas de Beus recognised the contrast between my dasymetric view of the election result and the map hung in the White House, snapped by *Reuters* journalist Trey Yingst in May 2017. De Beus's quick Photoshopped version of the picture, to the bottom right, juxtaposes not only the differences between the map types but also the differences between one version of the truth and another.

In his single tweet, de Beus illustrated the entire point of making the map and demonstrated the entire point of this book. It presents the same data in a different way. It leads to different insights, different interpretations, and a different perception.

Neither of the maps is right or wrong. They are different. Of course, we all have our own view on which one serves our needs, which we find most truthful, and which we prefer, but that's for us as individuals. Being able to assess the relative merits of different map types is also fundamental to our ability to understand how the message is being delivered. The title of Professor Mark Monmonier's classic book *How to Lie with Maps* (University of Chicago Press, 1991) might have been a good title for this book but he beat me to it.

The dasymetric dot density map demands to be seen in large format so I made it into a huge poster. If you'd like a copy, you can download it from esriurl.com/election2016.

Finally, are there 101 maps in this book? Did you believe the subtitle?

Redlands, CA
November 2020

One more map

The 2020 presidential election results as a ring chartmap

And with the comments on the previous page the epilogue was complete. Except, during the final editing stages, literally a day or so before the book was due to be handed off for printing I felt compelled to make one more map.

All the editorial work on the book was undertaken during September–November 2020, as the final weeks of the 2020 presidential election played out. Donald J. Trump was campaigning for a second term. His Democratic challenger was Joseph R. Biden Jr. As results emerged over many days, newsrooms, websites, TV, and individuals drew maps, graphs, charts, and plots to visualise the outcome. Many of these map types you've seen in this book of the 2016 results. Joe Biden won the 2020 presidential election. He flipped Wisconsin, Pennsylvania, and Michigan back to the Democrats, as well as Arizona and Georgia which have been Republican for many years. Biden won by 306 Electoral College votes to Trump's 232, the same margin by which Trump beat Clinton in 2016.

I was particularly drawn to a map made by Alan McConchie, lead cartographer at Stamen, the renowned data visualisation and cartography studio. Alan had made a proportional symbol map but with a twist. He'd overlaid proportional symbols of the number of votes for the winner with proportional symbols of the number of votes for the loser, for each state. The beauty of the design lay in the way in which the loser's symbol was white space. Visually, it knocked a hole through the winner's symbol leaving behind an outline which was symbolised using the colour of the winner's party. It created a map of the margin of victory in each state, and it was a map I hadn't thought of before, let alone made for this book. I contacted Alan, and he gave me permission to take the idea and remake it.

The map shows the excess votes, all scaled to the population total that mimicks the distribution of Electoral College votes. A state with a large population, such as Florida, but which has a narrow Republican win gets a thin red line. California, with a large population and a wide margin, gets a thick blue line. As Alan notes, in the Electoral College system all these excess votes are, in some respects, wasted. None of the large excesses for either candidate helped them in marginal states elsewhere.

As this entire book has emphasised, there's a rich variety of ways to represent empirical data. Even as I complete this book, new maps are emerging. And since the 2020 results differ from the 2016 election we all get to make new maps great again.

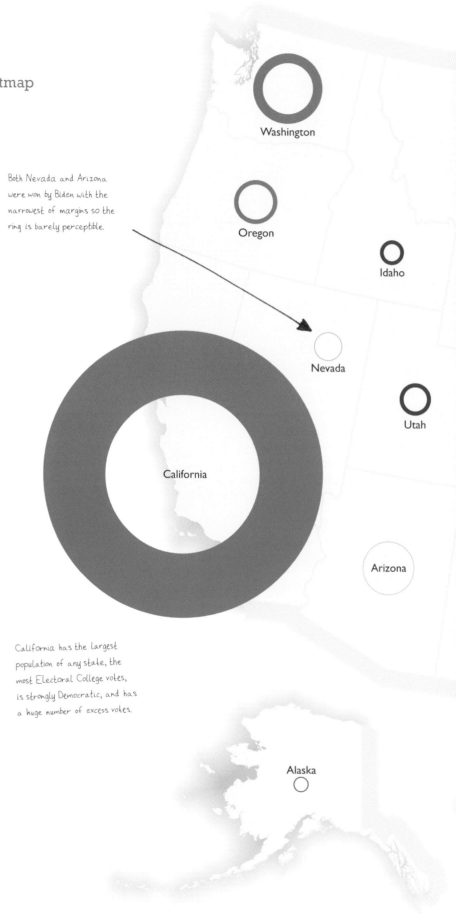

Both Nevada and Arizona were won by Biden with the narrowest of margins so the ring is barely perceptible.

California has the largest population of any state, the most Electoral College votes, is strongly Democratic, and has a huge number of excess votes.

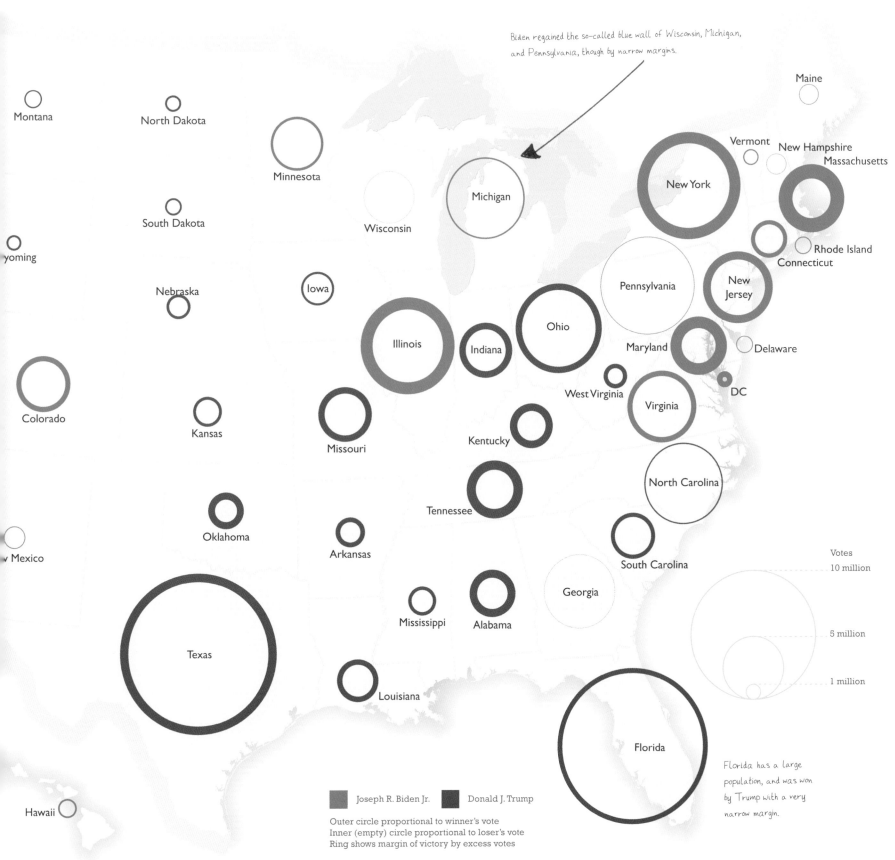

Biden regained the so-called blue wall of Wisconsin, Michigan, and Pennsylvania, though by narrow margins.

Maine

Montana

North Dakota

Vermont

New Hampshire

Massachusetts

Minnesota

New York

Michigan

South Dakota

Wisconsin

Rhode Island

Connecticut

yoming

Pennsylvania

New Jersey

Nebraska

Iowa

Ohio

Illinois

Indiana

Maryland

Delaware

Colorado

Kansas

West Virginia

Virginia

DC

Missouri

Kentucky

North Carolina

Oklahoma

Tennessee

Arkansas

South Carolina

Votes

10 million

Mexico

Georgia

Mississippi

Alabama

5 million

Texas

1 million

Louisiana

Florida

Florida has a large
population, and was won
by Trump with a very
narrow margin.

Hawaii

Joseph R. Biden Jr.   Donald J. Trump

Outer circle proportional to winner's vote
Inner (empty) circle proportional to loser's vote
Ring shows margin of victory by excess votes

Prior carte

Old is new again

Although all the maps in this book are new in the sense that they are newly created, there's really nothing new in this book. Most of the maps, graphs, charts, and plots in this book, and those seen more widely in the media, books, social media, and so on, use prior art. It's incredible to think of how graphical languages, ideas, and graphicacy more generally have evolved, yet most, if not all, of the techniques we're familiar with have a long and storied history. Many techniques were invented more than a century ago. Some have become commonplace because of their ubiquity, and even have become defaults in many software packages. Some techniques perhaps have not been as widely used but possibly from a lack of modern automation that makes the technique more challenging to design and produce.

This appendix lists the types of graphical representation that appear in this book. It describes the history and lineage of the techniques and points to some of the major milestones in the history of thematic cartography and data visualisation more widely. It's by no means comprehensive since it's specifically focussed on the techniques used in this book. It acts as a broad acknowledgment of the forerunners to modern interpretations of such techniques and highlights a few contemporary examples that have shown genuine innovation, and which provided specific inspiration for some of the maps in this book. Each technique is accompanied by a list of page numbers which reference the maps which have in full, or in part, made use of the technique described.

As the 33rd president of the United States, Harry S. Truman, once said, 'There is nothing new in the world except the history you do not know', and this is apposite to making maps. There's a lot of alleged reinvention that can be seen in contemporary cartography and data visualisation, but everyone stands on the shoulders of master mapmakers and cartographic innovators from days past, and they deserve credit for developing the techniques that our modern visual communication forms are firmly based on.

Looking at maps both historic and contemporary and learning from them is a remarkably good way to develop an understanding of visual communication. Part of the impetus for this book was to create an atlas of thematic mapping examples to showcase techniques. I encourage readers to explore prior art, look up some of the work, and let it help inspire your own thematic maps and data visualisation.

"There is nothing new in the world except the history you do not know."
– Harry S. Truman

3D symbols

3D symbology has been a mainstay of cartographic practice
for centuries. Although modern computer graphics support
the design and placement of volumetric symbols, maps and
3D perspective landscapes showing buildings, people, trees,
and animals rendered as 3D pictorial symbols have been
produced since early clay tablets in 2500 BC. In thematic maps,
3D symbology was a mainstay for mid-20th century American
cartographers such as Erwin Raisz and Richard Edes Harrison,
who made significant use of 3D volumetric symbology across
either planimetric or tilted maps.
p. 196

Anamorphic map

A forerunner to the more modern non-contiguous cartogram,
the first anamorphic map was created by French engineer
Émile Cheysson in 1888. It used several resized versions of
the outline of France to show a quantitative variable. Each
successively smaller map was positioned inside the other. The
map illustrated the decrease in travel time from Paris to other
parts of France over 200 years. For this specific use, isochrone
maps have become more widely used, but resizing and
deformation of map shapes is key to many modern cartograms.
p. 108

Area rectangles

In 1896, Jacques Bertillon created a map of Parisian
arrondissements, each with a rectangular symbol in
which overall population was measured by height and the
percentage of foreigners by width. The area is therefore
proportional to the absolute number of foreigners. This was a
forerunner to a range of techniques in which height and width
encode different variables. This might include cartograms and
treemaps, as well as maps that employ multivariate symbology
(e.g. Chernoff faces).
pp. 144, 216

Axonometric projection

This is a form of orthographic projection to support drawing
of objects in which the lines of sight are perpendicular to the
plane of projection and where the object is rotated around one
or more of its axes so that multiple sides are viewable. A 3D
population surface using an axonometric projection first was
published by German physicist Gustav Zeuner in 1869.
pp. 182, 186, 198, 200, 204

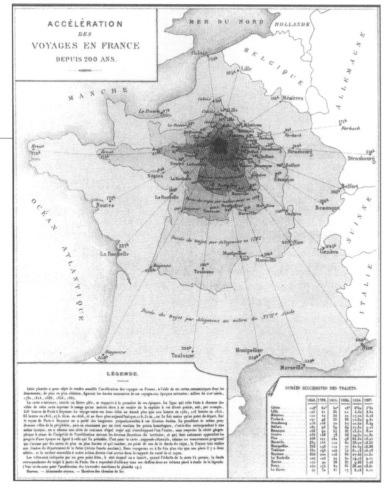

Travel time to Paris by Émile Cheysson, 1888

Bar chart

The bar chart was invented by William Playfair, a Scottish engineer and political economist in 1786, along with the line and area chart to visually represent economic data. The bar chart provides a quantitative graphical form that solves the problem of discrete visual comparison of data. Playfair is regarded by Harvard scholar Edward Tufte as one of the two great inventors of modern graphical design, the other being Swiss polymath Johann Heinrich Lambert whose work on addressing the special properties in map projections established that conformality and equal area were mutually exclusive. French cartographer and theorist Jacques Bertin illustrated a range of uses for bar charts on maps in his book *Sémiologie Graphique* (*Semiology of Graphics*) (Paris: Gauthier-Villars) in 1967.
pp. 130, 158, 204

Beeswarm plot

A beeswarm plot is essentially a form of dot plot, which has a long history in statistical graphics. English economist and logician William Stanley Jevons first used dot plots to graph the weight of British Sovereign coins by year. The beeswarm plot develops the basic method and positions dots in a 1D scatterplot that essentially superimposes a frequency-jittered strip chart on top of a box plot. The box plot itself was developed by American data visualisation specialist Mary Eleanor Spear in 1952 and subsequently enhanced as the box and whiskers plot by American mathematician John Tukey in 1970. By using dots, Tukey himself developed it into a textured dot strip, and American statistician and computer scientist Leland Wilkinson developed the algorithms to form the beeswarm plot using kernel densities to approximate a hand-drawn appearance in contrast to simply binning dots.
p. 140

Binning

Binning data into boundaries on thematic maps is as old as the choropleth map, which can be traced to Baron Pierre Charles Dupin in his 1826 map showing levels of illiteracy in France. In a contemporary sense, binning more commonly refers to aggregation of data variable occurrences within regular grids such as hexagons (hex-binning). Original data values are aggregated and replaced by a value representing some measure of their combined occurrence such as total or mean. It's effectively a spatial histogram.
p. 228

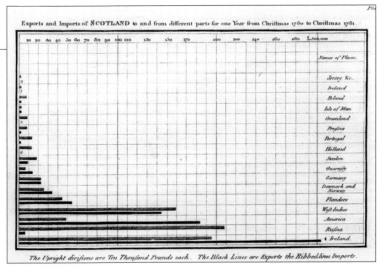

Bar chart of imports and exports of Scotland by William Playfair, 1786

Bivariate map

Bavarian statistician Georg Mayr used one of the earliest examples of a bivariate technique in his 1874 map of land use. It contained two variables, with the density of horses being classified into four categories, each with thickening vertical red lines as the density increased. The density of cattle also was categorised into four classes and shown with thickening green horizontal lines. Contemporary bivariate maps tend to use colour mixing to create solid fills for the discrete combinations of pairs of classified data variables, often for choropleth shading.
pp. 30, 40, 94, 228, 230

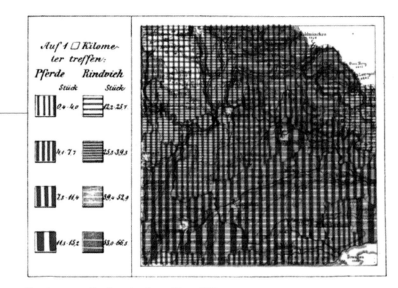

Bivariate map of land use by Georg Mayr, 1874

Block diagram

Also known as a *smooth 3D map*, in a block diagram, the height of the surface is proportional to the value of the variable being represented. The first use of a block diagram to display statistical data was published in 1879 by Italian mathematician Luigi Perozzo. It showed the absolute number of live male births from Swedish census data. It was essentially a 3D population pyramid and was referred to as a *stereogram* at the time. Perozzo's stereogram was certainly the forerunner of any 3D surface in which a data variable is encoded in the z-dimension. American cartographer George Jenks focussed much of his research, starting in the mid-1960s, on 3D map construction. One of his most famous maps was a 3D representation of the population density of central Kansas, US, using a fishnet-style block diagram, or smoothed statistical surface, produced by hand.
p. 186

Cartogram (compromise)

A compromise cartogram uses a geographical map as the background framework, and repeated symbols are positioned on top to represent the data variable. Without the geographical map, it would be a non-contiguous mosaic cartogram. Gabriel Dance and Nick Evershed of *The Guardian* developed the technique for the 2013 Australian election. The version in this book (p. 122) was inspired by the example made by Tom Pearson of the *Financial Times* (2016) as a way of showing the split of the Electoral College votes in the 2016 presidential election.
p. 122

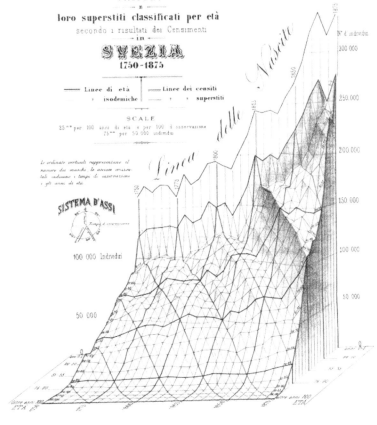

3D population pyramid of live male births by Luigi Perozzo, 1879

Cartogram (Demers)

The Demers cartogram is a technique for representing data for areas that eschews geography in preference for a geometric shape that represents the unit areas. It is ostensibly the same as a Dorling cartogram except it uses squares instead of circles to reduce the gaps between mapped objects. The Demers sacrifices distance to maintain contiguity. It was created by undergraduate student Steve Demers during his Bachelor of Arts in geography studies at the University of California, Santa Barbara, under the direction of research cartographer and Professor of Geography Keith Clarke in 2002. Demers also created a 20-foot span population cartogram of human figures on one of the campus walls. Each figure represented 10 million people.
p. 118

Cartogram (Dorling)

Developed by British geographer Danny Dorling in 1996, the Dorling cartogram is a technique for representing data for areas that replaces geographical shapes with a circle to represent the unit areas, sized by the data variable. A Dorling cartogram does not maintain shape, topology, or object centroids and is an abstract representation of the spatial pattern of the phenomenon being mapped. It can be likened to a proportional symbol map in which the symbols have been repositioned in a self-organising manner using cellular-automata methods to overcome the problems of overlaps caused by some non-contiguous cartogram techniques.
pp. 116, 124, 170, 174, 222

Cartogram (gridded)

A gridded cartogram takes a regularly sized and shaped polygon and tessellates it into a regular grid to form an overall shape that approximates the geographical area represented. It diminishes the geographical shape (sometimes severely) but attempts to maintain a more uniform topology that enables people to find familiar places without the need for perfect maintenance of adjacencies. Gridded cartograms generally use hexagons, squares, or diamonds that tessellate and use a single polygon to represent a unit area. They often are used as a framework to arrange other graphs, charts, and plots. Although not organised geographically, French statistician Toussaint Loua's shaded table (referred to as a *scalogram*) of 1873 was effectively a gridded cartogram that used squares, shaded by colour, to depict the relationship among variables for different areas.
pp.120, 152, 154, 156, 158, 160, 162, 164, 166, 188, 198, 200, 204, 226

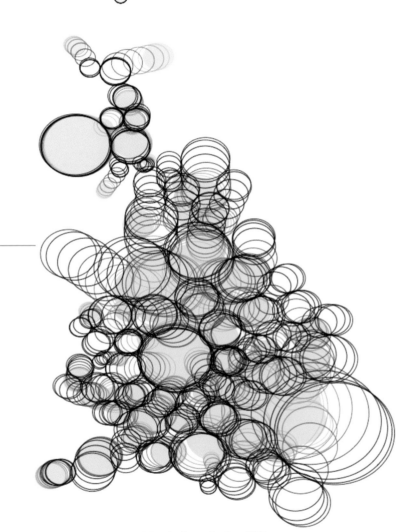

Population of Great Britain by Danny Dorling, 1996

Cartogram (mosaic)

A mosaic cartogram is like the gridded cartogram. It takes a regularly sized and shaped polygon and tessellates it into a regular grid to form an overall shape that approximates the geographical area represented. It diminishes the geographical shape (often less severely than a gridded cartogram) and attempts to maintain a more uniform topology. Mosaic cartograms generally use hexagons, squares, or diamonds that tessellate. They use a different number of polygons per unit area to represent some value of the data variable being mapped. An early mosaic display (though non-contiguous) was created by American economist and statistician Francis Amasa Walker for the 1874 *Statistical Atlas of the United States*. It used subdivided squares to show the relative proportions of variables per state. In 1905, the Imperial Statistical Office of the German Empire published what were referred to as *schematic maps* showing the 1904 harvest yields. It is recognised as an early form of mosaic cartogram based on tiled squares, with a different number of squares representing different regions and a choropleth shading scheme applied to each region.

pp. 112, 114, 122, 152, 154, 166, 198, 224

Cartogram (non-contiguous)

A non-contiguous cartogram begins with a geographical map from which all areas are scaled down independently according to a data value so that the desired size and area is created relative to all other areas. These areas retain the familiarity of geographical areas but sacrifice topology. French engineer Émile Cheysson published a map in 1888 that saw the shape of France resized and presented as a set of nested maps. Applying the technique across multiple countries would give a non-contiguous cartogram. Judy Olsen's 1976 projector method algorithm is perhaps the most widely used method of creating non-contiguous cartograms that preserve shape.

p. 108

Cartogram (contiguous)

Contiguous cartograms deform space by reshaping areas on the map while maintaining adjacencies. Originally referred to as *deformation cartograms* because the map is modified by pushing, pulling, and stretching the boundaries, the first computationally derived cartogram was made by the American-Swiss geographer and cartographer Waldo Tobler in 1973 using a rubber-sheeting method. Prior to Tobler's work,

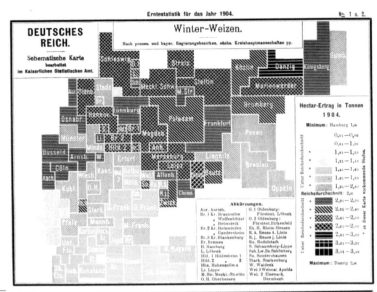

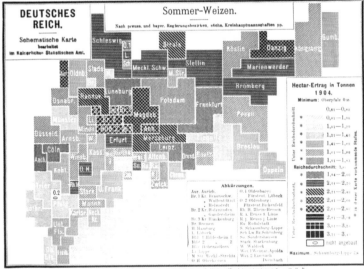

Harvest yields by the Imperial Statistical Office of the German Empire, 1905

these sorts of cartograms were made by hand. An early version of a cartogram that deformed size (though not shape) was published by French engineer Émile Cheysson in 1888. Many alternative algorithms exist with perhaps the most popular based on diffusion-based algorithms of applied mathematician and data scientist Michael Gastner and computational physicist Mark Newman in 2004. Gastner's later flow-based algorithm of 2018 (with Vivien Seguy and Pratyush More) has both optimised and modified the diffusion-based approaches.
p. 110

Chernoff faces
Chernoff faces were invented by American applied mathematician, statistician, and physicist Herman Chernoff in 1973 with the simple idea of creating cartoons of the human face where different data variables are represented by different facial features. Interpretation leant heavily on the notion that human beings are good at seeing and identifying small facial changes with minimum difficulty.
pp. 168, 170

Choropleth map
First developed by Baron Pierre Charles Dupin for his 1826 map showing levels of illiteracy in France, the map used an unclassed technique with shadings from black to white. As well as being the first choropleth map, it often is regarded as the first modern statistical map. It wasn't until 1938 that the actual term *choroplethe map* was introduced by the geographer John Kirtland Wright.
pp. 20, 22, 24, 26, 28, 36, 46, 80, 84, 178, 180, 182, 210

Chromastereoscopy
Chromastereoscopy is the generic name given to the patented Chromadepth™ system. It's a method of viewing based on differences in the diffraction of colour through prisms fitted into glasses. The technique was invented by US researcher Richard Steenblik in the early 1980s after he noticed that the bright colours on a video game seemed to lie in different depth planes. The technique of developing the optics for glasses that allowed for chromastereoscopic viewing wasn't perfected until 1992. The technique has been used in television by cable TV network Nickelodeon for the music video accompanying the 1998 single 'Psycho Circus' by rock band KISS and in map form by Scott Parker, who mapped the Grand Canyon using the technique in 2004.
p. 190

Illiteracy in France by Baron Pierre Charles Dupin, 1826

Circular line graph

Also known as a *polar line graph*, it's the equivalent of a rectangular line graph being wrapped into a circle. The horizontal axis of the graph becomes the circular axis of the circular graph and usually depicts time. If the circular axis depicts categories, the alternative radar chart would emerge. For both types, the quantitative data value being measured is shown on the radial axis. Émile Cheysson's maps for the *Albums of Statistics* for the French Ministry of Public Works often created new ways of visualising data by positioning innovative charts on maps. The 1883 map showing the population growth in all departments over 80 years is one such innovation. The circle atop each graph shows the population in 1841. The radial lines divide the circle into five-year intervals beginning in 1801 and continuing clockwise. The circular line graph traces how the population diverges from 1841 with red showing a decrease and grey showing an increase.
p. 162

Columns (vertical bars)

The bar chart was invented by Scottish engineer William Playfair in 1786 providing a quantitative graphical form that solves the problem of discrete visual comparisons of data. Positioning the base of a bar chart on a map and extruding the height of the bar vertically creates geographically located columns. Column charts have been positioned on maps at least since the late 1800s, including the maps made for the *Albums of Statistics* for the French Ministry of Public Works by Émile Cheysson. His 1881 map of the development of the railways used stacked columns. The 1883 map of expenditures for the first establishment of railways in the world illustrates the technique with multiple columns. In 3D, columns become cylinders, and the data variable is used to extrude them from either a digital elevation model as the surface or from a consistent base elevation. Hungarian-born American cartographer Erwin Raisz made extensive use of 3D columns in his thematic atlases in the mid-1900s.
pp. 192, 194, 198, 202

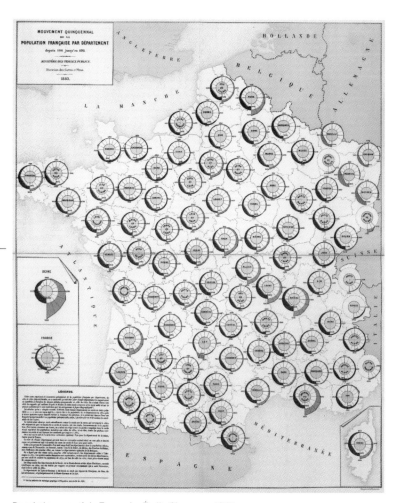

Population growth in France by Émile Cheysson, 1883

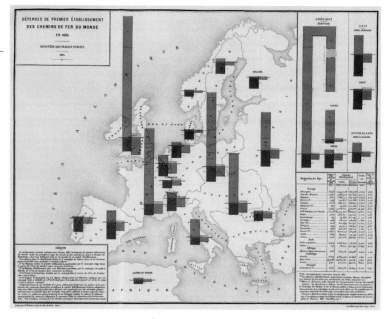

Map of expenditure of railways by Émile Cheysson, 1883

Dasymetric map

The first dasymetric map of population density was created by George Julius Poulett Scrope in 1833 although use of the term didn't emerge until 1911. Scrope never documented how he made the map, although the idea of using ancillary information to reallocate data to more meaningful spatial units clearly was established in his work. Many different methods have been developed to create a dasymetric map, with American-Swiss geographer and cartographer Waldo Tobler's pycnophylactic interpolation method finding popularity due to its use of a smooth density function extending to adjacent source zones while preserving the original count of each zone.
pp. 34, 68, 196, 228

Dials

A dial is generally a flat surface with markings used to display output—be that time or some other thing measured. Dials also can indicate direction in the form of a compass rose (sometimes known as a *wind rose*), and the ancient Greeks certainly maintained systems of points and winds based on celestial bodies used for orientation. In terms of cartography, many of the earliest uses of dials on maps relate to weather symbology. English polymath Francis Galton developed a series of glyphs with directional components for his 1861 meteorological charts, the first weather maps. The symbology for a meteorological station report acts as a dial with comprehensive information. Circular line graphs, polar area diagrams, and tree rings might all be a dial of some type.
pp. 96, 162, 164, 174

Directional arrows

The arrow is a well-used symbol in cartography. As a graphical mark, it immediately implies a direction and points to a destination. In 1504, it first was used by Portuguese cartographer Pedro Reinel as part of a fleur-de-lis to indicate north on a compass rose. In the 18th century, arrows began to be used to indicate flow of a phenomenon such as water, but it wasn't until 1903 when Hungarian-born historian Emil Reich used them to indicate army movements along lines that were used for indicating some aspect of the movement of people. In contemporary cartography, the arrow is used as a device to show movement of goods, people, ideas, or any other form of connectivity between places that might have qualitative or quantitative dimensions.
pp. 98, 100

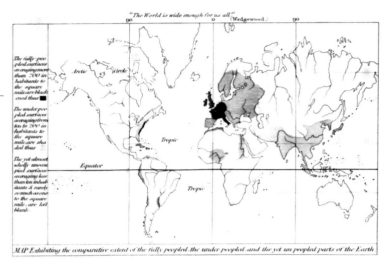

Population density by George Julius Poulett Scrope, 1833

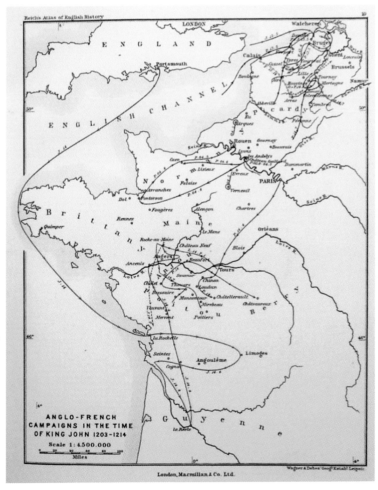

Army movements by Emil Reich, 1903

Dot density map

The use of dots on a map to show the location of data points often is attributed to epidemiologist John Snow's map of cholera in Soho, London, in 1854 (drawn by C. F. Cheffins). Each death was plotted against the street address. Yet the same technique was used by Valentine Seaman, an American physician whose two dot maps in 1795 pre-dated Snow to illustrate the distribution of yellow fever in New York. The first true dot density map was published in 1830 by French lithographer Armand Joseph Frère de Montizon showing the population of France by department where 1 dot represented 10,000 people.
pp. 66, 72, 74, 222

Distributive flow map

Although most of the lauded examples of distributive flow maps will lead directly to the work of French engineer Charles Joseph Minard in the mid-1850s, the first flow maps that used graduated widths to indicate quantity were published by British soldier Henry Drury Harness in 1837. Harness became an instructor at the Royal Military Academy at Woolwich and drew a range of maps of population and traffic for his Irish Railway Commission atlas.
pp. 100, 160

Halftone symbology

The invention of halftone symbology isn't easily credited although it's based on the appearance of the reprographic technique that uses dots of different sizes and spacing to simulate continuous tone images. British inventor William Henry Fox Talbot patented the use of textile screens to create halftone images in 1852 to support the more accurate printing of photographs in newspapers. In 1879, illustrator and printer Benjamin Henry Day Jr. produced a similar printing and photoengraving technique referred to as the Ben Day process, which created Ben Day dots. Although halftone uses dots of different sizes, Ben Day dots are the same size and distribution in contained areas. As technology progressed, digital halftone emerged in the 1970s through digital dot generators, image setters, and laser printers. Comic books from the 1950s onward made use of the Ben Day process for mass production in the four-colour printing process, and it was a hallmark of American artist Roy Lichtenstein. French cartographer and theorist Jacques Bertin published some excellent examples of the use of halftone for thematic data in *Semiology of Graphics* in 1967.
p. 214

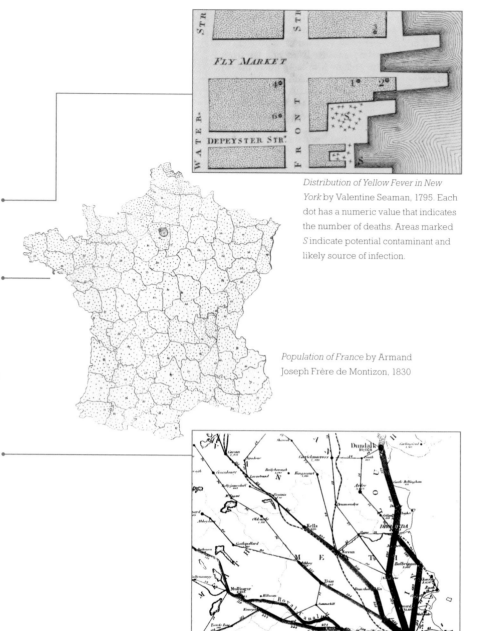

Distribution of Yellow Fever in New York by Valentine Seaman, 1795. Each dot has a numeric value that indicates the number of deaths. Areas marked *S* indicate potential contaminant and likely source of infection.

Population of France by Armand Joseph Frère de Montizon, 1830

Traffic flow in Ireland by Henry Drury Harness, 1837

Hillshades and shadows

Swiss painter, mathematician, surveyor, and cartographer Hans Conrad Gyger's *Grosse Landtafel des Kantons Zürich* (*Great Land Board of the Canton of Zurich*, 1664–67) was the first planimetric depiction of shaded relief. The 3D shading technique was so advanced, it was considered a military secret and hidden from his contemporaries. Lithographic printing techniques underpinned the development of tonal shades for choropleth maps in the mid-1800s, which also supported the development of shadows to accentuate symbol design. These became useful as graphical enhancements for 3D symbols in the mid-1900s in—for example, the statistical maps by Erwin Raisz. Shadows often are used for discrete symbols but only recently as full hillshade effects for statistical data.
pp. 32, 72, 92, 120, 194, 196

Histogram

Histograms have been used for centuries in many scientific fields to approximate and compress information. The term first was introduced by English mathematician and biostatistician Karl Pearson in the late 1800s to describe a common form of graphical representation in which columns mark as areas the frequency corresponding to the range of their base. A histogram is an approximate representation of the distribution of an array of numerical data. Bar charts almost certainly predate the development of the histogram. They first were used by William Playfair in 1786. The first use of a histogram is, unfortunately, unclear.
p. 144

Illuminated contour

Otherwise known as *relief contours* or *Tanaka contours*, this method of adding relief shading to contours originally was developed in the mid-1800s. Swiss scientific illustrator Ernst Heinrich Michaelis's *Passage du Splügen et de la Via Mala*, published in 1845, used varied line widths, most likely with a calligraphic pen to create an appearance of illumination and shadows. Limitations of printing techniques meant it wasn't possible to reproduce accurately the technique consistently in practice until the late 1950s. Professor Tanaka Kitiro developed the relief contour method by defining mathematical principles for their creation. The method varies line widths and uses black or white lines according to the angle of illumination and aspect of terrain. Most automated algorithms are based on the Tanaka method.
p. 92

Bivariate histogram showing the correlation between the heights of fathers (horizontal axis) and sons (vertical axis). This data was collected by Karl Pearson and Alice Lee between 1893 and 1898. The model was constructed around the same time, possibly under the supervision of Pearson. This example is also an early form of 3D modelling where height is used to encode a value.

Infographic

Infographic is not so much a map type but a word that defines a particular style of information visualisation. Although many maps, graphs, charts, and plots through history might be considered information graphics, they are not necessarily infographics. The word *infographic* first appeared in Éric de Grolier's *The Organization of Information Systems for Government and Public Administration*, published by UNESCO in 1979. The focus was on creating graphical representations of information, data, and knowledge succinctly, and it mattered little if it was a print or stenciled document, computer list, diagram, map, or photograph. Contemporary infographics tend to be collages of various components that each contribute a nugget of information to an overall theme. Alternatively, they often appear in news media as expressions of data through a unique visual metaphor. The inspiration for the infographic style on p. 142, which used interactive proportional balloons, was drawn from a similar approach by *The Guardian* and *Real Clear Politics* in their coverage of the 2012 presidential election campaign between Barack Obama and Mitt Romney (isbarackobamathepresident.com).
pp. 132, 142, 144

Extract of the *Plan de Paris* by Michel-Étienne Turgot, 1739

Interpolated surface shading

Representing a surface using variable shading was pioneered by the Belgium statistical and social theorist Adolphe Quetelet, who produced two maps in 1831 showing crimes against property and crimes against people using crayon shading. Quetelet abandoned the boundaries of the choropleth technique which others were using at the time in favour of an innovative attempt to show a smoothed continuous distribution with different tones of shading. Darker shading equals higher rates (of crime). In 1848, British cartographer and geographer August Petermann also made use of this technique in his map of cholera in the British Isles in the early 1830s and again in 1849, and then in 1852 to create maps of England and Wales that showed the distribution of the population based on the 1851 census of population survey. In contemporary computing, several algorithms provide ways to achieve the same and to create a rasterised digital elevation model (DEM) from input points. Nearest neighbour, kriging, inverse distance weighting, spline, and radial basis functions are all techniques that resolve the values of pixels across the DEM, which then are shaded continuously across a colour ramp according to the data value.
pp. 38, 76, 78, 80, 190

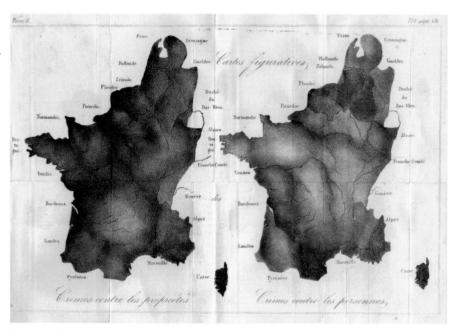

Crimes against property and crimes against people by Adolphe Quetelet, 1831

Isometric projection

Although used for centuries beforehand and originating in China, British chemist and natural philosopher William Farish formalised the method of the isometric version of the axonometric projection in his 1822 paper 'On Isometrical Perspective'. Farish recognised the need for accurate technical working drawings free of optical distortion caused by a perspective view of the world. The isometric projection visually represents 3D objects in two dimensions with equal foreshortening. An early example of its use in cartography is the detailed *Plan de Paris* published by Michel-Étienne Turgot in 20 sheets in 1739. It also was used to great effect by Hermann Bollmann in his *Bird's Eye View of New York* in 1962 and by Constantine Anderson covering much the same area in his 1985 map of midtown Manhattan. In thematic mapping, Hungarian-born American cartographer Erwin Raisz made use of 3D symbology in many of his atlases in the mid-1900s and the isometric perspective in his 1964 *Atlas of Florida*. pp. 120, 186, 198, 200, 204

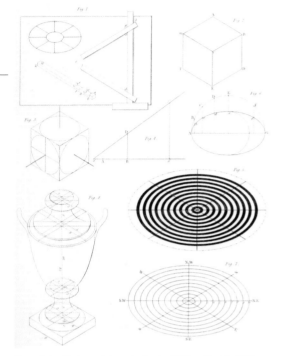

Formalising the isometric projection by William Farish, 1822

Isotype

In a forerunner to isotype, Irishman Michael George Mulhall formalised the pictogram in 1884, which was a graphic form of visualising phenomena using pictures that resembled a likeness of the phenomenon itself. Mulhall made pictograms for thematic data and sized them proportional to the data value. The Vienna Method of Pictorial Statistics was invented in 1924 by Austrian-born philosopher of science, sociologist, and political economist Otto Neurath, along with German designer and social scientist Marie Reidemeister (Neurath and Reidemeister married in 1941) and German artist Gerd Arntz. The term *isotype* (International System Of TYpographic Picture Education) was developed around 1935 after its inventors were forced to leave Vienna and flee to the Netherlands because of the rise of Austrian fascism. Isotype was used in the late 1930s and early 1940s in many publications, particularly in the United States. After the Neuraths fled to the United Kingdom from German invasion, they set up the isotype institute in 1942, and soon thereafter British wartime publications began to use isotype. Renowned graphic designer Herbert Bayer oversaw the production of the *World Geo-Graphic Atlas* for the US Container Corporation of America in 1853. Considered a masterpiece of cartographic design, it made substantial use of isotype in its thematic maps. pp. 70, 132

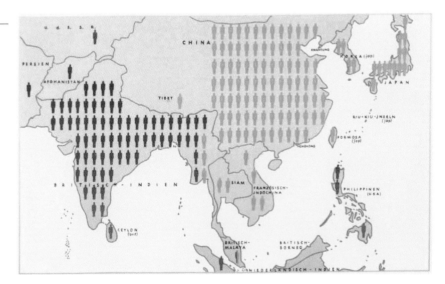

Extract from the 1930 *Statistical Atlas on Society and the Economy* by Otto Neurath and Gerd Arntz showing population in India and the Far East in 1930

Joy plot

In 2017, software engineer Jenny Bryan coined the term *joy plot* to describe a stacked series of area graphs that usually show categorical data on the y-axis and time on the x-axis. One of the defining features of a joy plot is overlap, as the peaks of individual graphs tend to puncture the base of the graph immediately above, creating a sense that the graph is somehow in front. Although not necessarily new, since they owe much to the development of area charts by Scottish engineer William Playfair in 1786, they took on new meaning with reference to the classic Joy Division 1979 *Unknown Pleasures* album cover, created by Peter Saville for Factory Records. They also were used in map form by French cartographer and theorist Jacques Bertin in his *Semiology of Graphics* in 1967. A joy plot is therefore a series of histograms, density plots, or time series for several segments, all aligned to the same horizontal scale. The name *joy plot* has been challenged more recently because of connotations with the origins of the band's name, Joy Division, which was the name given to groups of Jewish women in Nazi concentration camps kept for the sexual pleasure of soldiers. Given the history of the band's name, some prefer to refer to them as *ridgeline plots* instead.
p. 218

Example of a joy plot by Jacques Bertin, 1967

Labels as literal symbols

Labels have been a mainstay of cartographic practice for centuries. Labelling populated places or landscape features has been a necessary evil in cartography to enable the interpretation of maps and the translation of what is seen into a description. August Friedrich Wilhelm Crome is credited as the creator of the first thematic map of data for his 'new map of Europe which contains the strangest products and the most prestigious trading places along with the area content of all European countries in German square miles', published in 1782. Although the map contained some symbols, it predominantly used abbreviated letterforms that represented the occurrence of 56 products such as ores and industrial products.
p. 82

Line graph

Although Scottish engineer and political economist William Playfair largely is credited as the inventor of the line graph in 1786, Swiss polymath Johann Heinrich Lambert also began using line graphs to analyse empirical measurements in the late 1700s. He was one of the first to use graphical analyses

The use of abbreviated letterforms by August Friedrich Wilhelm Crome, 1782

to calculate derived information, such as estimating the rate of change in a data variable over time. Along with Playfair, Lambert is regarded by American Harvard scholar Edward Tufte as one of the two great inventors of modern graphical design. As this section of the book evidences, there are many others who have an equal claim.
pp. 130, 134, 156, 172

Masking

Broadly speaking, masking of map detail might refer to the omission of content or detail to focus on a separate aspect of the map. English astronomer, geophysicist, mathematician, meteorologist, and physicist Edmond Halley's weather map of 1686 showing prevailing ocean trade winds and his 1701 chart of compass variations in the Atlantic Ocean do just that by removing all land detail. Only the coastlines remain as the focus is entirely on the patterns of lines used to map the thematic feature. In contemporary cartography, making definition queries to isolate selected features based on specific criteria or turning layers off can achieve the same result. Some techniques such as dasymetric or value-by-alpha might have the appearance of applying a mask to parts of the map but those techniques rely on data reapportionment or the application of variable transparency to mute areas. Masking is, instead, binary.
pp. 36, 212

Multivariate point symbols

Multivariate symbology has a rich history in cartography. Matthew Fontaine Maury's *Wind and Current Chart of the North Atlantic*, 1850, encodes wind direction and strength, precipitation, and season via multivariate point symbols. Currents are shown as line features, encoding strength, temperature, season, and month. This map is two multivariate symbol layers, wind and currents, shown concurrently—an intersymbol coupling of intrasymbol encodings. Charles Joseph Minard's work provides many early examples of multivariate symbol encoding, later inspiring the work of Émile Cheysson, who created annual *Albums of Statistics* for the French Ministry of Public Works in the late 1800s.
pp. 64, 76, 78, 124, 168, 216

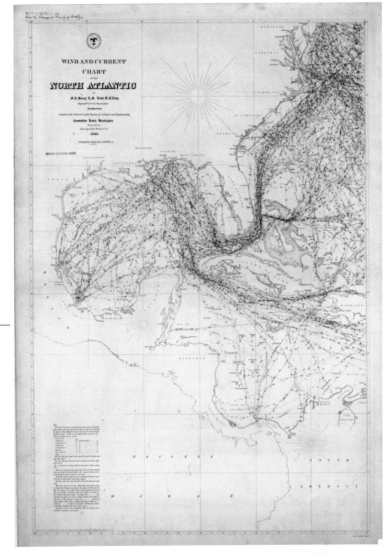

Wind and current chart of the North Atlantic by Matthew Fontaine Maury, 1850

Pattern fills

English astronomer, geophysicist, mathematician, meteorologist, and physicist Edmond Halley's use of dashed lines to mark prevailing wind on the first known weather map, in 1686, is perhaps also the first use of a pattern fill. The use of pattern fills to symbolise some characteristic of an area on maps saw extensive use due to dependence on copperplate engraving in the intaglio printing process. Map production was heavily reliant on intaglio printing from the demise of woodcut in the mid-16th century until the mid-19th century when lithography largely replaced engraving in the production of maps. Engraved maps were made by cutting lines and marks into a metal, usually copper. Patterns became a common method of showing differences between areas on the map, often on monochrome printed maps. In 1863, English polymath Francis Galton developed a large palette of pattern fill symbols to devise a comprehensive way of mapping weather. The map on p. 44 was inspired by French cartographer and theorist Jacques Bertin's use of a similar technique in his 1967 book, *Semiology of Graphics*. Not only did the pattern fill he developed illustrate qualitative difference but the pattern itself encoded proportions of three different variables.
pp. 44, 210

Map of the trade winds by Edmond Halley, 1686

Pictorial map

The earliest known printed map, *Rudimentum Novitorum*, was printed in Lübeck, Germany, in 1475. It was a T in O map (for Orbis Terrarum), essentially a pictorial map, and pictorial symbols had been used since early clay tablets, some 4,000 years earlier. The landscape is filled with representations of mountains, settlements, and people going about their business. Modern pictorial maps can look very much the same, filled with the characters to show a living landscape such as in the work of Uruguayan-born American cartoonist, comics artist, sculptor, illustrator, and cowboy Joseph Jacinto Mora, whose many maps of California provide excellent examples of the genre. Pictorial maps also have been used as satirical devices for centuries. They often lampoon characters or are used to comment about the social, economic, or political situation using the map as a graphical device and framework. This approach was prevalent in Britain during the second half of the 18th century, including the whimsical maps of Robert Dighton, 1793, and William Harvey's Geographical Fun maps, 1869. Comic and serio-comic maps developed into increasingly political (and sometimes prejudicial) visual metaphors. For example, Fred Rose began publishing maps in 1877 using

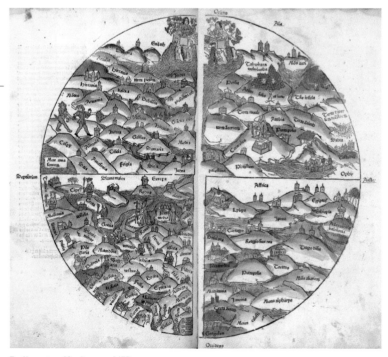

Rudimentum Novitorum, 1475

national stereotypes, animals, objects, and political figures, which became extremely popular. Johnson, Riddle & Co. produced a range of maps in the early 20th century depicting Europe. In 1914, *Hark! Hark! The Dogs Do Bark!* showed the continent at war with nations represented by various breeds of dog. The sailor is holding onto battleships via several dogheads as the Tsar of Russia approaches with a steamroller flattening everything in its path.
p. 230

Pie chart
Along with the bar chart, area chart, and line graph, Scottish engineer and political economist William Playfair also invented the pie chart, in 1801. In his publication titled *Statistical Breviary*, he illustrated the Turkish Empire's landholdings by slicing a circle into three wedges whose sizes were determined by land area. In colouring Asia's wedge green, Europe red, and Africa yellow, he not only displayed empirical proportions but differentiated the qualitative components by colour. French engineer Charles Joseph Minard was one of the first to position pie charts across a map, in 1858, to represent the proportions of cattle sent from around France to Paris for consumption.
pp. 64, 124, 130, 152, 164

Polar area chart
Although Florence Nightingale often mistakenly is credited as the inventor of the polar area chart, French lawyer (and amateur statistician) André-Michel Guerry created polar area charts in 1829 to show direction of the wind in eight sectors and births and deaths by hour of the day. His example used equal angular sectors with different radii and data encoded to the area of each segment, creating segments with different radii. Had Nightingale seen Guerry's work? Possibly. It's also possible she had seen them used by British epidemiologist William Farr, who was familiar with Guerry's work and had published polar area charts plotting mean temperature against mortality in London 1840–1850, published in 1852, some six years before Nightingale's more famous version. In fact, Nightingale collaborated with Farr on her versions so the influence may have been more direct. Polar area charts also are referred to colloquially as *coxcomb diagrams*, although this is a naming error that has been handed down through history.
pp. 162, 164

Hark! Hark! The Dogs Do Bark! by Johnson, Riddle & Co., 1914

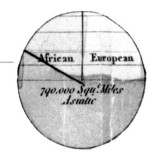

Proportions of the Turkish Empire by William Playfair, 1801

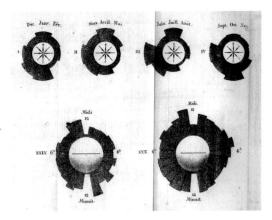

Polar area charts by André-Michel Guerry, 1829

Prism maps

3D depictions of geography have been around for centuries. Artistic bird's-eye views found popularity for mapping cities and small-extent landscapes. The development of 3D thematic or statistical maps is, by contrast, a relatively recent innovation. They also are referred to as a *stepped-relief map*, or *block map*, in which areas are extruded in height to a value they represent. In diagrammatic form, the *New York Journal* published a 3D representation of the forecast number of votes for candidates in the New York mayoral election in 1897. It showed the four candidates standing in front of a block of ballot papers with uniform length and width but variable height. The columns were extruded by the quantity of votes they were expected to receive. In map form, Erwin Raisz used the technique in his 1944 *Atlas of Global Geography*. He first produced a contiguous cartogram by tessellating squares to create rectangular shapes for countries, and then displayed it in an axonometric view with the countries extruded into blocks to show aspects such as death rate or hunger. In 1963, American cartographer George Jenks described the principle of extruding regions of a basemap in 3D according to a statistical variable but he noted that they currently must be drawn by hand using the same techniques as the preparation of physical geography block diagrams. Although not a map, French engineer Charles Joseph Minard's 1844 *tableau graphique* uses the spatial horizontal axis (along the Canal du Centre) effectively to create a profile through what would be a 3D prism map.
pp. 178, 180, 182, 184, 198

Profile lines

Also known as a *cross-section* or *transect*, profile lines have had a long history in geological mapping, in which the height of the profile line represents elevation, with geological features drawn underneath. For thematic data, Scottish engineer and political economist William Playfair's *Chart of Universal Commercial History* of 1805 uses a similar approach but with the value of the data variable used to plot the height of a line. Although often described as a mosaic plot, French engineer Charles Joseph Minard's 1844 tableau graphique uses a spatial horizontal axis, a profile along the Canal du Centre, to create a transect. It shows the transportation of commercial traffic through bars of varying width along the transect. Each bar is divided (a stacked bar chart), with height representing the amount of traffic. Consequently, the area of each segment shows the overall cost of transport.
p. 104

New York mayoral election prisms in the *New York Journal*, 1897

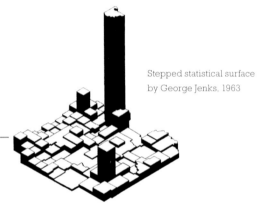

Stepped statistical surface by George Jenks, 1963

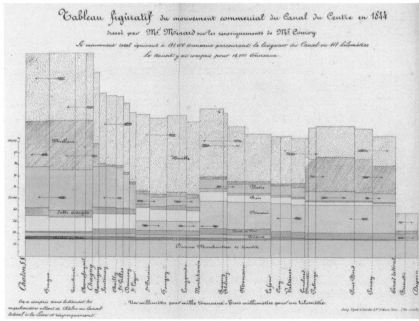

Charles Joseph Minard's 'tableau graphique' of the transport of commercial traffic along Canal du Centre, 1844

Proportional line symbols

The first use of proportional point symbols was by Scottish engineer and political economist William Playfair in 1801. The same concept first was applied to lines in 1837 by Henry Drury Harness on his map showing transportation in Ireland, in which the value of data was represented by a line whose thickness was proportional to the number of passengers (p. 252). French engineer Charles Joseph Minard made excellent use of proportional line symbology in his many flow maps in the mid-19th century, and on his famous map of Napoleon's 1812 march on Moscow, published in 1869. That map was, in fact, originally the bottom map on a sheet that contained another flow map, that of the campaign by military commander Hannibal some 2,000 years before that of Napoleon. Curiously, not all the lines on Minard's map of Napoleon's march were proportional. Although the legend says that 1 millimetre in line width is equivalent to 10,000 troops, at the point in which the map label indicates there are 10,000 troops, the line is 2 millimetres in width. Sometimes range-grading of symbols is used to symbolise extreme values on a proportional symbol map, and this is possibly what Minard did here to ensure overall clarity.
pp. 94, 220

Proportional point symbols

The first use of proportional symbols was made by Scottish engineer and political economist William Playfair in his book *Statistical Breviary*, published in 1801. He laid out an array of circles of different sizes, each representing the population of different cities through scaling the area of the circle to the data value. French engineer Charles Joseph Minard made numerous maps, graphs, and charts, but he was notable for merging the two and creating maps with charts and diagrams that sat on top. Published in 1858, his map of proportional symbols to show coal mine production was the first map to incorporate statistical diagrams with circles proportional to a data variable. A variation of the proportional symbol is the graduated symbol, in which data values are represented by range-graded symbols of fixed sizes. Each data value in a certain class is depicted using the same-sized symbol. Symbols also may be designed to have a pseudo-3D appearance in which the volume of the symbol is proportional to the data value being mapped. Hungarian-born American cartographer Erwin Raisz made use of 3D symbology in his 1944 *Atlas of Global Geography*.
pp. 54, 56, 58, 60, 62, 64, 72, 84, 142

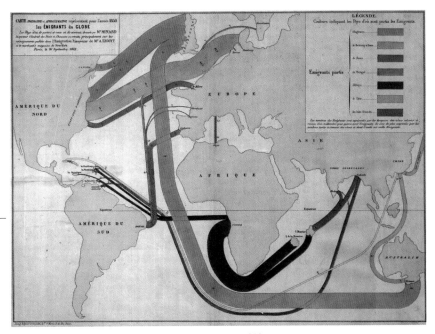

Worldwide emigrants of 1858 by Charles Joseph Minard, 1862

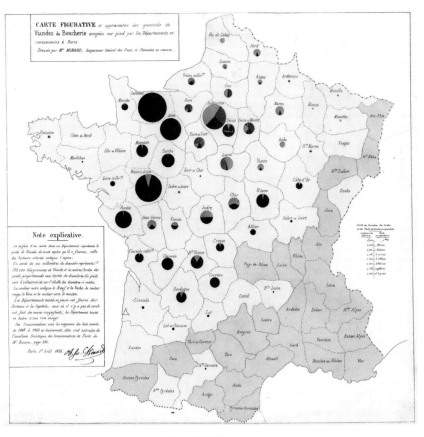

Coal mine production by Charles Joseph Minard, 1858

Sankey diagram

Sankey diagrams are named after Irish-born engineer Matthew Henry Phineas Riall Sankey, who used them in a figure showing the thermal efficiency of a steam engine in 1898. Although his name characterises the diagram, the technique has a longer history. Distributive flow charts already had been used on thematic maps to indicate transport flow in Ireland, published by British soldier Henry Drury Harness in 1837, and in French engineer Charles Joseph Minard's famous map of Napoleon's Russian campaign of 1812 (published in 1869). Both were a form of Sankey diagram, in map form.
p. 160

Scatterplot

In 1874, English polymath Francis Galton invented scatterplots to explore statistical correlation and regression. They are diagrams that use Cartesian coordinates to display values for two variables in a dataset. Values are displayed as a set of points with the position determined by one of the variables on the horizontal axis and the other variable on the vertical axis.
p. 130

Snake diagram

Linear diagrams long have been used to represent paths. In geographical terms, John Ogilby's strip maps in his *Britannia* road atlas in 1675 showed routes in a stylized form with key places along the route. The historical eruptions of Mount Vesuvius were mapped beautifully by Canadian-British traveller, geologist, writer, and artist John Auldjo in 1832. Lines also have been used in more abstract forms. For instance, American historian Walter Houghton created a linear diagram of intertwined lines to show the history of political parties in the United States in 1880. The general idea was updated by news media outlet FiveThirtyEight for 2016 presidential election coverage and again for 2020 coverage when they called it a *snake diagram*. The illustration on p. 134 was inspired by the FiveThirtyEight snake diagram.
p. 134

Space-time cube

The space-time cube originally was developed by Swedish geographer Torsten Hägerstrand in 1970. The cube's horizontal plane represents space in the x,y dimensions, and the vertical axis represents time. Any point in the 3D space

Historical eruptions of Mount Vesuvius by John Auldjo, 1832

locates a position in both space and time. Space-time cubes can be used to visualise point, line, or area features. If space is split into 3D bins, they also can hold properties that can be mapped, such as volume of a data variable at that specific position or the aggregation of finer resolution data within a 3D bin. Dutch Professor of Visual Analytics and Cartography Menno-Jan Kraak has been a pioneer in the resurgence of space-time cubes in interactive environments, particularly in 2014 to reimagine Charles Joseph Minard's classic map of Napoleon's march on Moscow.
pp. 146, 200

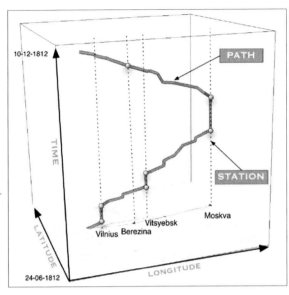

A space-time cube of Napoleon's 1812 march by Menno-Jan Kraak, 2014

Sparklines

Sparklines are typically small line graphs or bar charts, often used in-line to show general variation in a condensed way. Anglo-Irish novelist Lawrence Sterne used them as a typographical device to illustrate his narrative. They also were used in 1888 to indicate barometric pressure at various locations around the world for the 1883 eruption of Krakatoa. Professor Emeritus of Political Science, Statistics, and Computer Science Edward Tufte documented the style in 1983, calling it an 'intense continuous time-series'. The small charts often were used to show temperature or stock market prices. It wasn't until 2006 that Tufte introduced the term *sparkline* as 'small, high-resolution graphics embedded in a context of words, numbers, images'. A sparkline is a 'data-intense, design-simple, word-sized graphic'.
p. 172

Spatial treemap

A form of rectangular cartogram in which the map is schematicised with regions replaced by rectangles and the topology of the map is represented by the dual graph of the map. Perfect topology is hard to achieve, and there's often a trade-off between zero topological error and a perfect rectangular grid. Hungarian-born American cartographer Erwin Raisz introduced the idea of rectangular statistical cartograms in 1934. Several developments of the algorithms since have been published, and the idea of a spatially ordered treemap with a squared layout was proposed by UK academics Jason Dykes (Professor of Visualisation) and Jo Wood (Professor of Visual Analytics) in 2008. The spatial treemap on p. 138 uses a version of Dykes and Wood's algorithm.
p. 138

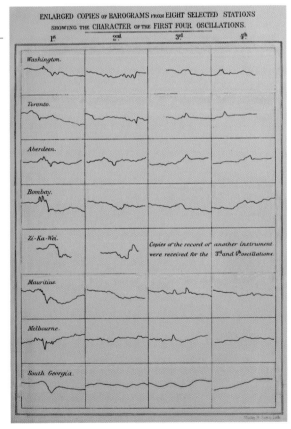

Global barometric pressure 'sparklines' for the 1883 eruption of Krakatoa, 1888

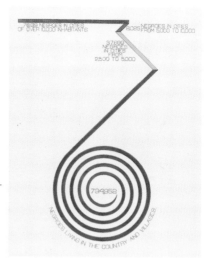

African Americans in Georgia by William Du Bois, 1900

Spiral point symbols

Spiral diagrams first were invented by American sociologist, socialist, historian, civil rights activist, pan-Africanist, author, writer, and editor William Edward Burghardt Du Bois. He developed the technique to visualise where the majority of Black people lived in Georgia in 1890 (published in 1900). The ingenious diagram was a way of positioning the different proportions of those who lived in urban areas against those who lived in rural areas on the same page. He also published charts that used concentric circles in the same publication, and both techniques were forerunners of spiral graphs. French cartographer and theorist Jacques Bertin formalised the structure of concentric ringed point symbols in his 1967 book *Semiology of Graphics*. American architects Joseph Russell Passonneau and Richard Saul Wurman used a similar encoding of multiple variables into circles in their 1967 atlas of American cities. These map types have been referred to colloquially as *tree-ring charts*.
pp. 76, 78, 166

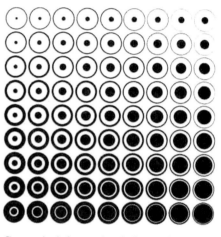

Concentric circle tree-rings by Jacques Bertin, 1967

Stacked chips

Stacking symbols in a pseudo-3D map to overcome problems of overlapping, or as a way of supporting counting, has a strong history in thematic mapping. Hungarian-born American cartographer Erwin Raisz used the concept to show a range of different thematic phenomena in his *Atlas of Global Geography* in 1944. For instance, import and export of goods were shown with stacks of the goods themselves, such as textiles, wood, paper, and cotton. Each layer in the stack represented a specific unit of value, so the higher the stack, the higher the value.
pp. 194, 200

Statistical contour

Also known as an *isoline* or *isopleth*, the first contour map of a data-based variable is credited to English astronomer, geophysicist, mathematician, meteorologist, and physicist Edmond Halley. He used lines of equal magnetic declination to create an isogonic map in 1701. The technique was extended to show areas of similar elevation with the same shading, which was lighter at higher elevations, developed by French publisher and cartographer Jean-Louis Dupain-Triel in 1782. French engineer Louis-Léger Vauthier produced one of the earliest known contour maps of thematic data in his 1874 map of population density in Paris.
pp. 88, 90, 92, 184

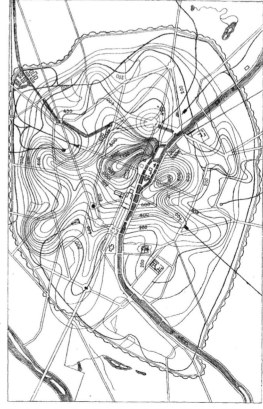

Population density in Paris by Louis-Léger Vauthier, 1874

Stepped contour

A 3D version of the statistical contour, in which each class interval between contours is extruded to a common height. The 2D approach can be traced back to French engineer Louis-Léger Vauthier's 1874 contoured maps of population density in Paris. American cartographer George Jenks' development of stepped surface maps in 1963 used extrusion to encode height into 3D on thematic maps. The same technique was discussed extensively by French cartographer and theorist Jacques Bertin in *Semiology of Graphics* in 1967.

p. 184

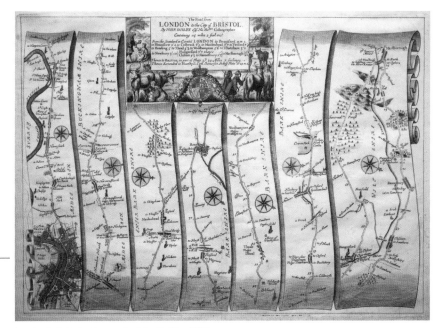

London to Bristol strip map by John Ogilby, 1675

Strip map

Scottish translator, impresario, and cartographer John Ogilby published the first road atlas, *Britannia*, in 1675. Comprehensive, it contained 100 plates of maps drawn at a consistent scale of one inch to the mile. It is known most widely for its use of an innovative scroll format, in which the map is split into narrow segments with each segment a continuation of the preceding one. The map on p. 102 is a pastiche of an Ogilby strip map.

p. 102

Small multiples

The first known use of small multiples to show a series of images as a coherent display was by Jesuit priest, physicist, and astronomer Christoph Scheiner in 1630. From his book *Rosa Ursina* (*The Sun*), he pioneered a way of representing the motion of sunspots across the sun's face. English-American photographer Eadweard Muybridge used a sequence of small multiples to develop his horses in motion graphic in 1886, pioneering animation and the development of the motion picture. American economist, statistician, journalist, educator, and military officer Francis Amasa Walker made substantial use of small multiples in the 1870 *Statistical Atlas of the United States* (published in 1874). Walker was superintendent of the US Census at the time. In fact, the atlas was innovative not only in the graphics' use of small multiples but also in other graphical techniques, such as mosaic charts, the use of side-by-side maps to show change over time, and diverging bar graphs showing demographic information tiled by state. The term *small multiples* wasn't, in fact, coined until 1990 by Edward Tufte.

pp. 48, 156, 158, 160, 162, 164, 166

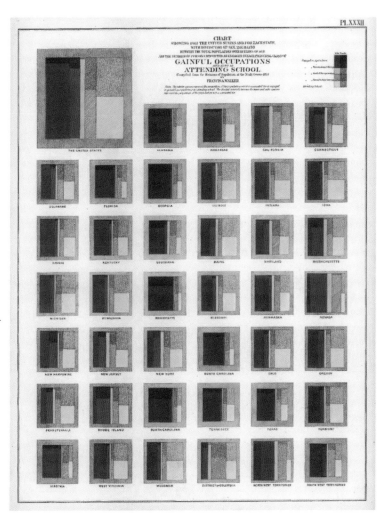

Employment in the United States by Francis Walker, 1874

Swing-o-meter

A specific graphical device most used in the reporting of elections by news media. The device was invented by Peter Milne and used in his reporting of the results in the southwest of the UK in 1955. The idea further was developed by David Butler, and the first physical model was used by the British Broadcasting Corporation in its television coverage of the 1959 general election. Digital swing-o-meters have become commonplace and are used to show the swing in percentage points in the vote toward one party or candidate and another. A limitation of the 2D version is that it restricts depiction of the swing to only two parties or candidates. A 3D swing-o-meter introduces different spaces, which can show how a swing might consist of a proportion from one party and a different proportion from another as the vote moves to a third party.
pp. 96, 174

Tables

Tabular formatting can be traced to antiquity. The invention of numerical tables for storing information is traceable to the early administrative archives of Sumer and Babylonia in Mesopotamia. Although many formal and informal styles exist, these tables from the late fourth millennium generally had vertical and horizontal rulings to separate categories of information. Detail was etched into clay tablets to account for fisheries, herded animals, grain production, and labour among other aspects of local life and the economy. Tables of empirical data began to appear in the early 1600s. They consisted of published tables of numbers and represented a branch of statistics which simply described facts using numbers. In addition to his profession as a haberdasher, Englishman John Graunt is regarded as the founder of demography, and possibly also the first epidemiologist, who pioneered his tables of mortality. Toward the late 1600s, tables were being used to report the first complete census (France, 1666), by English economist William Petty in his international statistical comparisons, and by English astronomer, geophysicist, mathematician, meteorologist, and physicist Edmond Halley in 1693.
p. 128

Timeline chart

Timelines come in many forms. The chronological map was created by French physician, botanist, writer, translator, and publisher Jacques Barbeu-Dubourg in 1753 to show

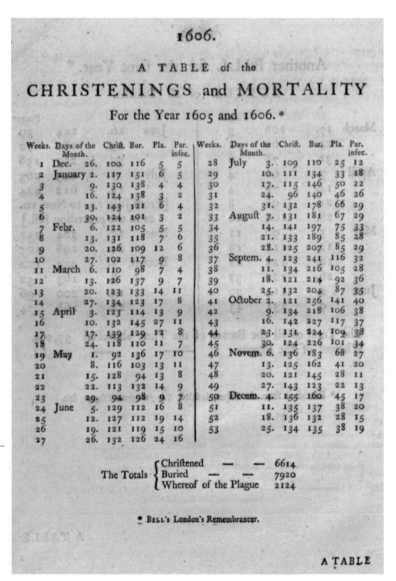

Table of mortality by John Graunt, 1606

an annotated timeline of history. It comprised a 54-foot scroll which illustrated events as well as names, grouped thematically and with symbols that showed character and profession. In 1765, English chemist, educator, scholar, and natural philosopher Joseph Priestly created an innovative timeline chart he called 'A Specimen of a Chart of Biography', in which individual bars represented the lifespan of a person, showing the lives and deaths of famous people in history. Longer bars meant a longer lifespan, and so the chart encoded a start date, end date, and span. He went on to create a further timeline in his 1769 'New Chart of History', which plotted the rise and fall of empires.
p. 130

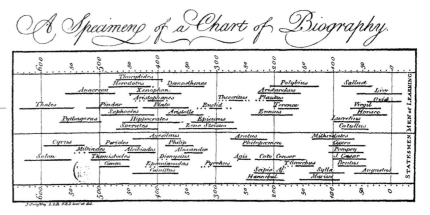

'A Specimen of a Chart of Biography' by Joseph Priestly, 1765

Treemap

A treemap is essentially a stacked column chart in which columns have different widths. French mathematician Charles Louis de Fourcroy is credited with the first use of geometric, proportional figures which he stacked into a square as an early example of a tableau graphique in 1782. Subsequently, German Georg von Mayr divided up a square in mosaic form to represent different dimensions of the dataset in 1874. Both were early forms of organising space in treemap form. More recently, in 1991, American computer scientist Ben Shneiderman developed one of the first computational techniques to nest rectangles to show some numerical measure in a space-constrained visualization of hierarchy. His 'slice and dice' method pioneered modern algorithms to support automated creation of treemaps although many alternatives now exist that order rectangles in different ways and which have different aspect ratios.
p. 136

Triangulated irregular network

With its origins in elevation modelling, a triangulated irregular network (TIN) is a representation of a continuous surface using vector-based triangular facets. It's a form of digital elevation model. TINs first were used by Swedish geographers Bengt-Erik Bengtsson and Stig Nordbeck to automate the drawing of contours through irregularly distributed datapoints in 1964 although the first use of triangles to explicitly represent a surface was by German geomorphologist K. Hormann in 1969.
p. 188

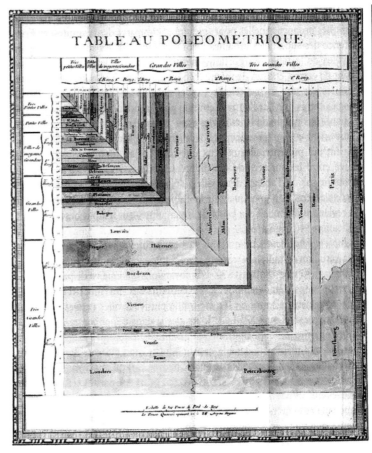

Demographic quantities as a tableau graphique by Charles Louis de Fourcroy, 1782

Trilinear plot

The trilinear plot first was used by American scientist Josiah Willard Gibbs in 1873 to show the relationship of three variables as a graph of x,y,z which sum to a constant. Otherwise known as a *ternary plot*, it graphically depicts the ratios of three variables as positions in an equilateral triangle. The use of trilinear plots in cartography is mostly as a trivariate choropleth map which seeks to map the relationship between three variables. It does so by applying three colours to each of the three sides of the plot and uses colour mixing theory mostly via subtractive mixing of cyan, magenta, and yellow (CMY) Because of its dependence on colour mixing, creation of a trivariate choropleth has been possible only by using computation. In 1981, Denis White produced a series of maps to illustrate land use in Massachusetts. Detailed categories were classified into the proportion of urban, open, and forest. The maps were created on an Advanced Electronic Designs 512 Imaging Terminal. In 1994, academic cartographer Cynthia Brewer set out a range of colour guidelines for cartography, including the concept of a 'three-variable balance scheme', which was effectively a simplified trilinear plot with internal areas classified into nine segments coloured according to CMY mixing.
p. 40

Unique values map

Mapping the difference between types across a map owes much to early geological mapping. In 1811, the geological environs of Paris were mapped by two Frenchmen, mining engineer George Cuvier and natural history professor Alexandre Brongniart. Hand-colouring showed different geological exposures planimetrically, as well as geological sections. In 1815, the first geological map of a large area was published by English geologist William Smith. British social researcher and reformer Charles Booth published his complete *Inquiry into Life and Labour in London* map in 1899, which is regarded as the first attempt at social cartography. The first sheets had been published a decade earlier with each building coloured to indicate the income and social class of its inhabitants. Although colour had been used on statistical maps prior to Booth, for instance, by French lithographer Armand Joseph Frère de Montizon on his map of political France in 1834, this monumental map was the first to apply it to a discrete thematic data variable so that spatial patterns could be explored.
pp. 16, 18, 48, 50, 104, 154, 174, 188, 192, 194, 196, 198, 200, 202, 204, 208, 210, 212

Extract from the *Inquiry into Life and Labour in London* by Charles Booth, 1899

Map of political France by Armand Joseph Frère de Montizon, 1834

Value-by-alpha

Value-by-alpha is a modification of the choropleth technique to visually equalise for population distribution by varying the transparency of an overprinted layer of population data. Value-by-alpha is heavily reliant on computational colour mixing of an alpha transparency layer with the underlying colours on a standard choropleth map. It was developed by cartographers Robert Roth, Andy Woodruff, and Zachary Johnson in 2010. Value-by-alpha modifies an underlying map to express some measure of a second variable. It's therefore a form of bivariate map, and its lineage can be traced to Bavarian statistician Georg Mayr's use of the bivariate technique in 1874.
pp. 30, 94, 228, 230

Violin plot

This method of plotting numerical data combines the box plot and kernel density plot. It incorporates a depiction of a smoothed kernel density plot on each side, which shows the probability density at different data values. A violin plot extends what can be interpreted from a box plot because it shows the entire data distribution. This can be useful when data is multimodal. The violin plot owes much to the development and experimentation of box plots by American statistician John Tukey in the mid- to late 1900s. It was formalised by statistician Jerry Hintze and Professor of Business Management Ray Nelson in 1998.
p. 140

Waffle grid

The use of multiple squares in a mosaic grid to show repetition of categorical data, and therefore some measure of quantity, is predicated on the work of Émile Cheysson, who created annual *Albums of Statistics* for the French Ministry of Public Works in the late 1800s. In particular, the map of traffic at French train stations from 1883 showed the proportions of freight and passengers in a regular square which were also proportional. American economist, statistician, businessman, inventor, and author Karl G. Karsten formalised the technique further in 1925 and referred to it as *hundred percent squares*. A 1932 map of deer breeding and dog breeding in the Russian Federation makes use of waffle grids, and 3D waffle grids were used by Hungarian-born American cartographer Erwin Raisz in his 1964 *Atlas of Florida*.
pp. 150, 152, 154, 198

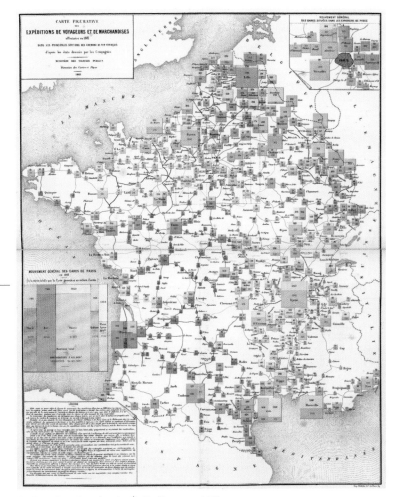

Traffic at French stations by Émile Cheysson, 1883

Deer and dog breeding in the Russian Federation, 1932

Credits

Except for the credits listed below, all maps in this book were originally designed and created by the author in their entirety using ArcGIS® Pro and data from ArcGIS® Living Atlas of the World. Graphs, charts, and plots were created using Adobe Illustrator. Some of the maps were finished using Adobe Illustrator. Book cover design and layout by the author.

p. ix (bottom right), 230, 231 (main map).
Photograph by Carlos Barria, copyright *Reuters*. Used with permission.

p. xvi
Extract from Maps and Sales Visualization by E. P. Hermann, 1922. Public domain.

p. xvii
Extract from Semiology of Graphics by Jacques Bertin, 1987. Courtesy of Esri Press. Used with permission.

p. xviii
Map of the popular vote of the 1880 US presidential election from the US Census Atlas, 1883. Public Domain.

p. xix
Result of the 1895 general election in the United Kingdom by The Times (London), 1895. Public domain.

pp. 64, 120, 170, 210 (main map), 212, 226
Designed in collaboration with John Nelson, Esri.

pp. 82, 116, 124, 170, 222
Designed in collaboration with Craig Williams, Esri.

pp. 94 (small map), 146, 188, 204
Designed in collaboration with Nathan Shephard, Esri.

p. 102 (main map)
Designed in collaboration with RJ Andrews, Info We Trust.

p. 113 (main map)
Designed in collaboration with Ben Flanagan, Esri UK.

pp. 136, 138, 150, 164
Designed in collaboration with Linda Beale, Esri.

p. 168
Designed in collaboration with Beichen Tian.

Except for the credits listed below, all the illustrations in the 'Epilogue' and 'Prior Carte' sections are in the public domain:

p. 234
Courtesy of Kenneth Field. Used with permission.

p. 241
Courtesy of Thomas de Beus. Used with permission.

p. 242
Designed in collaboration with Alan McConchie.

p. 249
Courtesy of Danny Dorling. Used with permission.

p. 252 (top and bottom)
Courtesy of David Rumsey Map Collection, David Rumsey Map Center, Stanford Libraries. Used with permission.

p. 257 (bottom)
Courtesy of the Otto and Marie Neurath Isotype collection, University of Reading, UK. Used with permission.

p. 258 (top), 265 (top), 266 (middle)
Courtesy of Esri Press. Used with permission.

p. 262 (middle)
Redrawn by author, based on the original by George Jenks, 1963.

p. 271
Courtesy of David Rumsey Map Collection, David Rumsey Map Center, Stanford Libraries. Used with permission.

p. 274
Illustration by Wesley Jones. Used with permission.

Happy mapping!